29th Annual Meeting of the North American Membrane Society (NAMS 2020)

AA005648

Held online due to COVID-19

Tempe, Arizona, USA
18-21 May 2020

ISBN: 978-1-7138-2788-7

Printed from e-media with permission by:

Curran Associates, Inc.
57 Morehouse Lane
Red Hook, NY 12571

Some format issues inherent in the e-media version may also appear in this print version.

Printed with permission by Curran Associates, Inc. (2021)

For permission requests, please contact North American Membrane Society (NAMS) at the address below.

North American Membrane Society (NAMS)
University of Arkansas
Ralph E. Martin Department of Chemical Engineering
3202 Bell Engineering Center
Fayetteville, AR 72701-1201
USA

Phone: +1 479 575-7926
Fax: +1 479 575-7926

names@uark.edu

Additional copies of this publication are available from:

Curran Associates, Inc.
57 Morehouse Lane
Red Hook, NY 12571 USA
Phone: 845-758-0400
Fax: 845-758-2633
Email: curran@proceedings.com
Web: www.proceedings.com

Oral Sessions - Monday

Presenting authors are indicated by an asterisk*. See the most up-to-date version of the program book at **www.membranes.org/nams-2020/**

Oral 1 - MEMBRANE FOULING I ..pg 1

12 pm – 1:40 pm, ROOM 1

Session Chairs

Steven Weinman
The University of Alabama

Dan Miller
Lawrence Berkeley National Laboratory

12:00 pm (Room 1, Monday)

Oral 1 - Electrochemical Prevention of Mineral Scale on Electrically Conducting Desalination Membranes

David Jassby (University of California, Los Angeles), Unnati Rao (University of California, Los Angeles), Bongyeon Jung* (University of California, Los Angeles).

12:20 pm (Room 1, Monday)

Oral 2 - Assessment of Oil Fouling By Oil-Membrane Interaction Energy

Henry Tanudjaja* (Nanyang Technological University), Jia Wei Chew (Nanyang Technological University).

12:40 pm (Room 1, Monday)

Oral 3 - Role of Instability Phenomena in Mitigating Membrane Fouling

Weiyi Li* (Southern University of Science and Technology), Xin Liu (Southern University of Science and Technology).

1:00 pm (Room 1, Monday)

Oral 4 - Do membrane hydrophobicity and hydrodynamic shear affect the initial deposition and pioneering colonization of anaerobes isolated from an anaerobic membrane bioreactor?

Yang Yang* (Ben-Gurion University of the Negev), Gideon Oron (Ben-Gurion University of the Negev), Moshe Herzberg (Ben-Gurion University of the Negev), Roy Bernstein (Ben-Gurion University of the Negev).

1:20 pm (Room 1, Monday)

Oral 5 - Nanofiltration of saline oil-water emulsions: Combined effect of salt concentration polarization and fouling by oil on flux performance

Charifa Hejase (Michigan State University), Vlad Tarabara* (Michigan State University).

Oral 2 – CARBON CAPTURE I ...pg. 3

12 pm – 1:40 pm, ROOM 2

Session Chairs

Winston Ho
Ohio State University

David Hopkinson
National Energy Technology Laboratory

Katherine Hornbostel
University of Pittsburg

12:00 pm (Room 2, Monday)

Oral 6 - Novel Facilitated Transport Membrane and Process for Post-Combustion Carbon Capture

Yang Han* (The Ohio State University), Kai Chen (The Ohio State University), Witopo Salim (The Ohio State University), Dongzhu Wu (The Ohio State University), Winston Ho (The Ohio State University).

12:20 pm (Room 2, Monday)

Oral 7 - Carbon Molecular Sieves with Superior H_2/CO_2 Separation Properties at Elevated Temperatures for Pre-combustion CO_2 Capture

Haiqing Lin* (The State University of New York at Buffalo), Hien Nguyen (The State University of New York at Buffalo), Maryam Omidvar (The State University of New York at Buffalo).

12:40 pm (Room 2, Monday)

Oral 8 - Carbon Capture Performance of Crosslinked Polyphosphazene Blend Membranes

Victor A Kusuma* (National Energy Technology Laboratory), Zi Tong (National Energy Technology Laboratory), Lingxiang Zhu (National Energy Technology Laboratory), Joshua McNally (Idaho National Laboratory), James Baker (National Energy Technology Laboratory), Christopher Orme (Idaho National Laboratory), David Hopkinson (National Energy Technology Laboratory), Frederick Stewart (National Energy Technology Laboratory).

1:00 pm (Room 2, Monday)

Oral 9 - Design and Techno-Economic Analysis of a Zeolite Membrane Reactor Intensified IGCC Process

Lie Meng (Arizona State University), Haoren Lu (Nexant, Inc.), Gerald Choi (Nexant, Inc.), Jerry Lin* (Arizona State University).

1:20 pm (Room 2, Monday)

Oral 10 - Synthesis of Sterically Hindered Polyvinylamine and Its Application in Facilitated Transport Membranes for CO2 Capture from Flue Gas

Ting-Yu Chen* (The Ohio State University), Xuepeng Deng (The Ohio State University), Li-Chiang Lin (The Ohio State University), Winston Ho (The Ohio State University).

Oral 3 – SEAWATER DESALINATION ...pg. 5

12 pm – 1:40 pm, ROOM 3

Session Chairs
Bill Phillip
University of Notre Dame

12:00 pm (Room 3, Monday)

Oral 11 - New Insights into Solute-Selectivity Relationship of Thin-Film Composite Polyamide Membranes

Xi Chen* (Columbia University), Chanhee Boo (Columbia University), Ngai Yin Yip (Columbia University).

12:20 pm (Room 3, Monday)

Oral 12 - Omniphobic PVDF-HFP nanofibrous membrane with Pine-needle-like hierarchical TiO2 nanostructures: Towards enhanced performance for membrane distillation

Weihua Qing* (New Jersey Institute of Technology), Xianhui Li (The University of Hong Kong), Chuyang Tang (University of Hong Kong / University of New South Wales).

12:40 pm (Room 3, Monday)

Oral 13 - Energy Barriers for Decoupled Cation and Anion Transport in Sub-nanometer Pores

Xuechen Zhou* (Yale University), Razi Epsztein (Technion), Jaehong Kim (Yale University), Menachem Elimilech (Yale University).

1:00 pm (Room 3, Monday)

Oral 14 - High Free Volume, Ion-containing Pentiptycene-based Polysulfone Membranes for Water Purification Application

Tao Wang* (University of Notre Dame), Ruilan Guo (University of Notre Dame), Feng Gao (University of Notre Dame), Bill Phillip (University of Notre Dame).

1:20 pm (Room 3, Monday)

Oral 15 - Zwitterionic Polysulfone Membranes for Pervaporation

Tejaswi Nori* (Arizona State University).

Oral 4 – EMERGING MATERIALS FOR LIQUID SEPARATION I ...pg. 8

12 pm – 1:40 pm, ROOM 4

Session Chairs
Hee-Jeung Oh
The Pennsylvania State University

Baoxia Mi
University of Berkeley, California

Piran Kidambi
Vanderbilt University

12:00 pm (Room 4, Monday)

Oral 16 - Charge-modified polysulfones as a platform for membrane separations

Matthew Green* (Arizona State University), Yi Yang (Arizona State University), Mani Modayil Korah (Arizona State University), Hoda Shokrollahzadeh Behbahani (Arizona State University), Tejaswi Nori (Arizona State University).

12:20 pm (Room 4, Monday)

Oral 17 - Diffusion-Induced in situ Growth of Covalent Organic Frameworks for Composite Membranes

Priyanka Manchanda* (King Abdullah University of Science and Technology), Suzana Nunes (King Abdullah University of Science and Technology).

12:40 pm (Room 4, Monday)

Oral 18 - Lyotropic Liquid Crystals with Water-Continuous Nanostructure as Thin-film Composite Nanofiltration Membranes

Yizhou Zhang* (University of Pennsylvania), Chinedum Osuji (University of Pennsylvania).

1:00 pm (Room 4, Monday)

Oral 20 - Two-Dimensional Covalent Organic Frameworks with Various Backbones and Pore Surface Engineering as Selective Layers of Thin-Film Composite Membranes

Phuoc Duong (University of Wyoming), Valerie Kuehl (University of Wyoming), Veronica Spaulding (University of Wyoming), Jiashi Yin (University of Wyoming), John Hoberg (University of Wyoming), Bruce Parkinson (University of Wyoming), Katie Li-Oakey* (University of Wyoming).

Oral 5 – INORGANIC MATERIALS...pg .10

12 pm – 1:40 pm, ROOM 5

Session Chairs
Kumar Varoon Agrawal
Ecole polytechnique fédérale de Lausanne

Luis Francisco Villalobos
Ecole polytechnique fédérale de Lausanne

12:00 pm (Room 5, Monday)

Oral 21 - Synthesis of polycrystalline ZIF-8 membranes in few minutes for CO2/N2 and CO2/CH4 separation

Kumar Varoon Agrawal* (Ecole polytechnique fédérale de Lausanne), Jian Hao (Ecole polytechnique fédérale de Lausanne), Deepu Babu (Ecole polytechnique fédérale de Lausanne).

12:20 pm (Room 5, Monday)

Oral 22 - Zeolite membrane separator for safe Li-ion batteries

Kishen Rafiz (Arizona State University), Jerry Lin* (Arizona State University).

12:40 pm (Room 5, Monday)

Oral 23 - Preparation and nanofiltration properties of thin film meso-porous oxide membranes via the sonochemical precipitation method

Minghui Qiu* (Nanjing Tech University), Zhihao Jin (Nanjing Tech University), Hendrik Verweij (The Ohio State University), Yiqun Fan (Nanjing Tech University).

1:00 pm (Room 5, Monday)

Oral 24 - Optimization of the performance of the photocatalytic TiO2 membranes fabricated by suspension plasma spray (SPS) process

Elnaz Alebrahim* (Concordia University), Md. Saifur Rahaman (Concordia University), Christian Moreau (Concordia University).

1:20 pm (Room 5, Monday)

Oral 25 - Few-nanometer thick organosilica membranes for high-temperature H2/CO2 separation

Haiqing Lin* (The State University of New York at Buffalo), Lingxiang Zhu (Energy National Energy and Technology Laboratory), Liang Huang (The State University of New York at Buffalo).

Oral Sessions - Tuesday

Presenting authors are indicated by an asterisk*. See the most up-to-date version of the program book at **www.membranes.org/nams-2020/**

Oral 6 - MEMBRANE FOULING II ..pg. 13

11:00 am – 12:40 pm, ROOM 1

Session Chairs
Steven Weinman
The University of Alabama

Dan Miller
Lawrence Berkeley National Laboratory

11:00 am (Room 1, Tuesday)

Oral 26 - Exploring and modeling the effect of pattern geometry on nanofiltration membrane fouling

Anna Malakian* (Clemson University), Lucas Messick (Clemson University), Tara Spitzer (Clemson University), Scott Husson (Clemson University).

11:20 am (Room 1, Tuesday)

Oral 27 - Oscillation induced aggregation to reduce colloidal fouling in microfiltration

Hamdy Abo Zaid (Ben-Gurion University of the Negev), Levi Gheber (Ben-Gurion University of the Negev), Jack Gilron* (Ben-Gurion University of the Negev).

11:40 am (Room 1, Tuesday)

Oral 28 - Living Filtration Membranes Demonstrate Anti-Biofouling Properties

Carson Bechtel (Montana Technological University), Katherine Zodrow* (Montana Technological University).

12:00 pm (Room 1, Tuesday)

Oral 29 - Covalently grafting graphene oxide onto UF membranes to improve antifouling properties

Xiaoyi Chen* (State University of New York at Buffalo), Erda Deng (State University of New York at Buffalo), Dongwon Park (State University of New York at Buffalo), Blaine Pfeifer (State University of New York at Buffalo), Haiqing Lin (State University of New York at Buffalo).

12:20 pm (Room 1, Tuesday)

Oral 30 - Laser-Induced Graphene for Charged Membrane Applications Enabled by Sequential Infiltration Synthesis

David Bergsman* (Massachusetts Institute of Technology), Beza Getachew (Massachusetts Institute of Technology), Jeffrey Grossman (Massachusetts Institute of Technology).

12:40 pm (Room 1, Tuesday)

Oral 151 - Mineral scale formation and dissolution in RO operation in cyclic modes of operation

Yoram Cohen (University of California, Los Angeles), Anditya Rahardianto* (University of California, Los Angeles), Yeunha Kim (University of California, Los Angeles), Tae Lee (University of California, Los Angeles), Muhammad Bilal (University of California, Los Angeles).

Oral 7 – CARBON CAPTURE II ...pg. 16

11:00 am – 12:40 pm, ROOM 2

Session Chairs
Winston Ho
Ohio State University

David Hopkinson
National Energy Technology Laboratory

Katherine Hornbostel
University of Pittsburg

11:00 am (Room 2, Tuesday)

Oral 31 - High flux CO2-selective single-layer graphene membranes: synthesis and scale-up

Kumar Varoon Agrawal* (Ecole polytechnique fédérale de Lausanne), Shiqi Huang (Ecole polytechnique fédérale de Lausanne).

11:20 am (Room 2, Tuesday)

Oral 32 - Tuning the transport properties of CANAL ladder polymers by alkyl substitutions and backbone conformation modification

Francesco Maria Benedetti * (Massachusetts Institute of Technology), Holden Lai (Stanford University), Jun Myun Ahn (Stanford University), Zexin Jin (Stanford University), Albert Wu (Massachusetts Institute of Technology), Maria Grazia De Angelis (University of Bologna), Yan Xia (Stanford University), Zachary Smith (Massachusetts Institute of Technology).

11:40 am (Room 2, Tuesday)

Oral 33 - Subambient Carbon Capture using Next Generation Hollow Fiber Membrane Modules at the National Carbon Capture Center

David Hasse* (Air Liquide), Shilu Fu (Air Liquide), Shdhir Kulkarni (Air Liquide), Alex Augustine (Air Liquide), Trapti Chaubey (Air Liquide), Dean Kratzer (Air Liquide), Noemi Collado (Air Liquide).

12:00 pm (Room 2, Tuesday)

Oral 34 - Scalable charge-modified polysulfone IL support membrane morphologies for CO2 capture in space

Mani Modayil Korah* (Arizona State University), Matthew Green (Arizona State University), Yi Yang (Arizona State University).

12:20 pm (Room 2, Tuesday)

Oral 35 - Ultra-thin supported graphene oxide membrane for CO2 capture

Yi Zhou* (Ohio State University).

Oral 8 – EMERGING MATERIALS FOR LIQUID SEPARATION II ...pg. 18

11:00 pm – 12:40 pm, ROOM 3

Session Chairs

Hee-Jeung Oh
The Pennsylvania State University

Baoxia Mi
University of Berkeley, California

Piran Kidambi
Vanderbilt University

11:00 am (Room 3, Tuesday)

Oral 36 - Ultrathin Two-Dimensional Covalent Organic Framework (COF) Membranes for Molecular Separations

David Shaffer* (University of Houston), Rahul Shevate (University of Houston).

11:20 am (Room 3, Tuesday)

Oral 37 - Toward Systematic Tuning of Thin-Film Composite Membrane Selectivity

Mackenzie Anderson* (University of California, Los Angeles).

11:40 am (Room 3, Tuesday)

Oral 38 - Dry-Wet Phase Inversion Synthesis Role on Graphene Oxide-Polysulfone Membrane for Membrane Distillation

Lucy M Camacho* (Texas A&M University-Kingsville), Samuel Olatunji (Texas A&M University-Kingsville).

12:00 pm (Room 3, Tuesday)

Oral 39 - Phosphorene- Based Antifouling Membranes: Synthesis, Fabrication and Applications

Joyner Eke* (University of Kentucky), Isabel Escobar (University of Kentucky).

12:20 pm (Room 3, Tuesday)

Oral 40 - Understanding Virus Prefilters in Improving the Filtration Performance of the

Planova BioEX Virus Filter for Monoclonal Antibodies

Solomon Isu* (University of Arkansas), AAA ZZZ (Pennsylvania State University), Xianghong Qian (University of Arkansas), Andrew Zydney (Pennsylvania State University), Ranili Wickramasinghe (University of Arkansas).

Oral 9 – OSMOTICALLY DRIVEN PROCESSES ...pg. 21

11:00 pm – 12:40 pm, ROOM 4

Session Chairs

Tony Straub
University of Colorado Boulder

Andrea Achilli
The University of Arizona

11:00 am (Room 4, Tuesday)

Oral 41 - Reverse-fluxed Draw Solutes Removal and Microalgae Production in an Osmotic Photobioreactor System For Complete Water Recovery from Secondary Effluent

Zixuan Wang* (Washington University in St. Louis), Zhen He (Washington University in St. Louis), Yi-Ying Lee (University of Maryland Center for Environmental Science and University of Maryland Baltimore Country), David Scherr (Virginia Polytechnic Institute and State University), Yantao Li (University of Maryland Center for Environmental Science and University of Maryland Baltimore County).

11:20 am (Room 4, Tuesday)

Oral 42 - Zwitterionic Hydrogels with Polyamide-Skin Layers Mitigating Concentration Polarization for Forward Osmosis

Thien Tran* (University at Buffalo, The State University of New York), Shiwei Pan (Wanhua Chemical Group Co., Ltd), Xiaoyi Chen (The State University of New York at Buffalo), Adrienne Blevins (University of Colorado Boulder), Yifu Ding (University of Colorado Boulder), Haiqing Lin (The State University of New York at Buffalo).

11:40 am (Room 4, Tuesday)

Oral 43 - A Simplified Modeling Framework to Investigate Osmotically Driven Processes at the System-scale

Zachary Binger* (The University of Arizona), Andrea Achilli (The University of Arizona).

12:00 pm (Room 4, Tuesday)

Oral 44 - Experimental Demonstration of Power Generation from Fertilizer via Pressure Retarded Osmosis

Pouyan Pourmovahed (Oakland University), Jonathan Maisonneuve* (Oakland University).

12:20 pm (Room 4, Tuesday)

Oral 45 - Mitigation of Biological Fouling in a Forward Osmosis Membrane Bioreactor

Derrick Satterfield (University of Nevada, Reno), Jack Griffin (University of Nevada, Reno), Terin George (University of Nevada, Reno), Sage Hiibel* (University of Nevada, Reno).

Oral 10 – WATER REUSE...pg. 23

11:00 pm – 12:40 pm, ROOM 5

Session Chairs

Vicky Karanikola
Ecole polytechnique fédérale de Lausanne

Jack Girlon
Ben-Gurion University of the Negev

11:00 am (Room 5, Tuesday)

Oral 46 - Desalination for a Circular Water Economy

Megan Mauter* (Stanford University).

11:20 am (Room 5, Tuesday)

Oral 47 - The impact of pre-chlorination on microfiltration fouling during municipal wastewater effluent filtration for water reuse

Kunal Gupta* (Texas A&M University), Shankar Chellam (Texas A&M University).

11:40 am (Room 5, Tuesday)

Oral 48 - Understanding colloidal fouling formation during inline coagulation/ultrafiltration process

Bianca Souza Chaves* (The University of Arizona), Andrea Achilli (The University of Arizona).

12:00 pm (Room 5, Tuesday)

Oral 49 - Combination of AnMBR with outdoor microalgae cultivation for industrial wastewater treatment, resource recovery and near-zero waste

Roy Bernstein* (Ben-Gurion University at Negev).

12:20 pm (Room 5, Tuesday)

Oral 50 - Membrane Surface Characterization Upon Cleaning of a Heavily Fouled Reverse Osmosis Membrane from an Advanced Reclamation Facility

Bilal Abada* (Texas A&M University), Shankar Chellam (Texas A&M University).

Oral Sessions - Wednesday

Presenting authors are indicated by an asterisk*. See the most up-to-date version of the program book at **www.membranes.org/nams-2020/**

Oral 11 – POLYMERIC AND MIXED-MATERIALS – GAS SEPARATION I ...pg 26

11:00 am – 12:40 pm, ROOM 1

Session Chairs
Ben Sundell
Armaco

Zachary Smith
Massachusetts Institute of Technology

Xiaoli Ma
University of Wisconsin-Milwaukee

11:00 am (Room 1, Wednesday)

Oral 51 - Effect of thermal treatment on the structure and gas transport properties of a triptycene-based polybenzoxazole exhibiting configurational free volume

Ryan D. Crist (University of Oklahoma), Zihan Huang (University of Notre Dame), Ruilan Guo (University of Notre Dame), Michele Galizia* (University of Oklahoma).

11:20 am (Room 1, Wednesday)

Oral 52 - Optimizing Plasticization Benefits of Polyimides for Membrane-based Natural Gas Sweetening

Yang Liu* (Georgia Institute of Technology), William Koros (Georgia Institute of Technology), Gongping Liu (Nanjing Tech University), Mohamed Eddaudi (King Abdullah University of Science and Technology), Zhijie Chen (King Abdullah University of Science and Technology).

11:40 am (Room 1, Wednesday)

Oral 53 - New Facilitated Transport Membranes for Hydrogen Purification from Coal-Derived Syngas

Yang Han* (The Ohio State University), Xuepeng Deng (The Ohio State University), Li-Chiang Lin (The Ohio State University), Winston Ho (The Ohio State University).

12:00 pm (Room 1, Wednesday)

Oral 54 - High Performance Gas Separation Membranes from CANAL Ladder Polymerization

Yan Xia (Stanford University), Holden Lai* (Stanford University).

12:20 pm (Room 1, Wednesday)

Oral 55 - Carbon Molecular Sieve Membranes for Petrochemical Gas Separations

Surendar Venna* (Dow Inc.), Thomas Fitzgibbons (Dow Inc.), Derrick Flick (Dow Inc.), James Heard (Dow Inc.), Jay (Junqiang) Liu (The Dow Chemical Company), Hali McCurry (Dow Inc.), Nikki Montanez (Dow Inc.), Gerard Rogers (Dow Inc.), Li Tang (Dow Inc.), Cole Witham (Dow Inc.), Abhishek Roy (Dow Inc.).

Oral 12 – BIOINSPIRED AND BIOMIMETIC MATERIALS...pg. 28

11:00 am – 12:40 pm, ROOM 2

Session Chairs
Yuexiao Shen
Texas Tech University

Patrick Saboe
National Renewable Energy Laboratory

11:00 am (Room 2, Wednesday)

Oral 56 - Artificial Water Channels- toward biomimetic membranes for desalination

Mihail Barboiu* (Institut Europeen des Membranes).

11:20 am (Room 2, Wednesday)

Oral 57 - Nature-inspired ion conducting polymers for energy conversion and storage devices

Shudipto K Dishari* (University of Nebraska-Lincoln).

11:40 am (Room 2, Wednesday)

Oral 58 - Ultra-High Ionic Exclusion Through Carbon Nanomembranes

Yang Yang (Bielefeld University), Roland Hillmann (Bielefeld University), Yubo Qi (Bielefeld University), Riko Korzetz (Bielefeld University), Niklas Biere (Bielefeld University), Daniel Emmrich (Bielefeld University), Michael Westphal (Bielefeld University), Björn Büker (Bielefeld University), Andreas Hütten (Bielefeld University), André Beyer (Bielefeld University), Dario Anselmetti (Bielefeld University), Armin Gölzhäuser* (Bielefeld University).

12:00 pm (Room 2, Wednesday)

Oral 59 - High Density Membrane Protein-Polymer Nanosheets-Based Biomimetic Membranes

Yu-Ming Tu* (The University of Texas at Austin), Woochul Song (The University of Texas at Austin), Benny Freeman (The University of Texas at Austin), Manish Kumar (The University of Texas at Austin).

12:20 pm (Room 2, Wednesday)

Oral 60 - Biomimetic carbon nanotube and AAO systems utilizing electro-dynamic interfaces

Bruce Hinds* (University of Washington).

Oral 13 – MICROPOROUS / CATALYTIC MATERIALS ...pg. 30

11:00 am – 12:40 pm, ROOM 3

Session Chairs
Ayse Asatekin
Tufts University

Will Tarpeh
Stanford University

Vlad Tarabara
Michigan State University

11:00 am (Room 3, Wednesday)

Oral 61 - Membranes with Integrated Nanomaterials: Material Science Advances and Applications

Dibakar Bhattacharyya* (University of Kentucky), Hongyi (Derek) Wan (University of Kentucky), Anthony Saad (University of Kentucky), Ronald Vogler (University of Kentucky), Trisha Nickerson (University of Kentucky), Ashish Aher (University of Kentucky), Saiful Islam (University of Kentucky).

11:20 am (Room 3, Wednesday)

Oral 62 - Efficient Ammonia Decomposition in a Catalytic Membrane Reactor to Enable Hydrogen Storage and Utilization

Zhenyu Zhang (CSM), Javishk Shah (CSM), J. Douglas Way (CSM), Colin Wolden* (Colorado School of Mines).

11:40 am (Room 3, Wednesday)

Oral 63 - Tuning Ultrafiltration Membrane Performance via Surface Graft Polymerization of Acrylic Acid Enabled by Air Plasma Activation

Yian Chen* (University of California, Los Angeles), Soomin Kim (University of California, Los Angeles), Yoram Cohen (University of California, Los Angeles).

12:00 pm (Room 3, Wednesday)

Oral 64 - Monitoring protein fouling on patterned membranes by light microscopy and simulation

Anna Malakian* (Clemson University), Bowen Ling (Stanford University), Ilenia Battiato (Stanford University), Scott Husson (Clemson University).

12:20 pm (Room 3, Wednesday)

Oral 65 - Influences of Microwave Irradiation on Performances of Membrane Filtration and Catalytic Degradation of Perfluorooctanoic Acid (PFOA)

Fangzhou Liu* (New Jersey Institute of Technology).

Oral 14 – HIGH SALINITY STREAMS, BRINE MINIMIZATION AND ZLD ...pg. 33

11:00 am – 12:40 pm, ROOM 4

Session Chairs
Kerri Hickenbottom
The University of Arizona

Jon Brant
University of Wyoming

11:00 am (Room 4, Wednesday)

Oral 66 - Minimal and zero liquid discharge with reverse osmosis using low-salt-rejection membranes

Zhangxin Wang* (Yale University), Menachem Elimelech (Yale University), Akshay Deshmukh (Yale University), Yuhao Du (Yale University).

11:20 am (Room 4, Wednesday)

Oral 67 - Evaluation of Integrated Electrocoagulation-Microfiltration and Direct Contact Membrane Distillation Processes for Treating Produced Water

Mahmood Jebur* (University of Arkansas), Ranil Wickramasinghe (University of Arkansas), Mahdi Malmali (Texas Tech University), Xianghong Qian (University of Arkansas), Yunxia Hu (Tianjin Polytechnic University), Yuhe Cao (University of Arkansas), Yu-Hsuan Chiao (University of Arkansas).

11:40 am (Room 4, Wednesday)

Oral 68 - A Novel Membrane Distillation-Crystallization Approach for Inland Brine Treatment

Evangelos Balis (University of Nevada, Reno), Jack Griffin (University of Nevada, Reno), Sage Hiibel* (University of Nevada, Reno).

12:00 pm (Room 4, Wednesday)

Oral 69 - Critical Analysis of Membrane Distillation at High Salinity

Mukta Hardikar* (The University of Arizona), Andrea Achilli (The University of Arizona).

12:20 pm (Room 4, Wednesday)

Oral 70 - Testing mineral scaling propensities of fluorinated and non-fluorinated polyamide thin film composite membranes

Sankaranarayanan Ayyakudi Ravichandran* (University of Colorado Boulder), Josue Velasco (University of Colorado Boulder), Saied Delagah (U.S. Department of Interior), John Pellegrino (University of Colorado).

Oral 15 – PROCESS INTENSIFICATION AND INTEGRATION ...pg. 35

11:00 am – 12:40 pm, ROOM 5

Session Chairs
Hannah Murnen
Compact Membrane Systems

Ed Sanders
Air Liquide

11:00 am (Room 5, Wednesday)

Oral 71 - Energy Efficiency of Electro-Driven Brackish Water Desalination: Electrodialysis Significantly Outperforms Membrane Capacitive Deionization

Mohan Qin* (University of Wisconsin-Madison), Sohum Patel (Yale University), W. Shane Walker (The University of Texas at El Paso), Menachem Elimelech (Yale University).

11:20 am (Room 5, Wednesday)

Oral 72 - CO2-Selective Membrane for Enhancing H2 Utilization in Solid Oxide Fuel Cells

Kai Chen* (The Ohio State University), Witopo Salim (Membrane Technology & Research Inc.), Yang Han (The Ohio State University), Mike Gasda (Bloom Energy Corporation), Winston Ho (The Ohio State University).

11:40 am (Room 5, Wednesday)

Oral 73 - Flow Patterns through densely packed hollow fibers: a CFD study for membrane distillation

Albert Kim* (University of Hawaii), Hyeon-Ju Kim (Korea Research Institute of Ships and Ocean Engineering), Deok-Soo Moon (Korea Research Institute of Ships and Ocean Engineering).

12:00 pm (Room 5, Wednesday)

Oral 74 - Analysis of Membrane Processes for CO2 Removal and H2 Reuse for Solid Oxide Fuel Cells

Kai Chen* (The Ohio State University), Yang Han (The Ohio State University), Mike Gasda (Bloom Energy Corporation), Winston Ho (The Ohio State University).

12:20 pm (Room 5, Wednesday)

Oral 75 - Experimental evaluation of an optimized concentration gradient battery

Fei Liu* (University of North Carolina at Chapel Hill), Ryan Kingsbury (University of North Carolina at Chapel Hill), Mikayla Armstrong (University of North Carolina at Chapel Hill, Orlando Coronell (University of North Carolina at Chapel Hill).

Oral 16 – POLYMERIC AND MIXED-MATERIALS – GAS SEPARATION II ...pg. 38

1:00 pm – 3:00 pm, ROOM 1

Session Chairs
Ben Sundell
Armaco

Zachary Smith
Massachusetts Institute of Technology

Xiaoli Ma
University of Wisconsin-Milwaukee

1:00 pm (Room 1, Wednesday)

Oral 76 - Microporous Pentiptycene-based Polymers with Heterocyclic Rings for High Performance Gas Separation Membranes

Zihan Huang* (University of Notre Dame), Claire Yin (University of Notre Dame), Gregory Kline (University of Notre Dame), Ruilan Guo (University of Notre Dame).

1:20 pm (Room 1, Wednesday)

Oral 77 - High-Performance Gas Separation Membranes Based on Poly(benzimidazole)

Alexander Bridge* (The University of Texas at Austin), Joshua Moon (The University of California Santa Barbara), Joan Brennecke (The University of Texas at Austin), Benny Freeman (The University of Texas at Austin).

1:40 pm (Room 1, Wednesday)

Oral 78 - Polybenzimidazole-Derived Carbon Molecular Sieve Hollow Fiber Membranes with Tailored Oxygen Selective Transport

Jong Geun Seong* (Los Alamos National Laboratory), John Matteson (Los Alamos National Laboratory), Jeremy Lewis (Los Alamos National Laboratory), John Baca (Los Alamos National Laboratory), Alexander Josephson (Los Alamos National Laboratory), Troy Holland (Los Alamos National Laboratory), Joel Kress (Los Alamos National Laboratory), Kathryn Berchtold (Los Alamos National Laboratory), Rajinder Pal Singh Singh (Los Alamos National Laboratory).

2:00 pm (Room 1, Wednesday)

Oral 79 - Tailoring structural and functional features of high-performance ionenes and ionic composites designed for membrane-based gas separations

Kathryn O'Harra* (University of Alabama), Irshad Kammakakam (University of Alabama), Jason Bara (University of Alabama).

2:20 pm (Room 1, Wednesday)

Oral 80 - Azo-UiO-66 MOF for low-energy CO2 capture adsorbents and mixed matrix membranes for CO2/N2 separation

Bradley Ladewig* (Karlsruhe Institute of Technology), Nicholaus Prasetya (Imperial College London).

2:40 pm (Room 1, Wednesday)

Oral 81 - Physical aging of sub-micron PBI membranes at elevated temperatures

Melanie Merrick* (The University of Texas at Austin), Benny Freeman (The University of Texas at Austin).

Oral 17 – MEMBRANE SYNTHESIS AND CASTING I ...pg. 41

1:00 pm – 3:00 pm, ROOM 2

Session Chairs
Christine Duval
Case Western Reserve University

1:00 pm (Room 2, Wednesday)

Oral 82 - Polyamide Nanofiltration Membrane with Highly Uniform Sub-nanometre Pores for Sub-1Å Precision Separation

Yuanzhe Liang* (Vanderbilt University).

1:20 pm (Room 2, Wednesday)

Oral 83 - Synthesis of Ultrahigh MW Polyvinylamine for Incorporation in Membranes for CO2 Capture

Kai Chen* (The Ohio State University), Yang Han (The Ohio State University), Winston Ho (The Ohio State University).

1:40 pm (Room 2, Wednesday)

Oral 84 - Surface Patterning of Microporous Membranes using Thermally Induced Phase Separation under Confinement

Shouhong Fan* (University of Colorado at Boulder), Yifu Ding (University of Colorado Boulder).

2:00 pm (Room 2, Wednesday)

Oral 85 - Membrane with cross-linked zwitterionic nanopores achieves sub-nanometer separations

Samuel J Lounder* (Tufts University), Ayse Asatekin (Tufts University).

2:20 pm (Room 2, Wednesday)

Oral 86 - Highly Permeable and Selective Crosslinked Polymer Membranes for Energy Efficient Gas Separation

Si Li* (University of Notre Dame), Gregory Kline (University of Notre Dame), Ruilan Guo (University of Notre Dame).

2:40 pm (Room 2, Wednesday)

Oral 152 - Lyotropic liquid crystal templating for making ultrafiltration membranes

Sahar Qavi (New Mexico State University), Aaron Lindsay (New Mexico State University), Millicent Firestone (Los Alamos National Laboratory), Reza Foudazi* (New Mexico State University).

Oral 18 – DOWNSTREAM BIOPROCESSING ... pg. 43

1:00 pm – 3:00 pm, ROOM 3

Session Chairs
Prity Bengani-Lutz
Repligen

Onur Kas
UCB

James McGrath
University of Rochester

1:00 pm (Room 3, Wednesday)

Oral 87 - Membrane adsorbers for medical isotope purification

Christine E Duval* (Case Western Reserve University), Maura Sepesy (Case Western Reserve University), Priyanka Suresh (Case Western reserve University).

1:20 pm (Room 3, Wednesday)

Oral 88 - New Multimodal Anion-Exchange Membranes for Polishing of Biologics

Joshua Osuofa* (Clemson University), Daniel Henn (Purilogics, LLC), Anna Forsyth (Purilogics, LLC), Jinxiang Zhou (Purilogics, LLC), Scott Husson (Clemson University).

1:40 pm (Room 3, Wednesday)

Oral 89 - Fouling and Retention Behavior of BioOptimal MF-SL Microfilter during Cell Culture Harvesting

Ranil Wickramasinghe (University of Arkansas), Xianghong Qian* (University of Arkansas), Da Zhang (University of Arkansas).

2:00 pm (Room 3, Wednesday)

Oral 90 - Development of nanopocket membranes for tangential flow analyte capture (TFAC) of extracellular vesicles

Thomas Gaborski* (Rochester Institute of Technology), Mehdi Dehghani (Rochester Institute of Technology), Shayan Gholizadeh (Rochester Institute of Technology).

2:20 pm (Room 3, Wednesday)

Oral 91 - Catch and Display Diagnostics: Capturing Non-protein Biologics on Ultrathin Silicon Membranes Using Size and Affinity-based Separations

Michael Klaczko* (University of Rochester), Kilean Lucas (University of Rochester), William Houlihan (University of Rochester), Julia Kuebel (University of Rochester), Jonathan Flax (University of Rochester), Richard Waugh (University of Rochester), James McGrath (University of Rochester).

Oral 19 – MEMBRANE CHARACTERIZATION ...pg. 45

1:00 pm – 3:00 pm, ROOM 4

Session Chairs
Santiago Romero
University of Edinburgh

1:00 pm (Room 4, Wednesday)

Oral 92 - Tailoring the Structure and Performance of Model Polyamide-based Membranes for Desalination

William Mulhearn (NIST), Peter Beaucage (NIST), Ryan Nieuwendaal (NIST), Christopher Soles (NIST), Christopher Stafford* (NIST).

1:20 pm (Room 4, Wednesday)

Oral 93 - Applying Transition-State Theory to Model Solute Transport in Membranes with Sub-nanometer Pores

Razi Epsztein* (Technion), Ryan M. Duchanois (Yale University), Cody L. Ritt (Yale University), Menachem Elimelech (Yale University).

1:40 pm (Room 4, Wednesday)

Oral 94 - Unraveling the Clay-Membrane Interactions during An Osmotically Driven Process via Optical Coherence Tomography

Xin Liu* (Southern University of Science and Technology), Weiyi Li (Southern University of Science and Technology).

2:00 pm (Room 4, Wednesday)

Oral 95 - Local density and free volume inhomogeneities govern transport properties in reverse osmosis membranes

Michael Geitner* (The Pennsylvania State University), Tyler Culp (The Pennsylvania State University), Abhishek Roy (Dow), Mou Paul (Dow), Steve Jons (DuPont), Jeffrey Wilbur (DuPont Water Solutions), Manish Kumar (The University of Texas at Austin), Enrique Gomez (The Pennsylvania State University).

2:20 pm (Room 4, Wednesday)

Oral 96 - Probing the Structure and Dynamics of Membranes under Realistic Operating Conditions, using Operando Spectroscopy

Casey O'Brien* (University of Notre Dame).

Oral 20 – MEMBRANES FOR ELECTROCHEMICAL APPLICATIONS I ...pg. 47

1:00 pm – 3:00 pm, ROOM 5

Session Chairs

Geoff Geise
University of Virginia

Orlando Coronell
University of North Carolina at Chapel Hill

Chris Arges
Louisiana State University

Shudipto Dishari
University of Nebraska

1:00 pm (Room 5, Wednesday)

Oral 97 - Modelling and validation of concentration dependence of ion exchange membrane permselectivity: significance of convection and Manning's counter-ion condensation theory

Ryan Kingsbury* (University of North Carolina at Chapel Hill), Orlando Coronell (University of North Carolina at Chapel Hill).

1:20 pm (Room 5, Wednesday)

Oral 98 - The Influence of Ion Association on Ion Solubility in Ion Exchange Membranes

Rahul Sujanani* (The University of Texas at Austin), Oscar Nordness (The University of Texas at Austin), Joshua Saunders (The University of Texas at Austin), Joan Brennecke (The University of Texas at Austin), Lynn Katz (The University of Texas at Austin), Benny Freeman (The University of Texas at Austin).

1:40 pm (Room 5, Wednesday)

Oral 99 - Functionalized Ceramic Ion Exchange Membranes for Electrodialysis

Gregory M Newbloom* (Membrion, Inc.), Rachel Malone (Membrion, Inc.), Ryan Kingsbury (Membrion, Inc.), Aditya Salunkhe (Membrion, Inc.).

2:00 pm (Room 5, Wednesday)

Oral 100 - Tradeoff in membranes for artificial photosynthesis: Ion transport and product crossover

Sarah M Dischinger* (Lawrence Berkeley National Laboratory), Dan Miller (Lawrence Berkeley National Laboratory), Blaine Carter (Lawrence Berkeley National Laboratory), Shubham Gupta (Lawrence Berkeley National Laboratory).

2:20 pm (Room 5, Wednesday)

Oral 101 - Highly Selective Ion Separations with Pressure-driven Flow through Charged Nanoporous Membranes

Chao Tang* (University of Notre Dame), Andriy Yaroshchuk (Polytechnic University of Catalonia), Merlin Bruening (University of Notre Dame).

Oral Sessions - Thursday

Presenting authors are indicated by an asterisk*. See the most up-to-date version of the program book at **www.membranes.org/nams-2020/**

Oral 21 – MOLECULAR AND PROCESS MODELING ...pg. 50

11:00 am – 12:40 pm, ROOM 1

Session Chairs
David Warsinger
Purdue University
Shihong Lin
Vanderbilt University

11:00 am (Room 1, Thursday)

Oral 102 - Effect of packing nonuniformity at the fiber bundle-case interface on flow distribution and separation performance for hollow fiber membrane modules

Lili Sun (University of Toledo), Atabong Etiendem (University of Toledo), Glenn Lipscomb* (University of Toledo, Toledo).

11:20 am (Room 1, Thursday)

Oral 103 - Estimating fluid pressure gradients within crosslinked aromatic polyamide using molecular dynamics

Riley Vickers* (University of North Carolina at Chapel Hill,) Timothy Weigand (University of North Carolina at Chapel Hill), Casey Miller (University of North Carolina at Chapel Hill), Orlando Corone!! (University of North Carolina at Chapel Hill).

11:40 am (Room 1, Thursday)

Oral 104 - High-throughput computational prediction of the cost of carbon capture using mixed matrix membranes

Janice Steckel* (U.S. Department of Energy), Christopher Wilmer (University of Pittsburgh), Samir Budhathoki (DOE/NETL), Olukayode Ajayi (Brewer Science.

12:00 pm (Room 1, Thursday)

Oral 105 - Opportunities for high productivity and selectivity desalination via osmotic distillation with improved membrane design

Anthony Straub (University of Colorado Boulder), Sangsuk Lee* (University of Colorado Boulder).

12:20 pm (Room 1, Thursday)

Oral 106 - Ion association as key element in RO and NF modeling

Viatcheslav Freger* (Technion – IIT)

Oral 22 – MEMBRANE SYSTHESIS AND CASTING II ...pg. 52

11:00 am – 12:40 pm, ROOM 2

Session Chairs
Christine Duval
Case Western Reserve University

11:00 am (Room 2, Thursday)

Oral 107 - Atomically Thin Graphene Membranes for Desalination and Molecular Separations

Piran Kidambi* (Vanderbilt University).

11:20 am (Room 2, Thursday)

Oral 108 - A Comparison of PolarClean, Gamma-Valerolactone and their Mixture as Bio-derived solvents for Polysulfone Membrane Fabrication

Xiaobo Dong* (University of Kentucky), Tequila Harris (Georgia Tech), Isabel Escobar (University of Kentucky).

11:40 am (Room 2, Thursday)

Oral 109 - Electrospun Pretreatment Membranes

Husain Mithaiwala* (Arizona State University), Matthew Green (Arizona State University).

12:00 pm (Room 2, Thursday)

Oral 110 - Aqueous phase separation of responsive copolymers for sustainable and mechanically stable membranes

Wouter Nielen* (University of Twente), Joshua D. Willott (University of Twente), Wiebe M. de Vos (University of Twente).

12:20 pm (Room 2, Thursday)

Oral 111 - Stepwise synthesis of oligoamide film on porous supports for preparing new types of membranes

Paramita Manna (Ben-Gurion University of the Negev), Roy Bernstein (Ben Gurion University), Ron Kasher* (Ben Gurion University of the Negev).

Oral 23 – MEMBRANE DISTILLATION AND PERVAPORATION ...pg. 54

11:00 am – 12:40 pm, ROOM 3

Session Chairs
Lee Vane
United States Environmental Protection Agency

11:00 am (Room 3, Thursday)

Oral 112 - Elucidating the Trade-off between Membrane Wetting Resistance and Water Vapor Flux in Membrane Distillation

Chenxi Li (University of British Columbia), Xuesong Li (Tongji University), Xuewei Du (Colorado State University), Ying Zhang (University of British Columbia), Tiezheng Tong (Colorado State University), Arun Kota (North Carolina State University), Jongho Lee* (University of British Columbia).

11:20 am (Room 3, Thursday)

Oral 113 - Cost optimization of gap membrane distillation

Timothy V Bartholomew* (National Energy Technology Laboratory), Alexander Dudchenko (Stanford University), Nicholas Siefert (National Energy Technology Laboratory), Meagan Mauter (Stanford University).

11:40 am (Room 3, Thursday)

Oral 114 - Modeling Pilot-Scale Air Gap Membrane Distillation as a Special Case of Permeate Gap Membrane Distillation

Mukta Hardikar* (The University of Arizona), Phakdon (University of Arizona), Itzel Marquez (Central Michigan University), Eduardo Saez (University of Arizona), Andrea Achilli (The University of Arizona).

12:00 pm (Room 3, Thursday)

Oral 115 - Economic Evaluation of Membrane Distillation as a Competitive Brackish Water Desalination Technology

Haamid S Usman* (Concordia University), Md. Saifur Rahaman (Concordia University), Khaled Touati (Concordia University).

12:20 pm (Room 3, Thursday)

Oral 116 - Conducting thermal energy to the membrane/water interface for the enhanced desalination of hypersaline brines using membrane distillation

Jingbo Wang* (University of California, Los Angeles), Yiming Liu (University of California, Los Angeles), Unnati Rao (University of California, Los Angeles), Navid Ebrahimi (University of California, Los Angeles), Eric Hoek (University of California, Los Angeles), Tzahi Cath (Colorado School of Mines), Nils Tilton (Colorado School of Mines), Craig Turchi (NREL), Yongho Sungtaek Ju (University of California, Los Angeles), David Jassby (University of California, Los Angeles).

Oral 24 – MODULE MODELING AND DESIGN ...pg. 57

11:00 am – 12:40 pm, ROOM 4

Session Chairs
David Ladner
Clemson University

Nils Tilton
Colorado School of Mines

Grigorios Panagakos
National Energy Technology Lab

11:00 am (Room 4, Thursday)

Oral 117 - Rough or wiggly? Membrane topology and morphology for fouling control

Bowen Ling* (Stanford University), David Ladner (Clemson University), Ilenia Battiato (Stanford University).

11:20 am (Room 4, Thursday)

Oral 118 - Characterization of laminar, transitional, and turbulent flow regime in membrane modules

Alexander Dudchenko* (Stanford University), Meagan Mauter (Stanford University).

11:40 am (Room 4, Thursday)

Oral 119 - 3D Numerical simulations of temperature polarization in vacuum membrane distillation systems with active membrane heating

Mark Dudley* (Colorado School of Mines), Nils Tilton (Colorado School of Mines), David Jassby (UCLA), Eric Hoek (UCLA), Craig Turchi (NREL), Tzahi Cath (Colorado School of Mines), Michael Heeley (Colorado School of Mines).

12:00 pm (Room 4, Thursday)

Oral 120 - Direct numerical simulations of unsteady mixing in direct contact membrane distillation systems with different spacer blockages and vertical offsets

Jincheng Lou (Colorado School of Mines), Jacob Johnston (Colorado School of Mines), Denis Martinand (University of Aix Marseille), Nils Tilton* (Colorado School of Mines).

12:20 pm (Room 4, Thursday)

Oral 121 - Hydrodynamic load exerted on a moving bundle of hollow fibers using constraint dissipative hydrodynamics: hydrorattle simulation

Albert S. Kim* (University of Hawaii at Manoa), Kwang Jin Lee (Kolon Industries, Inc.), Moo Seok Lee (Kolon Industries, Inc.), Hyeon-Ju Kim (Korea Research Institute of Ships and Ocean Engineering), Jung-Hyun Moon (Korea Research Institute of Ships and Ocean Engineering).

Oral 25 – MEMBRANES FOR ELECTROCHEMICAL APPLICATIONS II ...pg. 59

11:00 am – 12:40 pm, ROOM 5

Session Chairs
Geoff Geise
University of Virginia

Orlando Coronell
University of North Carolina at Chapel Hill

Chris Arges
Louisiana State University

Shudipto Dishari
University of Nebraska

11:00 am (Room 5, Thursday)

Oral 122 - Understanding water-splitting thermodynamics and kinetics in micropatterned bipolar membranes

Subarna Kole* (Louisiana State University), Christopher Arges (Louisiana State University).

11:20 am (Room 5, Thursday)

Oral 123 - Co-permeation behavior of methanol and acetate in polyether-based cation exchange membranes

Jung Min Kim* (Auburn University), Bryan Beckingham (Auburn University).

11:40 am (Room 5, Thursday)

Oral 124 - Carbon nanotube thin film deposition on hollow fiber membranes: addressing the scalability challenges of conductive membranes

Melissa J Larocque* (McMaster University), David Latulippe (McMaster University), Charles De Lannoy (McMaster University).

12:00 pm (Room 5, Thursday)

Oral 125 - Selective ion conducting membranes for non-aqueous redox flow battery applications

Patrick McCormack (University of Virginia), Hongxi Luo (University of Virginia), Gary Koenig (University of Virginia), Geoffrey M Geise* (University of Virginia).

12:20 pm (Room 5, Thursday)

Oral 126 - Graphene/Nafion Based Membrane Structure for Crossover Mitigation in Energy Storage and Conversion Systems

Saheed A Bukola* (National Renewable Energy Laboratory), Zhaodong Li (National Renewable Energy Laboratory), Christopher Antunes (National Renewable Energy Laboratory), Jason Zack (National Renewable Energy Laboratory), Glenn Teeter (National Renewable Energy Laboratory), Jeffrey Blackburn (National Renewable Energy Laboratory), Bryan Pivovar (National Renewable Energy Laboratory).

Oral 26 – ORGANIC SOLVENT SEPARATIONS ...pg. 61

1:00 pm – 3:00 pm, ROOM 1

Session Chairs

Ryan Lively
Georgia Institute of Technology

Steve White
Membrane Technology and Research

Michele Galizia
University of Oklahoma

Neel Rangnekar
ExxonMobil

1:00 pm (Room 1, Thursday)

Oral 127 - Molecular origin of flux non-linearity in Organic Solvent Nanofiltration

Kelly P. Bye (University of Oklahoma), Michele Galizia* (University of Oklahoma).

1:20 pm (Room 1, Thursday)

Oral 128 - Thin-film nanocomposite membranes with enhanced permeability and selectivity for organic solvent forward osmosis

Bofan Li* (National University of Singapore), Susilo Japip (National University of Singapore), Tai-Shung Chung (National University of Singapore).

1:40 pm (Room 1, Thursday)

Oral 129 - Studies in separation of organic solvent mixtures during reverse osmosis and nanofiltration by a perfluorodioxole copolymer membrane

John Chau (New Jersey Institute of Technology), Kamalesh Sirkar* (New Jersey Institute of Technology).

2:00 pm (Room 1, Thursday)

Oral 130 - Complex aromatic hydrocarbon mixture OSRO separation using defect-engineered Torlon® hollow fiber membranes

Hye Youn Jang (Georgia Institute of Technology), Ryan Lively (Georgia Institute of Technology).

2:20 pm (Room 1, Thursday)

Oral 131 - Organic solvent nanofiltration (OSN) membranes for separation of close-boiling mixtures and high-boilers

Petrus Cuperus* (SolSep BV), Ingrid Wienk (SolSep BV).

2:40 pm (Room 1, Thursday)

Oral 150 - Polymer Membrane-based Liquid Hydrocarbon Fractionation

Ronita Mathias* (Georgia Institute of Technology), Kirstie Thompson (Georgia Institute of Technology), Daeok Kim (Imperial College London), Jihoon Kim (Imperial College London), Neel Rangnekar (ExxonMobil), JR Johnson (ExxonMobil), Scott Hoy (ExxonMobil Research and Engineering), Irene Bechis (Imperial College London), Andrew Tarzia (Imperial College London), Kim Jelfs (Imperial College London), Benjamin McCool (ExxonMobil Research and Engineering), Andrew Livingston (Imperial College London), M.G. Finn (Georgia Institute of Technology), Ryan Lively (Georgia Institute of Technology).

Oral 27 – CONTAMINANT REMOVAL FROM WATER ...pg 64

1:00 pm – 3:00 pm, ROOM 2

Session Chairs
Prakhar Prakash
Chevron

1:00 pm (Room 2, Thursday)

Oral 132 - Adsorption of organic micropollutants to polymer surfaces probed by second harmonic scattering laser spectroscopy

Will Cole (University of California Berkeley), Hoyun Wei (University of California Berkeley), Son Nguyen (University of California Merced), Charles Harris (University of California Berkeley), Dan Miller* (LBNL), Richard Saykally (University of California Berkeley).

1:20 pm (Room 2, Thursday)

Oral 133 - Laser-induced graphene polymer composite membranes as electrically active filters for contaminant removal

Christopher J Arnusch* (Ben Gurion University of the Negev).

1:40 pm (Room 2, Thursday)

Oral 134 - Printing zwitterionic copolymer thin film composite (TFC) membranes: Enabling Tunability of Thickness and Remarkable Permeances for Nanofiltration Applications

Xin Qian* (University of Connecticut), Jeffrey McCutcheon (University of Connecticut), Ayse Asatekin (Tufts University), Samuel Lounder (Tufts University), Tulasi Ravindran (University of Connecticut).

2:00 pm (Room 2, Thursday)

Oral 135 - Enhancing antimicrobial properties of Chitosan/ Graphene Oxide membrane for waste water treatment application

Raheleh Daneshpour* (University of Arkansas), Lauren Greenlee (University of Arkansas).

2:20 pm (Room 2, Thursday)

Oral 136 - Early-stage membrane fouling detection for improving membrane cleaning efficiency

Thomas Schäfer* (Polymat, University of the Basque Country), Iliane Rafaniello (POLYMAT, University of the Basque Country).

Oral 28 – INDUSTRIAL APPLICATIONS ...pg. 66

1:00 pm – 3:00 pm, ROOM 3

Session Chairs
Dibakar Bhattacharyya
University of Kentucky

CJ Kurth
Solecta Membranes

1:00 pm (Room 3, Thursday)

Oral 137 - Produced water desalination using high temperature membranes

Cheng Chen* (Chevron), Prakhar Prakash (Chevron), Xiaofei Huang* (Hydranautics), Rich Franks (Hydranautics).

1:20 pm (Room 3, Thursday)

Oral 138 - Membrane based upgrading of Biogas: Using Realtime Controls and Selective Distribution of Membrane Types to Effectively Respond to Varying Gas Flow and Gas Quality

Steven K Pedersen* (Evonik Canada Inc.), Erik Hoving (Evonik Cooperation).

1:40 pm (Room 3, Thursday)

Oral 139 - Industrial Applications of Membranes in the Membrane Science, Engineering and Technology (MAST) Center at the University of Arkansas

Ranil Wickramasinghe* (University of Arkansas).

2:00 pm (Room 3, Thursday)

Oral 140 - High Throughput Development of Carbon Molecular Sieve Membranes and Adsorbents for Olefin/Paraffin Separations

Jay (Junqiang) Liu* (The Dow Chemical Company).

Oral 29 – AWARDS SESSION ...pg. 68

1:00 pm – 3:00 pm, ROOM 4

Session Chairs
Caleb Funk
DuPont

Bruce Hinds
University of Washington

1:00 pm (Room 4, Thursday)

Oral 141 - 3D Printed Adsorber for Capturing Chemotherapy Drugs before They are Released in the Body

Hee Jeung Oh* (Pennsylvania State University), Mariam Aboian (Yale University), Michael Yi (University of California, Berkeley), Jacqueline Maslyn (University of California, Berkeley), Whitney Loo (University of California, Berkeley), Xi Jiang (Lawrence Berkeley National Laboratory), Dilworth Parkinson (Lawrence Berkeley National Laboratory), Mark Wilson (University of California, San Francisco), Teri Moore (UCSF), Colin Yee (UCSF), Grecory Robbins (Carbon, Inc.), Florian Barth (Carbon, Inc.), Joseph DeSimone (Carbon, Inc.), Steven Hetts (University of California, San Francisco), Nitash Balsara (University of California, Berkeley).

1:20 pm (Room 4, Thursday)

Oral 142 - Mechanical properties of thin-film composite membranes and the roles they play on transport in osmotic processes

Jaime Idarraga-Mora* (Clemson University), Anthony Childress (Clemson University), Parker Friedel (Clemson University), Michael Lemelin (Clemson University), Alton O'Neal (Clemson University), Morgan Pfeiler (Clemson University), Steven Weinman (The University of Alabama), Apparao Rao (Clemson University), David Ladner (Clemson University), Scott Husson (Clemson University).

1:40 pm (Room 4, Thursday)

Oral 143 - Engineering Selective Desalination Membranes by Controlling Functional Group Configuration

Hongxi Luo* (University of Virginia), Kevin Chang (University of Virginia), Kevin Bahati (University of Virginia), Geoffrey Geise (University of Virginia).

2:00 pm (Room 4, Thursday)

Oral 144 - More than wetting resistance: How membrane surface wettability regulates water vapor permeability and mineral scaling in membrane distillation

Tiezheng Tong* (Colorado State University), Xuewei Du (Colorado State University), Wei Wang (North Carolina State University), Arun Kota (North Carolina State University), Yiming Yin (Colorado State University), Kofi Christie (Vanderbilt University), Shihong Lin (Vanderbilt University).

2:20 pm (Room 4, Thursday)

Oral 145 - Comprehensive study of dissolved methane harvesting using omniphobic membrane contactor

Abhishek Dutta* (University of British Columbia).

2:40 pm (Room 4, Thursday)

Oral 146 - Quantification of the CO2-induced plasticization of glassy polymer and CMS film membranes from the behavior of CH4 diffusivity, and how observations of multicomponent sorption and permeation rectified recurring

Giuseppe Genduso* (King Abdullah University of Science and Technology), Ingo Pinnau (KAUST).

Oral 30 – MEMBRANES FOR ELECTROCHEMICAL APPLICATIONS III ...pg. 72

1:00 pm – 3:00 pm, ROOM 5

Session Chairs

Geoff Geise
University of Virginia

Orlando Coronell
University of North Carolina at Chapel Hill

Chris Arges
Louisiana State University

Shudipto Dishari
University of Nebraska

1:00 pm (Room 5, Thursday)

Oral 147 - Tailoring Non-Aqueous Electrolyte Concentration enables Optimum Ion Transport in Pentablock Terpolymer Cation Exchange Membranes

Michelle Lehmann* (University of Tennessee), Jameson Tyler (University of Tennessee), Ethan Self (Oak Ridge National Laboratory), Jagjit Nanda (Oak Ridge National Laboratory), Tomonori Saito (Oak Ridge National Laboratory), Thomas Zawodzinski (University of Tennessee).

1:20 pm (Room 5, Thursday)

Oral 148 - Powering up Biomedical Devices by Harnessing Energy from Reverse Electrodialysis using Sodium Concentrations from the Flow of Human Blood

Efecan Pakkaner* (University of Arkansas), Jessica Orton (University of Arkansas), Jamie Hestekin (University of Arkansas), Christa Hestekin (University of Arkansas).

1:40 pm (Room 5, Thursday)

Oral 149 - Ionic conductivity of ion exchange membranes: Measurement techniques and salt concentration dependence

José Carlos (University of Michigan, Ann Arbor),
Jovan Kamcev* (University of Michigan, Ann Arbor).

Poster Sessions - Monday

Presenting authors are indicated by an asterisk*. See the most up-to-date version of the program book at **www.membranes.org/nams-2020/**

Poster 1: Materials ...pg. 74

2:00 pm – 3:00 pm

Session Chairs
Boya Xiong
The University of Alabama

Beza Getachew
Massachusetts Institute of Technology

Yuexiao Shen
Texas Tech University

Poster 1 - Influence of aliphatic and aromatic fluorine groups on the gas permeability and morphology of fluorinated polyimides

Albert Wu* (MIT), James Drayton (MIT), Katherine Mizrahi Rodriguez (MIT), Zach Smith (MIT).

Poster 2 - Interlayer Spacing and Separation Performance of Graphene Oxide Membranes in Organic Solvent

Sunxiang Zheng* (University of California, Berkeley), Baoxia Mi (University of California, Berkeley).

Poster 3 - Fluoride- and Hydroxide-Containing Facilitated Transport Membranes for CO2 Removal From Solid Oxide Fuel Cells

Kai Chen* (The Ohio State University), Witopo Salim (Membrane Technology & Research Inc.), Yang Han (The Ohio State University), Mike Gasda (Bloom Energy Corporation), Winston Ho (The Ohio State University).

Poster 4 - Advancing NF Membranes: Role of Surface Charge and Support

Francisco Leniz* (University of Kentucky), Dibakar Bhattacharyya (University of Kentucky), Isabel Escobar (University of Kentucky).

Poster 5 - The development of honeycomb-graphene oxide coated porous polymeric membranes

Feihu Wang* (Ben-Gurion University of the Negev).

Poster 7 - Activated Carbon as a Photothermal Absorber for Solar Driven Air-Gap Membrane Distillation

Maryam AlQaydi* (Khalifa University), Arwa Alshareif (Khalifa University), Ibrahim Mustafa (Khalifa University), Faisal AlMarzooqi (Khalifa University), Hassan Arafat (Khalifa University).

Poster 8 - Application of 3D Printed Minimal Surface Area Spacers on PVDF Membranes for Treating Fracking Wastewater via Membrane Distillation

Shahriar Habib* (The University of Alabama), Steven Weinman (The University of Alabama).

Poster 9 - Molecularly Engineered 6FDA-based Polyimide Membranes for Sour Natural Gas Separation

Zhongyun Liu* (Georgia Institute of Technology), Yang Liu (Georgia Institute of Technology), Wulin Qiu (Georgia Institute of Technology), William Koros (Georgia Institute of Technology).

Poster 10 - Membrane with cross-linked zwitterionic nanopores achieves sub-nanometer separations

Samuel Lounder* (Tufts University), Ayse Asatekin (Tufts University).

Poster 11 - Tuning Pore Size and Robustness of Membranes Formed by Scalable Self-Assembly of Random Copolymer Micelles

Luca Mazzaferro* (Tufts University), Ilin Sadeghi (Massachusetts Institute of Technology), Ayse Asatekin (Tufts University).

Poster 12 - Electron tomography reveals the 3D microstructure of virus filtration membranes

Kaitlyn Brickey (Pennsylvania State University), Andrew Zydney (Penn State University), Enrique Gomez (Penn State University).

Poster 13 - Tuning the binding strength of membrane adsorbers in radiochromatography

Priyanka Suresh* (Case Western reserve University), Christine Duval (Case Western Reserve University).

Poster 14 - Resilient Hollow Fiber Nanofiltration Membranes Fabricated from Copolymers

Michael Dugas* (University of Notre Dame).

Poster 17 - Effect of Metal Ions Dissociated in Cross-linked Poly(Ethylene Oxide) (XLPEO) on Physical Properties and Gas Transport Characteristicss

Taliehsadat Alebrahim* (The State University of New York at Buffalo), Alisa Chakraborty (The State University of New York at Buffalo), Haiqing Lin (The State University of New York at Buffalo).

Poster 19 - High Density Membrane Protein-Polymer Nanosheets-Based Biomimetic Membranes

Yu-Ming Tu* (The University of Texas at Austin), Woochul Song (University of Texas at Austin), Benny Freeman (The University of Texas at Austin), Manish Kumar (University of Texas at Austin).

Poster 20 - Polyvinyl Alcohol (PVA) Hydrogel Underlayer to Support CVD Graphene on Ultrafiltration Membranes as a Support for Water Desalination

Mansour Saberi* (Clemson University), Scott Husson (Clemson University), Stephen Creager (Clemson University).

Poster 21 - Ion Transport Improvement by Surface Modified Beads

Jamie Hestekin (University of Arkansas), Leticia Santos de Souza* (University of Arkansas), John Moore (University of Arkansas), Christa Hestekin (University of Arkansas).

Poster 22 - Surface Nano-structuring with Tethered Poly(acrylic acid) Chains for Tuning Ultrafiltration Membrane Performance

Yian Chen* (University of California, Los Angeles), Yoram Cohen (University of California, Los Angeles).

Poster 23 - Membrane Capacitive Deionization using Quaternized Polymer Blends

Robert McNair* (University of Manchester), Robert Dryfe (University of Manchester), Gyorgy Szekely (King Abdullah University of Science and Technology).

Poster 24 - Ion-Stabilized Organic Solvent Nanofiltration Membranes from PIM/PBI Polymer Blends

Gergo Ignacz* (King Abdullah University of Science and Technology), Gyorgy Szekely (King Abdullah University of Science and Technology).

Poster 25 - Synthesis and Gas Transport Properties of CO2-philic Anionic Poly(ionic liquid) Composite Membranes

Irshad Kammakakam* (University of Alabama), Jason Bara (University of Alabama).

Poster 26 - Synthesis and Characterization of Chlorinated RO Membrane Model Compounds – Insights into the Role of N-Cl Species in the Chlorination of Polyamide Membranes

Dean Welsh* (DuPont / FilmTec Corporation), Rachel Ehlert (DuPont), Kyoungmoo Koh (DuPont), Mou Paul (Dow), Tom Peterson (Dow), Nipon Pothayee (Dow), Mark Rickard (DuPont), Abhishek Roy (DOW), David Wilson (Dow).

Poster 27 - Spray-Coated Graphene Oxide Hollow Fibers for Nanofiltration

Gheorghe Falca* (King Abdullah University of Science and Technology)), Lakshmeesha Upadhyaya (King Abdullah University of Science and Technology), Valentina Elena Musteata (King Abdullah University of Science and Technology), Suzana Nunes (King Abdullah University of Science and Technology).

Poster 28 - Laser-Induced Graphene for Charged Membrane Applications Enabled by Sequential Infiltration Synthesis

David Bergsman* (Massachusetts Institute of Technology), Beza Getachew (MIT), Jeffrey Grossman (Massachusetts Institute of Technology).

Poster 31 - Nafion based mosaic nanofiltration membranes with tuneable charge and permeability

Michael Zelner* (Technion - IIT), Philipp Jahn (Universität Duisburg-Essen), Mathias Ulbricht (Universität Duisburg-Essen), Viatcheslav Freger (Technion - IIT).

Poster 32 - Graphene Oxide (GO) Nanocomposite Anion Exchange Membranes for Improved Electrodialysis

Levente Cseri* (University of Manchester), Peter M. Budd (University of Manchester), Gyorgy Szekely (King Abdullah University of Science and Technology).

Poster 33 - Large-Area Composite-Membrane based on Ultra-Thin Carbon Nanomembranes

Nikolaus Meyerbroeker* (CNM Technologies GmbH), Polina Angelova (CNM Technologies GmbH), Henning Vieker (CNM Technologies GmbH), Albert Schnieders* (CNM Technologies GmbH).

Poster 34 - Biomimetic carbon nanotube-containing membranes for water purification

Mauricio Dantus (Technion), Yun-Chiao Yao (Lawrence Livermore National Laboratory), Aleksandr Noy (Lawrence Livermore National Laboratory), Viatcheslav Freger (Technion - IIT).

Poster 35 - Preparation of organic solvent and thermal resistant polytriazole membranes with enhanced mechanical properties

Stefan Chisca* (KAUST), Gheorghe Falca (KAUST), Valentina Elena Musteata (King Abdullah University of Science and Technology), Suzana Nunes (King Abdullah University of Science and Technology).

Poster 36 - Hydrogen-sieving single-layer graphene membranes obtained by crystallographic and morphological optimization of catalytic copper foil

Mojtaba Rezaei* (Ecole polytechnique fédérale de Lausanne), Kumar Varoon Agrawal (Ecole polytechnique fédérale de Lausanne).

Poster 37 - Nanoporous, Ionic, Lyotropic Liquid Crystal Polymer Membranes: permanent pore modification, characterization of ion-exchange properties, and fabrication of ultrathin films

Michael McGrath* (University of Colorado at Boulder).

Poster 38 - Porous, high surface area polymers with controlled pore size <20 nm, via 1-step controlled radical polymerization

Mahati Chintapalli* (PARC).

Poster 39 - Oxone®-Mediated TEMPO-Oxidized Cellulose Nanomaterials Form I and Form II

John Moore* (University of Arkansas).

Poster 40 - 6FDA-based Ionic Polyimides-Ionic Liquid Composites for Advanced Gas Separation

Grayson Dennis* (University of Alabama).

Poster 136 - Single-layer graphene membranes by crack-free transfer for gas separation with sub-angstrom resolution

Shiqi Huang* (Ecole polytechnique fédérale de Lausanne), Jing Zhao (Nanjing Tech University), Kumar Varoon Agrawal (Ecole polytechnique fédérale de Lausanne).

Poster 137 - Centimeter-Scale Gas-Sieving Nanoporous Single-Layer Graphene Membrane

Wan-Chi Lee (Ecole polytechnique fédérale de Lausanne), Kumar Varoon Agrawal (Ecole polytechnique fédérale de Lausanne).

Poster 138 - Synthetic saponite clays as additives for reducing aging effects in PIM-1 membranes

Federico Begni (Università degli Studi del Piemonte Orientale), Geo Paul (Università degli Studi del Piemonte Orientale), Elsa Lasseuguette* (University of Edinburgh), Enzo Mangano (University of Edinburgh), Chiara Bisio (Università degli Studi del Piemonte Orientale), Maria-Chiara Ferrari (U. Edinburgh), Giorgio Gatti (Università degli Studi del Piemonte Orientale).

Poster 139 - Free volume manipulation of a 6FDA-HAB polyimide using a solid-state protection/deprotection strategy

Sharon Lin* (Massachusetts Institute of Technology), Taigyu Joo (Massachusetts Institute of Technology), Francesco Maria Benedetti (Massachusetts Institute of Technology), Laura Chen (Massachusetts Institute of Technology), Albert Wu (MIT), Katherine Mizrahi Rodriguez (MIT), Qihui Qian (Massachusetts Institute of Technology), Cara Doherty (CSIRO), Zach Smith (MIT).

Poster 140 - Facile and time-efficient carboxylic acid functionalization of PIM-1: effect on molecular packing and gas separation performance

Katherine Mizrahi Rodriguez* (MIT), Albert Wu (MIT), Qihui Qian (Massachusetts Institute of Technology), Gang Han (MIT), Sharon Lin (Massachusetts Institute of Technology), Hyunhee Lee (MIT), Won Seok Chi (Chonnam National University), Francesco Maria Benedetti (Massachusetts Institute of Technology), Cara Doherty (CSIRO), Zach Smith (MIT).

Poster 141 - Strategy for acquiring high CO2 permeance performance for dense ceramic-carbonates membranes by total conductivity modification

Oscar Ovalle* (Arizona State University), Jerry Lin (Arizona State University).

Poster 149 - Synthesis and in situ functionalization of microfiltration membranes via high internal phase emulsion templating

Ryan Zowada* (New Mexico State Univeristy), Muchu Zhou (New Mexico State University), Reza Foudazi (New Mexico State University), Anna Malakian (Clemson University).

Poster Sessions - Tuesday

Presenting authors are indicated by an asterisk*. See the most up-to-date version of the program book at **www.membranes.org/nams-2020/**

Poster 2: Processes ...pg. 92

1:00 pm – 2:00 pm

Session Chairs

Boya Xiong
Massachusetts Institute of Technology

Beza Getachew
Massachusetts Institute of Technology

Yuexiao Shen
Texas Tech University

Poster 41 - Polyelectrolyte complexation induced Aqueous Phase Separation for the next generation of sustainable membranes

Muhammad Irshad Baig* (University of Twente), Elif Nur Durmaz (University of Twente), Joshua D. Willott (University of Twente), Wiebe M. de Vos (University of Twente).

Poster 42 - Impact of Post-Synthetic Modification Routes on Filler Structure and Performance in Metal–Organic Framework Based Mixed-matrix Membranes

Qihui Qian* (Massachusetts Institute of Technology).

Poster 43 - Nanofiltration Membranes by Salinity Gradient Induced Aqueous Phase Separation

Elif Nur Durmaz* (University of Twente), Muhammad Irshad Baig (University of Twente), Joshua D. Willott (University of Twente), Wiebe M. de Vos (University of Twente).

Poster 44 - Co-permeation behavior of methanol and acetate in polyether-based cation exchange membranes

Jung Min Kim* (Auburn University), Bryan Beckingham (Auburn University).

Poster 45 - Radio frequency induction heated membranes in vacuum membrane distillation

Arezou Anvari* (Temple University), Avner Ronen (Temple University).

Poster 46 - Using Inorganic Salts in Forward Osmosis to Remove Water from High Osmotic Pressure Mineral Processing Effluents

Noel Devaere* (University of Toronto), Vladimiros Papangelakis (University of Toronto).

Poster 47 - Radio Frequency Induction Heated Membranes in Vacuum Membrane Distillation

Arezou Anvari* (Temple University), Avner Ronen (Temple University).

Poster 48 - Membrane Processes for CO2 Removal and H2 Reuse for Solid Oxide Fuel Cells: Process Design and Techno-Economic Analysis

Kai Chen* (The Ohio State University), Yang Han (The Ohio State University), Mike Gasda (Bloom Energy Corporation), Winston Ho (The Ohio State University).

Poster 49 - Conducting thermal energy to the membrane/water interface for the enhanced desalination of hypersaline brines using membrane distillation

Jingbo Wang* (University of California, Los Angeles), Yiming Liu (University of California, Los Angeles), Unnati Rao (UCLA), Navid Ebrahimi (University of California, Los Angeles), Eric Hoek (UCLA), Tzahi Cath (Colorado School of Mines), Nils Tilton (Colorado School of Mines), Craig Turchi (NREL), Yongho Sungtaek Ju (University of California, Los Angeles), David Jassby (UCLA).

Poster 50 - Membrane Surface Characterization Upon Cleaning of a Heavily Fouled Reverse Osmosis Membrane from an Advanced Reclamation Facility

Bilal Abada* (Texas A&M University), Shankar Chellam (Texas A&M University).

Poster 51 - Biomimetic Analysis of crown ethers as ion channels for rapid and selective Li+ transport

Chenhao Yao* (University of Texas at Austin), Manish Kumar (University of Texas at Austin), Nathaniel Lynd (The University of Texas at Austin).

Poster 52 - Multiphysics Simulation of CO2 Removal from Seawater using Hollow Fiber Membrane Contactors

Joanna Rivero (University of Pittsburgh).

Poster 53 - Reactive membranes to prevent fouling by generating in situ microbubbles

Weiliang Bai* (University of Texas at Austin), Manish Kumar (University of Texas at Austin), Navid Saleh (Univeristy of Texas at Austin).

Poster 54 - The Influence of Ion Association on Ion Solubility in Ion Exchange Membranes

Rahul Sujanani* (The University of Texas at Austin), Oscar Nordness (The University of Texas at Austin), Joshua Saunders (The University of Texas at Austin), Joan Brennecke (The University of Texas at Austin), Lynn Katz (The University of Texas at Austin), Benny Freeman (The University of Texas at Austin).

Poster 55 - Enhanced water evaporation through graphene nanopores via the Kelvin effect

Yuhang Fang* (Purdue University), Prabudhya Roy Chowdhury (Purdue University), Xiulin Ruan (Purdue University), David Warsinger (Purdue University).

Poster 56 - Batch counterflow reverse osmosis for energy efficient desalination at high recoveries

Abhimanyu Das* (Purdue University), David Warsinger (Purdue University).

Poster 57 - The Effect of Microplastic Shape, Size, and Concentration on Membrane Performance in Municipal Wastewater Treatment

Ryan LaRue* (McMaster University), Blake Patterson (McMaster University), Todd Hoare (McMaster University), David Latulippe (McMaster University).

Poster 58 - Does modifying commercial reverse osmosis membranes with solvent pre-treatment and additional polymerization enhance water permeability and salt rejection?

Mikayla Armstrong* (University of North Carolina at Chapel Hill), Orlando Coronell (University of North Carolina at Chapel Hill).

Poster 59 - Thin-film composite gas separation membranes prepared by interfacial polymerization of macrocyclic molecules

Woochul Song* (The University of Texas at Austin), Jaesung Park (The University of Texas at Austin), Benny Freeman (The University of Texas at Austin), Manish Kumar (The University of Texas at Austin).

Poster 60 - Development of Novel Processes for the Synthesis of Membranes with Special Wettability

Siamak Nejati* (University of Nebraska-Lincoln).

Poster 61 - Dissipative Particle Dynamics Simulation of PVDF Membrane Formation via Non-Solvent Induced Phase Separation Process

Suphanat Aphinyan (University of Arkansas), Eric Ledieu (University of Arkansas), David Ford (University of Arkansas), Xianghong Qian (University of Arkansas).

Poster 62 - Influences of Microwave Irradiation on Performances of Membrane Filtration and Catalytic Degradation of Perfluorooctanoic Acid (PFOA)

Fangzhou Liu* (NJIT).

Poster 63 - Performance of Combining Ozonation, Ceramic Membrane Filtration with Biological Activated Carbon Filtration for the Advanced Treatment of Hypersaline Petrochemical Wastewater

Jinting Hu* (Tsinghua Shenzhen International Graduate School).

Poster 64 - Cinchona-decorated cyclodextrin organocatalyst for asymmetric synthesis in a continuous-flow membrane reactor

Peter Kisszekelyi* (Budapest University of Technology and Economics) Abdulaziz Alammar (University of Manchester), Jozsef Kupai (Budapest University of Technology and Economics), Peter Huszthy (Budapest University of Technology and Economics) Barabas Julia (Budapest University of Technology and Economics), Tibor Holtzl (Budapest University of Technology and Economics), Lajos Szente (Cyclolab, Ciklodextrin Kutató-Fejlesztő Kft), Carlo Bawn (University of Manchester), Ralph Adams (University of Manchester), Gyorgy Szekely (King Abdullah University of Science and Technology).

Poster 65 - Membrane Separation of Nitrogen and Phosphorus Nutrients with Downstream Recovery as Struvite Fertilizer

Zahra Anari* (University of Arkansas), Lauren Greenlee (University of Arkansas).

Poster 66 - Effects of resin bead chemistries on the selective removal of industrially relevant metal ions using wafer-enhanced electrodeionization

Humeyra Ulusoy Erol* (University of Arkansas), Jamie Hestekin (University of Arkansas), Christa Hestekin (University of Arkansas).

Poster 67 - Ceramic membrane centered hybrid processes for water treatment

WANYI FU* (Tsinghua University), Xihui Zhang (Tsinghua University).

Poster 69 - Membrane charge weakly affects ion transport in reverse osmosis

Mikhail Stolov* (Technion - Israel Institute of Technology).

Poster 70 - Experimental measurement of single- and mixed-gas permeation through photo-responsive gas membranes

Bradley Ladewig* (Karlsruhe Institute of Technology), Jinju Zhang (Karlsruhe Institute of Technology), Nicole Jung (Karlsruhe Institute of Technology).

Poster 71 - Radical filtration for Water Purification

Shuyana Heredia* (University of Twente).

Poster 72 - membrane-toolkit: a Python package for fast, accurate, automated experimental data management

Ryan Kingsbury* (University of North Carolina at Chapel Hill).

Poster 73 - In-situ measurement of the mechanical properties of a foulant layer at a membrane surface

Jose Agustin Epstein* (Technion - Israel Institute of Technology), Guy Ramon (Technion - Israel Institute of Technology).

Poster 74 - On the reliability and comparability of organic solvent nanofiltration reports

Hai Anh Le Phuong* (The University of Manchester), Christopher F. Blanford (The University of Manchester), Gyorgy Szekely (King Abdullah University of Science and Technology).

Poster 76 - Experimental Design Study of Reinforced Hollow Fiber Membrane Casting

Walter Kosar* (Arkema Inc.), Gregory O'Brien (Arkema Inc.), Michele Crane (Arkema Inc.).

Poster 77 - Determination of Carbon Nanomembrane Permeability Coefficients via Radioactive Tracer Experiments

Raphael Dalpke (Bielefeld University), Anna Dreyer (Bielefeld University), Riko Korzetz* (Bielefeld University), André Beyer (Bielefeld University), Karl-Josef Dietz (Bielefeld University), Armin Gölzhäuser (Bielefeld University).

Poster 79 - Biocatalytic membrane reactor using immobilized phosphotriesterase for the degradation of pesticides

Giuseppe Vitola (National Research Council of Italy, Insitute on membrane Technology, CNR-ITM), Rosalinda Mazzei (National Research Council of Italy, Insitute on membrane Technology, CNR-ITM), Lidietta GIORNO* (National Research Council of Italy - Institute on Membrane Technology (CNR-ITM)).

Poster 80 - Membrane Filtration of Poly(dT60) Single-Stranded DNA

Hossein Nouri Alavijeh* (Clarkson University), Ruth E. Baltus (Clarkson University).

Poster 81 - Scaling Resistance in Nanophotonics-Enabled Solar Membrane Distillation

Francois Perreault* (Arizona State University).

Poster 82 - Generalized model for the prediction of the permeability of mixed-matrix using impermeable fillers of diverse geometry

Haoyu Wu* (University of Ottawa), Boguslaw Kruczek (University of Ottawa), Jules Thibault (U. Ottawa).

Poster 83 - Factors controlling the structure formation in isoporous hollow fiber membranes

Kirti Sankhala* (Helmholtz-Zentrum Geesthacht, Geesthacht), D. C. Florian Wieland (Helmholtz-Zentrum Geesthacht, Geesthachtge), Joachim Koll (Helmholtz-Zentrum Geesthacht, Geesthacht), Maryam Radjabian (Helmholtz-Zentrum Geesthacht, Geesthacht), Clarissa Abetz (Helmholtz-Zentrum Geesthacht), Volker Abetz (Helmholtz-Zentrum Geesthacht).

Poster 84 - Solubility of Rare Earth Sulfates and Chlorined toward Electrochemical Membrane Recovery: Effect of pH

Mohammad U. Shafiq (Texas A&M University-Kingsville), Lucy Camacho* (Texas A&M University-Kingsville).

Poster 142 - Improved batch reverse osmosis configuration for better energy efficiency

Sandra Cordoba* (Purdue University), Abhimanyu Das (Purdue University), David Warsinger (Purdue University).

Poster 143 - Microalgae Filtration using Electrochemically Reactive Ceramic Membrane: Filtration Performances, Fouling Kinetics and Foulant Layer Characteristics

Qingquan Ma* (NJIT).

Poster 144 - Green solvent PVC ultrafiltration membrane

Baicang Liu* (Sichuan University), Wancen Xie (Sichuan University).

Poster 145 - Porous single-layer graphene membranes for carbon capture

Kuang-Jung Hsu* (Ecole polytechnique fédérale de Lausanne), Shiqi Huang (Ecole polytechnique fédérale de Lausanne), Guangwei He (Ecole polytechnique fédérale de Lausanne), Kumar Varoon Agrawal (Ecole polytechnique fédérale de Lausanne).

Poster 146 - Bimodal Reverse Osmosis and Pressure Retarded Osmosis Framework

Akshay Rao* (Purdue University), Stephen Coan (Purdue University), Luke Wrede (Purdue University), Owen Li (Purdue University), George Elias (Purdue University), sandra cordoba (Purdue University), Michael Roggenburg (Purdue University), David Warsinger (Purdue University), Luciano Castillo (Purdue University).

Poster 147 - Effect of packing nonuniformity at the fiber bundle-case interface on flow distribution and separation performance for hollow fiber membrane modules

Lili Sun* (University of Toledo), Atabong Etiendem (University of Toledo), Glenn Lipscomb (University of Toledo, Toledo).

Poster 148 - Techniques for Measuring Gas and Vapor Solubility and Transport in Membranes

Mark Roper* (Hiden Isochema Ltd), Darren Broom (Hiden Isochema Ltd), Michael Benham (Hiden Isochema Ltd).

Poster 3: Applications ...pg. 111

2:00 pm – 3:00 pm

Session Chairs

Boya Xiong
Massachusetts Institute of Technology

Beza Getachew
Massachusetts Institute of Technology

Yuexiao Shen
Texas Tech University

Poster 85 - Super water permeable PVDF UF membrane for fine separation

Shun Shimura* (Toray Industries, Inc.), Hiroki Eimura (Toray Industries, Inc.), Masayuki Hanakawa (Toray Industries, Inc.), Masahiro Kimura (Toray Industries, Inc.).

Poster 86 - Highly Permeable Polyethersulfone Substrates with Bicontinuous Structure for CO2-Selective Composite Membranes

Ruizhi Pang* (The Ohio State University), Kai Chen (The Ohio State University), Yang Han (The Ohio State University), Winston Ho (The Ohio State University).

Poster 87 - Nutrient recovery from synthetic livestock wastewater effluent using electroactive membranes

Kartikeya Kekre* (Temple University), Arezou Anvari (Temple University), Avner Ronen (Temple University).

Poster 88 - High Aspect Ratio Nanoparticles with Controlled Interfacial Transport for Improved Reverse Osmosis Performance

Ethan Smith* (Virginia Tech), Stephen Martin (Virginia Tech).

Poster 89 - GO-based Membrane as Barrier against Toxic Vapors/Gases vs. Membrane-Supported MOFs

Cheng Peng (New Jersey Institute of Technology), Zafar Iqbal (New Jersey Institute of Technology), Kamalesh Sirkar* (New Jersey Institute of Technology), Gregory Peterson (U.S. Army Edgewood Chemical Biological Center).

Poster 90 - Sterically Hindered Poly(N-methyl-N-vinylamine) Membranes for CO2 Capture from Flue Gas

Ting-Yu Chen* (The Ohio State University), Xuepeng Deng (The Ohio State University), Li-Chiang Lin (The Ohio State University), Winston Ho (The Ohio State University).

Poster 91 - The impact of pre-chlorination on microfiltration fouling during municipal wastewater effluent filtration for water reuse

Kunal Gupta* (Texas A&M University), Shankar Chellam (Texas A&M University).

Poster 92 - Rapid Concentration and Isotopic Screening of Plutonium from Aqueous Systems Using Functionalized Membranes

James Foster* (Clemson University), Scott Husson (Clemson University), Timothy DeVol (Clemson University), Brian Powell (Clemson University).

Poster 93 - Amine-Containing CO2-Selective Membrane and Process for Carbon Capture from Flue Gas

Yang Han* (The Ohio State University), Kai Chen (The Ohio State University), witopo salim (The Ohio State University), dongzhu wu (The Ohio State University), Winston Ho (The Ohio State University).

Poster 94 - Facilitated Transport Membranes with Tunable Amine–CO2 Chemistry for Hydrogen Purification

Yang Han* (The Ohio State University), Xuepeng Deng (The Ohio State University), Li-Chiang Lin (The Ohio State University), Winston Ho (The Ohio State University).

Poster 95 - High-Performance Gas Separation Membranes Based on Poly(benzimidazole)

Alexander Bridge* (The University of Texas at Austin), Joshua Moon (California Nanosystems Institute, The

University of California Santa Barbara), Joan Brennecke (The University of Texas at Austin), Benny Freeman (The University of Texas at Austin).

Poster 96 - Development and Testing of Pervaporation Desalination Membranes

Elisabeth Thomas* (Arizona State University), Mary Laura Lind (Arizona State University).

Poster 97 - Synthesis of charged PEG macromonomers and their application towards antifouling thin-film composite membranes

Swati Sundararajan* (Ben-Gurion University of the Negev), Ron Kasher (Ben Gurion University of the Negev).

Poster 99 - Membrane Adsorbers for Medical Isotope Production

Maura Sepesy* (Case Western Reserve University), Benjamin Fugate (Case Western Reserve University), Christine Duval (Case Western Reserve University).

Poster 100 - Controlling Membrane Pore Structure of Polyelectrolyte Multilayer Nanofiltration Membranes for Selective Ion Removal

Ryan DuChanois* (Yale University).

Poster 101 - Uranium concentration using reactive polymer thin films and thin-film composite membranes for spectroscopic analyses

Abenazer Darge* (Clemson University), Yugantar Gera (Clemson University), Timothy DeVol (Clemson University), Scott Husson (Clemson University).

Poster 102 - Macrocyclic Nanofiltration Membranes for the Removal of Per- and Polyfluoroalkyl Substances in Drinking Water

Elham Abaie (Texas Tech University), Limeimei XU (Texas Tech University), Yuexiao Shen (Texas Tech University).

Poster 103 - Ultrafiltration for the purification of nucleic acid-based therapeutics

Ivan Manzano* (Penn State University), Andrew Zydney (Penn State University).

Poster 105 - UF membranes modified by covalently grafted graphene oxide with improved antifouling properties

Xiaoyi Chen* (State University of New York at Buffalo), Erda Deng (State University of New York at Buffalo), Dongwon Park (State University of New York at Buffalo), Blaine Pfeifer (State University of New York at Buffalo), Haiqing Lin (The State University of New York at Buffalo).

Poster 106 - Defining new performance metrics for the application of NF membranes in 'tough-to-treat' industrial wastewater applications

Abhishek Premachandra* (McMaster University), Nicole Perna (McMaster University), Susan O'Brien (McMaster University), Jacob McGivern (McMaster University), David Latulippe (McMaster University).

Poster 107 - Printing polyepoxyether thin film composite (TFC) membranes: Achieving membrane chemical robustness with tunable permeance for Nanofiltration Applications

Xin Qian* (University of Connecticut), Jeffrey McCutcheon (University of Connecticut), Rhea Verbeke (KU Leuven), Ivo Vankelecom (KU Leuven).

Poster 108 - A New Concept for Generating Mechanical Work from Gas Permeation

Sarah Moussaddy (Oakland University), Jonathan Maisonneuve* (Oakland University).

Poster 109 - Influences of Membrane Morphology and Biocompatibility on Hemodialysis Treatment Outcomes

Heloisa Westphalen (University of Saskatchewan), Amira abdelrasoul* (University of Saskatchewan).

Poster 110 - Highly permeable polymeric membrane fabrication through interfacial

polymerization of pillar[5]arene artificial water channels

Matthew Skiles (University of Texas at Austin), Woochul Song (University of Texas at Austin), Diana Cintron (University of Texas at Austin), Manish Kumar (University of Texas at Austin).

Poster 111 - Sorption of ternary gas mixtures of CO2, CH4 and C2H6 in PIM-1, and role of solubility in multicomponent gas separation

Francesco Maria Benedetti* (University of Bologna), Eleonora Ricci (University of Bologna), Antonella Noto (University of Bologna), Maria Grazia De Angelis (University of Bologna), Tim Merkel (Membrane Technology & Research Inc.), Jianyong Jin (The University of Auckland).

Poster 112 - Tailoring structural and functional features of high-performance ionenes and ionic composites designed for membrane-based gas separations

Kathryn O'Harra* (University of Alabama), Irshad Kammakakam (University of Alabama), Jason Bara (University of Alabama).

Poster 113 - High-temperature H2 purification from atom-thick g-C3N4 nanosheets hosting ordered nanopores

Luis Francisco Villalobos* (Ecole polytechnique fédérale de Lausanne), Kumar Varoon Agrawal (Ecole polytechnique fédérale de Lausanne), Mostapha Dakhchoune (Ecole polytechnique fédérale de Lausanne).

Poster 114 - Interpenetrating Polymer Networks for Organic Solvent Nanofiltration Membranes

Gergo Ignacz* (King Abdullah University of Science and Technology), Gyorgy Szekely (King Abdullah University of Science and Technology).

Poster 115 - Enhancing antimicrobial properties of Chitosan/ Graphene Oxide membrane for waste water treatment application

Raheleh Daneshpour* (University of Arkansas), Lauren Greenlee (University of Arkansas).

Poster 116 - Hindering Lattice Flexibility Of Metal Organic Frameworks For CO2-sieving

Deepu Babu* (Ecole polytechnique fédérale de Lausanne), Kumar Varoon Agrawal (Ecole polytechnique fédérale de Lausanne).

Poster 117 - Graphene-based nanocomposite membranes with a mussel inspired polydopamine coating for produced water treatment

Gyorgy Szekely (King Abdullah University of Science and Technology), Abdulaziz Alammar* (University of Manchester).

Poster 118 - Enhanced Degradation of 1,4-dioxane by Photo-Fenton Reactive Ceramic Membrane

SHAN XUE* (NJIT).

Poster 119 - Zwitterionic-Containing Ultra-thin Hydrogel Selective Layer for Fouling-Resistant Ultrafiltration Membranes

Alice Oliveira Aguiar* (Tufts University), Hyunmin Yi (Tufts University), Ayse Asatekin (Tufts University).

Poster 120 - Amphiphilic tercopolymer membranes for water vapor removal in air dehumidification

Faheem Akhtar* (KAUST), Kim Choon Ng (King Abdullah University of Science and Technology (KAUST)), Klaus-Viktor Peinemann (KAUST).

Poster 121 - Synthesis of sodalite precursor nanosheets and facile assembly for hydrogen purification

Mostapha Dakhchoune* (Ecole polytechnique fédérale de Lausanne), Kumar Varoon Agrawal (Ecole polytechnique fédérale de Lausanne), Luis Francisco Villalobos (Ecole polytechnique fédérale de Lausanne).

Poster 122 - Novel Electrospun Nanoparticle Nanofiber Composites for CO2 separation

Amogh Meshram* (Arizona State University), Bin Mu (Arizona State University).

Poster 123 - Novel Nanofiltration Membranes for Isolation of Pharmaceutical Compounds

Gazelle Vaseghi* (Compact Membrane Systems), Kenneth Pennisi (Compact Membrane Systems), Hannah Murnen (Compact Membrane Systems), Sudip Majumdar (Compact Membrane Systems).

Poster 124 - Nexar Coated Hollow fibers for membrane dehumidification

Lakshmeesha Upadhyaya* (KAUST).

Poster 126 - A novel RGO-CCNT-based catalytic membrane with co-enhanced permeability and catalytic activity for high-efficiency degradation of charged antibiotics

Wenli Jiang* (University of California Berkeley), Sunxiang Zheng (University of California, Berkeley), Monong Wang (UC Berkeley), Baoxia Mi (UC Berkeley).

Poster 127 - Energy and performance optimization for electrodialysis water treatment

Zahra Abbasian Chaleshtari (New Mexico State University), Abdulhameed Alalwani (New Mexico State University), Reza Foudazi* (New Mexico State University).

Poster 128 - Enhancement of Nanofiltration Membranes through Chemical Bonding of Lignosulfonic Acid

Phillip Sandman* (University of Kentucky).

Poster 129 - TiO2 Nanowires Based System for Urea Photodecomposition and Dialysate Regeneration

Guozheng Shao* (Univ. of Wash.), Yushi Zang (University of Washington), Bruce Hinds (Univ. of Washington).

Poster 130 - Improving Membrane Performance for CO2 Capture With Ultrahigh MW Polyvinylamine

Kai Chen* (The Ohio State University), Yang Han (The Ohio State University), Winston Ho (The Ohio State University).

Poster 131 - Haloorganic Sorption and Degradation by Temperature Responsive Membranes

Rollie Mills* (University of Kentucky).

Poster 133 - The impact of monochloramines and dichloramines on reverse osmosis membranes in wastewater potable reuse process trains: A pilot-scale study

HYEJIN LEE (McMaster University), Amin Halali (McMaster University), Siva Sarathy (Trojan Technologies), Charles-François de Lannoy* (McMaster).

Poster 134 - Synchrotron-based X-ray Spectroscopy to Investigate the Fouling Mechanism on Reverse Osmosis Membranes

Valerie Niemann* (Stanford University), Hans-Georg Steinrück (Paderborn University), Michael Toney (SLAC National Accelerator Laboratory), William Tarpeh (Stanford University), Sharon Bone (SLAC National Accelerator Laboratory).

Poster 135 - Detection of biofouling on gold-coated MF membranes by in-situ electrical impedance spectroscopy

Nan Zhang* (McMaster University), HYEJIN LEE (McMaster University), Charles-François de Lannoy (McMaster University).

Poster 4: Undergraduate Competition ...pg. 133

2:00 pm – 3:00 pm

Session Chairs
Boya Xiong
Massachusetts Institute of Technology

Beza Getachew
Massachusetts Institute of Technology

Yuexiao Shen
Texas Tech University

UG 1 - Enhancing the Energy Efficiency of Desalination via Batch Reverse Osmosis

Katie Brodersen* (Purdue University).

UG 2 - Continuous covalent organic framework membranes for dye/salt separation

Miguel Jaimes* (UW-Milwaukee).

UG 3 - Functionalization of Polyvinylidene Fluoride Membranes for Ion Separation Applications

Ronald Vogler * (University of Kentucky), Saiful Islam (University of Kentucky), Dibakar Bhattacharyya (U. Kentucky).

UG 4 - Ammonia and Water Flux in Membrane Distillation Using Spacers

Nobuyo Watanabe* (Barnard College of Columbia University).

UG 5 - A Liquid-liquid Hollow-Fiber Membrane contactor device to remove oceanic CO2

Christopher Snodgrass* (University of Pittsburgh).

UG 6 - Predicting Octanol/Water Partition Coefficients from Molecular Structure

Vy (Jasmine) Tran* (Miami University).

UG 7 - Evaluating Electrocoagulation Process for Treating Produced Water

Alix-Cynthia-Ineza Karangwa* (University of Arkansas).

UG 8 - Influence of molecular interactions, membrane swelling and plasticization on pure and mixed fluid transport in OSN membranes

Kelly Bye* (University of Oklahoma).

UG 9 - Reduced graphene oxide membranes: properties and potential water applications

Trisha Nickerson* (University of Kentucky), Ashish Aher (U. Kentucky), Dibakar Bhattacharyya (U. Kentucky).

UG 11 - Perylene-Polyimide membranes

Aryan Louise Tan* (UCLA).

UG 12 - Effect of mechanical strain on the transport properties of thin-film composite membranes used in osmotic processes

Alton O'Neal* (Clemson University).

UG 13 - Advanced Water Treatment for Direct Potable Reuse

Ciara Lugo* (University of Arizona).

REVISED SECTION FOR IMMEDIATE REFERENCE ON TUESDAY, MAY 19th, 2020

Revised Schedule of Poster Sessions

Processes Poster Session 1PM EDT – 3 PM EDT
Location: Virtual Hallway 1- Mary Laura

Applications Poster Session 1PM EDT – 3 PM EDT
Location: Virtual Hallway 2- Isabel

Materials Poster Session (Take 2) 1PM EDT – 2 PM EDT
Location: Virtual Hallway 3- Jamie

UG Poster Session 2 PM EDT – 3 PM EDT
Location: Virtual Hallway 3- Jamie

Processes Poster Session (Virtual Hallway 1), Host: Mary Laura Lind, Time: 1:00 pm – 3:00 pm EDT

#	First name	Last name	iPoster Title
41	Muhammad Irshad	Baig	Polyelectrolyte complexation induced Aqueous Phase Separation for the next generation of sustainable membranes
42	Qihui	Qian	Impact of Post-Synthetic Modification Routes on Filler Structure and Performance in Metal–Organic Framework Based Mixed-matrix Membranes
43	Elif Nur	Durmaz	Nanofiltration Membranes by Salinity Gradient Induced Aqueous Phase Separation
44	Jung Min	Kim	Co-permeation behavior of methanol and acetate in polyether-based cation exchange membranes
45	Arezou	Anvari	Radio frequency induction heated membranes in vacuum membrane distillation
46	Noel	Devaere	Using Inorganic Salts in Forward Osmosis to Remove Water from High Osmotic Pressure Mineral Processing Effluents
47	Arezou	Anvari	Radio Frequency Induction Heated Membranes in Vacuum Membrane Distillation
48	Kai	Chen	Membrane Processes for CO2 Removal and H2 Reuse for Solid Oxide Fuel Cells: Process Design and Techno-Economic Analysis
49	Jingbo	Wang	Conducting thermal energy to the membrane/water interface for the enhanced desalination of hypersaline brines using membrane distillation
50	Bilal	Abada	Membrane Surface Characterization Upon Cleaning of a Heavily Fouled Reverse Osmosis Membrane from an Advanced Reclamation Facility
51	Chenhao	Yao	Biomimetic Analysis of crown ethers as ion channels for rapid and selective Li+ transport
52	Joanna	Rivero	Multiphysics Simulation of CO2 Removal from Seawater using Hollow Fiber Membrane Contactors
53	Weiliang	Bai	Reactive membranes to prevent fouling by generating in situ microbubbles
54	Rahul	Sujanani	The Influence of Ion Association on Ion Solubility in Ion Exchange Membranes
55	Yuhang	Fang	Enhanced water evaporation through graphene nanopores via the Kelvin effect
56	Abhimanyu	Das	Batch counterflow reverse osmosis for energy efficient desalination at high recoveries
57	Ryan	LaRue	The Effect of Microplastic Shape, Size, and Concentration on Membrane Performance in Municipal Wastewater Treatment
58	Mikayla D	Armstrong	Does modifying commercial reverse osmosis membranes with solvent pre-treatment and additional polymerization enhance water permeability and salt rejection?
59	Woochul	Song	Thin-film composite gas separation membranes prepared by interfacial polymerization of macrocyclic molecules
60	Siamak	Nejati	Development of Novel Processes for the Synthesis of Membranes with Special Wettability
61	Eric	Ledieu	Dissipative Particle Dynamics Simulation of PVDF Membrane Formation via Non-Solvent Induced Phase Separation Process
62	Fangzhou	Liu	Influences of Microwave Irradiation on Performances of Membrane Filtration and Catalytic Degradation of Perfluorooctanoic Acid (PFOA)
63	Jinting	Hu	Performance of Combining Ozonation, Ceramic Membrane Filtration with Biological Activated Carbon Filtration for the Advanced Treatment of Hypersaline Petrochemical Wastewater
64	Peter	Kisszekelyi	Cinchona-decorated cyclodextrin organocatalyst for asymmetric synthesis in a continuous-flow membrane reactor
65	Zahra	Anari	Membrane Separation of Nitrogen and Phosphorus Nutrients with Downstream Recovery as Struvite Fertilizer
66	Humeyra B	Ulusoy Erol	Effects of resin bead chemistries on the selective removal of industrially relevant metal ions using wafer-enhanced electrodeionization.
67	WANYI	FU	Ceramic membrane centered hybrid processes for water treatment
69	Mikhail	Stolov	Membrane charge weakly affects ion transport in reverse osmosis
70	Bradley P.	Ladewig	Experimental measurement of single- and mixed-gas permeation through photo-responsive gas membranes
71	Shuyana A.	Heredia	Radical filtration for Water Purification
72	Ryan S	Kingsbury	membrane-toolkit: a Python package for fast, accurate, automated experimental data management
73	Jose Agustin	Epstein	In-situ measurement of the mechanical properties of a foulant layer at a membrane surface
74	Hai Anh	Le Phuong	On the reliability and comparability of organic solvent nanofiltration reports
76	Walter	Kosar	Experimental Design Study of Reinforced Hollow Fiber Membrane Casting
77	Riko	Korzetz	Determination of Carbon Nanomembrane Permeability Coefficients via Radioactive Tracer Experiments
79	Lidietta	GIORNO	Biocatalytic membrane reactor using immobilized phosphotriesterase for the degradation of pesticides
80	Hossein	Nouri Alavijeh	Membrane Filtration of Poly(dT60) Single-Stranded DNA
81	Francois	Perreault	Scaling Resistance in Nanophotonics-Enabled Solar Membrane Distillation
82	Haoyu	Wu	Generalized model for the prediction of the permeability of mixed-matrix using impermeable fillers of diverse geometry
83	Kirti	Sankhala	Factors controlling the structure formation in isoporous hollow fiber membranes
84	Lucy M	Camacho	Solubility of Rare Earth Sulfates and Chlorined toward Electrochemical Membrane Recovery: Effect of pH
142	Sandra Patricia	Cordoba	Improved batch reverse osmosis configuration for better energy efficiency
143	Qingquan	Ma	Microalgae Filtration using Electrochemically Reactive Ceramic Membrane: Filtration Performances, Fouling Kinetics and Foulant Layer Characteristics
144	Baicang	Liu	Green solvent PVC ultrafiltration membrane
145	Kuang-Jung	Hsu	Porous single-layer graphene membranes for carbon capture
146	Akshay K.	Rao	Bimodal Reverse Osmosis and Pressure Retarded Osmosis Framework
147	Lili	Sun	Effect of packing nonuniformity at the fiber bundle-case interface on flow distribution and separation performance for hollow fiber membrane modules
148	Mark	Roper	Techniques for Measuring Gas and Vapor Solubility and Transport in Membranes

Applications Poster Session (Virtual Hallway 2), Host: Isabel Escobar, Time: 1:00 pm – 3:00 pm EDT

#	First name	Last name	iPoster Title
85	Shun	Shimura	Super water permeable PVDF UF membrane for fine separateion
86	Ruizhi	Pang	Highly Permeable Polyethersulfone Substrates with Bicontinuous Structure for CO2-Selective Composite Membranes
87	Kartikeya	Kekre	Nutrient recovery from synthetic livestock wastewater effluent using electroactive membranes
88	Stephen M	Martin	High Aspect Ratio Nanoparticles with Controlled Interfacial Transport for Improved Reverse Osmosis Performance
89	Cheng	Peng	GO-based Membrane as Barrier against Toxic Vapors/Gases vs. Membrane-Supported MOFs
90	Ting-Yu	Chen	Sterically Hindered Poly(N-methyl-N-vinylamine) Membranes for CO2 Capture from Flue Gas
91	Kunal	Gupta	The impact of pre-chlorination on microfiltration fouling during municipal wastewater effluent filtration for water reuse
92	James C	Foster	Rapid Concentration and Isotopic Screening of Plutonium from Aqueous Systems Using Functionalized Membranes
93	Yang	Han	Amine-Containing CO2-Selective Membrane and Process for Carbon Capture from Flue Gas
94	Yang	Han	Facilitated Transport Membranes with Tunable Amine–CO2 Chemistry for Hydrogen Purification
95	Alexander	Bridge	High-Performance Gas Separation Membranes Based on Poly(benzimidazole)
96	Elisabeth	Thomas	Development and Testing of Pervaporation Desalination Membranes
97	Swati	Sundararajan	Synthesis of charged PEG methacrylate macromonomers and their application towards antifouling thin-film composite membranes
99	Maura	Sepesy	Membrane Adsorbers for Medical Isotope Production
100	Ryan	DuChanois	Controlling Membrane Pore Structure of Polyelectrolyte Multilayer Nanofiltration Membranes for Selective Ion Removal
101	Abenazer W	Darge	Uranium concentration using reactive polymer thin films and thin-film composite membranes for spectroscopic analyses
102	Elham	Abaie	Macrocyclic Nanofiltration Membranes for the Removal of Per- and Polyfluoroalkyl Substances in Drinking Water
103	Ivan	Manzano	Ultrafiltration for the purification of nucleic acid-based therapeutics
105	Xiaoyi	Chen	UF membranes modified by covalently grafted graphene oxide with improved antifouling properties
106	Abhishek	Premachandra	Defining new performance metrics for the application of NF membranes in 'tough-to-treat' industrial wastewater applications
107	Xin	Qian	Printing polyepoxyether thin film composite (TFC) membranes: Achieving membrane chemical robustness with tunable permeance for Nanofiltration Applications
108	Jonathan	Maisonneuve	A New Concept for Generating Mechanical Work from Gas Permeation
109	Heloisa	Westphalen	Influences of Membrane Morphology and Biocompatibility on Hemodialysis Treatment Outcomes
110	Matthew	Skiles	Highly permeable polymeric membrane fabrication through interfacial polymerization of pillar[5]arene artificial water channels
111	Francesco Maria	Benedetti	Sorption of ternary gas mixtures of CO2, CH4 and C2H6 in PIM-1, and role of solubility in multicomponent gas separation
112	Kathryn E	O'Harra	Tailoring structural and functional features of high-performance ionenes and ionic composites designed for membrane-based gas separations
113	Luis Francisco	Villalobos	High-temperature H2 purification from atom-thick g-C3N4 nanosheets hosting ordered nanopores.
114	Gergo	Ignacz	Interpenetrating Polymer Networks for Organic Solvent Nanofiltration Membranes
115	Raheleh	Daneshpour	Enhancing antimicrobial properties of Chitosan/ Graphene Oxide membrane for waste water treatment application
116	Deepu	Babu	Hindering Lattice Flexibility Of Metal Organic Frameworks For CO2-sieving
117	Gyorgy	Szekely	Graphene-based nanocomposite membranes with a mussel inspired polydopamine coating for produced water treatment
118	SHAN	XUE	Enhanced Degradation of 1,4-dioxane by Photo-Fenton Reactive Ceramic Membrane
119	Alice	Oliveira Aguiar	Zwitterionic-Containing Ultra-thin Hydrogel Selective Layer for Fouling-Resistant Ultrafiltration Membranes
120	Faheem Hassan	Akhtar	Amphiphilic tercopolymer membranes for water vapor removal in air dehumidification
121	Mostapha	Dakhchoune	Synthesis of sodalite precursor nanosheets and facile assembly for hydrogen purification
122	Amogh	Meshram	Novel Electrospun Nanoparticle Nanofiber Composites for CO2 separation
123	Gazelle	Vaseghi	Novel Nanofiltration Membranes for Isolation of Pharmaceutical Compounds
124	Lakshmeesha	Upadhyaya	Nexar Coated Hollow fibers for membrane dehumidification
126	Wenli	Jiang	A novel RGO-CCNT-based catalytic membrane with co-enhanced permeability and catalytic activity for high-efficiency degradation of charged antibiotics
127	Zahra	Abbasian	Energy and performance optimization for electrodialysis water treatment
128	Phillip	Sandman	Enhancement of Nanofiltration Membranes through Chemical Bonding of Lignosulfonic Acid
129	Guozheng	Shao	TiO2 Nanowires Based System for Urea Photodecomposition and Dialysate Regeneration
130	Kai	Chen	Improving Membrane Performance for CO2 Capture With Ultrahigh MW Polyvinylamine
131	Rollie G	Mills	Haloorganic Sorption and Degradation by Temperature Responsive Membranes
133	Charles-François	de Lannoy	The impact of monochloramines and dichloramines on reverse osmosis membranes in wastewater potable reuse process trains: A pilot-scale study
134	Valerie	Niemann	Synchrotron-based X-ray Spectroscopy to Investigate the Fouling Mechanism on Reverse Osmosis Membranes
135	Nan	Zhang	Detection of biofouling on gold-coated MF membranes by in-situ electrical impedance spectroscopy
150	Hao Tang	Tang	Highly efficient plasmonic membrane activation of peroxide for alcohol oxidation
151	Seyede Fateme	BANIHASHEMI	Synthesis of γ-alumina supported ZIF-8 membranes for enhanced propylene/propane separation

Materials Poster Session (Virtual Hallway 3), Host: Jamie Hestekin, Time: 1:00 pm – 2:00 pm EDT			
#	First name	Last name	iPoster Title
1	Albert	Wu	Influence of aliphatic and aromatic fluorine groups on the gas permeability and morphology of fluorinated polyimides
2	Sunxiang	Zheng	Interlayer Spacing and Separation Performance of Graphene Oxide Membranes in Organic Solvent
3	Kai	Chen	Fluoride- and Hydroxide-Containing Facilitated Transport Membranes for CO2 Removal From Solid Oxide Fuel Cells
4	Francisco	Leniz	Advancing NF Membranes: Role of Surface Charge and Support
5	Feihu	Wang	The development of honeycomb-graphene oxide coated porous polymeric membranes
7	Maryam Khalfan	AlQaydi	Activated Carbon as a Photothermal Absorber for Solar Driven Air-Gap Membrane Distillation
8	Shahriar	Habib	Application of 3D Printed Minimal Surface Area Spacers on PVDF Membranes for Treating Fracking Wastewater via Membrane Distillation
9	Zhongyun	Liu	Molecularly Engineered 6FDA-based Polyimide Membranes for Sour Natural Gas Separation
10	Samuel J	Lounder	Membrane with cross-linked zwitterionic nanopores achieves sub-nanometer separations
11	Luca	Mazzaferro	Tuning Pore Size and Robustness of Membranes Formed by Scalable Self-Assembly of Random Copolymer Micelles
12	Kaitlyn P	Brickey	Electron tomography reveals the 3D microstructure of virus filtration membranes
13	Priyanka	Suresh	Tuning the binding strength of membrane adsorbers in radiochromatography
14	Michael	Dugas	Resilient Hollow Fiber Nanofiltration Membranes Fabricated from Copolymers
15	Tao	Wang	High Free Volume, Ion-containing Pentiptycene-based Polysulfone Membranes for Water Purification Application
16	Justin	Easa	Evolution of surface and bulk carbon species derived from propylene and their influence on the interaction of hydrogen with palladium
17	Taliehsadat	Alebrahim	Effect of Metal Ions Dissociated in Cross-linked Poly(Ethylene Oxide) (XLPEO) on Physical Properties and Gas Transport Characteristicss
19	Yu-Ming	Tu	High Density Membrane Protein-Polymer Nanosheets-Based Biomimetic Membranes
20	Mansour	Saberi	Polyvinyl Alcohol (PVA) Hydrogel Underlayer to Support CVD Graphene on Ultrafiltration Membranes as a Support for Water Desalination
21	Leticia	Santos	Ion Transport Improvement by Surface Modified Beads
22	Yian	Chen	Surface Nano-structuring with Tethered Poly(acrylic acid) Chains for Tuning Ultrafiltration Membrane Performance
23	Robert	McNair	Membrane Capacitive Deionization using Quaternized Polymer Blends
24	Gergo	Ignacz	Ion-Stabilized Organic Solvent Nanofiltration Membranes from PIM/PBI Polymer Blends
25	Irshad	Kammakakam	Synthesis and Gas Transport Properties of CO2-philic Anionic Poly(ionic liquid) Composite Membranes
26	Dean M	Welsh	Synthesis and Characterization of Chlorinated RO Membrane Model Compounds – Insights into the Role of N-Cl Species in the Chlorination of Polyamide Membranes
27	Gheorghe	Falca	Spray-Coated Graphene Oxide Hollow Fibers for Nanofiltration
28	David-S	Bergsman	Laser-Induced Graphene for Charged Membrane Applications Enabled by Sequential Infiltration Synthesis
31	Michael	Zelner	Nafion based mosaic nanofiltration membranes with tuneable charge and permeability
32	Levente	Cseri	Graphene Oxide (GO) Nanocomposite Anion Exchange Membranes for Improved Electrodialysis
33	Albert	Schnieders	Large-Area Composite-Membrane based on Ultra-Thin Carbon Nanomembranes
34	Mauricio	Dantus	Biomimetic carbon nanotube-containing membranes for water purification
35	Stefan	Chisca	Preparation of organic solvent and thermal resistant polytriazole membranes with enhanced mechanical properties
36	Mojtaba	Rezaei	Hydrogen-sieving single-layer graphene membranes obtained by crystallographic and morphological optimization of catalytic copper foil
37	Michael J.	McGrath	Nanoporous, Ionic, Lyotropic Liquid Crystal Polymer Membranes: permanent pore modification, characterization of ion-exchange properties, and fabrication of ultrathin films
38	Mahati	Chintapalli	Porous, high surface area polymers with controlled pore size <20 nm, via 1-step controlled radical polymerization
39	John P	Moore	Oxone®-Mediated TEMPO-Oxidized Cellulose Nanomaterials Form I and Form II
40	Grayson	Dennis	6FDA-based Ionic Polyimides-Ionic Liquid Composites for Advanced Gas Separation
136	Shiqi	Huang	Single-layer graphene membranes by crack-free transfer for gas separation with sub-angstrom resolution
137	Wan-Chi	Lee	Centimeter-Scale Gas-Sieving Nanoporous Single-Layer Graphene Membrane
138	Elsa	Lasseuguette	Synthetic saponite clays as additives for reducing aging effects in PIM-1 membranes
139	Sharon	Lin	Free volume manipulation of a 6FDA-HAB polyimide using a solid-state protection/deprotection strategy
140	Katherine Mizrahi	Rodriguez	Facile and time-efficient carboxylic acid functionalization of PIM-1: effect on molecular packing and gas separation performance
141	Oscar	Ovalle	Strategy for acquiring high CO2 permeance performance for dense ceramic-carbonates membranes by total conductivity modification
149	Muchu	Zhou	Synthesis and in situ functionalization of microfiltration membranes via high internal phase emulsion templating

Undergrad. Poster Session (Virtual Hallway 3), Host: Jamie Hestekin, Time: 2:00 pm – 3:00 pm EDT			
#	First name	Last name	iPoster Title
UG 1	Katie	Brodersen	Enhancing the Energy Efficiency of Desalination via Batch Reverse Osmosis
UG 2	Miguel	Jaimes	Continuous covalent organic framework membranes for dye/salt separation
UG 3	Ronald	Vogler	Functionalization of Polyvinylidene Fluoride Membranes for Ion Separation Applications
UG 4	Nobuyo	Watanabe	Ammonia and Water Flux in Membrane Distillation Using Spacers
UG 5	Christopher	Snodgrass	A Liquid-liquid Hollow-Fiber Membrane contactor device to remove oceanic CO2
UG 6	Vy (Jasmine)	Tran	Predicting Octanol/Water Partition Coefficients from Molecular Structure

UG 7	Alix Cynthia Ineza	Karangwa	Evaluating Electrocoagulation Process for Treating Produced Water
UG 8	Kelly	Bye	Influence of molecular interactions, membrane swelling and plasticization on pure and mixed fluid transport in OSN membranes
UG 9	Trisha	Nickerson	Reduced graphene oxide membranes: properties and potential water applications
UG 11	Aryan Louise	Tan	Perylene-Polyimide membranes
UG 12	Alton	O'Neal	Effect of mechanical strain on the transport properties of thin-film composite membranes used in osmotic processes
UG 13	Ciara	Lugo	Advanced Water Treatment for Direct Potable Reuse

NORTH AMERICAN MEMBRANE SOCIETY

29TH ANNUAL MEETING

2020 ONLINE MAY 18TH – 21ST

EMERGING MEMBRANE MATERIALS, PROCESSES, AND APPLICATIONS

@2020NAMS

www.membranes.org/nams-2020/

Program Book

Conference chairs

Manish Kumar, The University of Texas at Austin
Mary Laura Lind, Arizona State University

Edited by

Woochul Song, The University of Texas at Austin

Acknowledgement
Image for program book cover from
Prof. Christopher G. Arges
Cain Department of Chemical Engineering, Louisiana State University

North American Membrane Society

NAMS 2020 online - Program at a Glance

Date	Time (EDT*)	Event
Monday May 18	11:00 AM - 12:00 PM	Plenary Session
	12:00 PM – 1:40 PM (Parallel Oral Sessions)	Membrane Fouling I
		Carbon Capture I
		Seawater Desalination
		Emerging Materials for Liquid Separations I
		Inorganic Materials
	2:00 PM – 3:00 PM	Poster Session 1: Materials
Tuesday May 19	11:00 AM – 1:00 PM (Parallel Oral Sessions)	Membrane Fouling II
		Carbon Capture II
		Emerging Materials for Liquid Separations II
		Osmotically Driven Processes
		Water Reuse
	1:00 PM – 2:00 PM	Poster Session 2: Processes
	2:00 PM – 3:00 PM (Parallel Poster Sessions)	Poster Session 3: Applications
		Poster Session 4: Undergraduate Competition
Wednesday May 20	11:00 AM – 12:40 PM (Parallel Oral Sessions)	Polymeric and Mixed-Materials – Gas Separation I
		Bioinspired and Biomimetic Membranes
		Microporous / Catalytic Membranes
		High Salinity Streams, Brine Mineralization and ZLD
		Process Intensification and Integration
	1:00 PM – 3:00 PM (Parallel Oral Sessions)	Polymeric and Mixed-Materials – Gas Separation II
		Membrane Synthesis and Casting I
		Downstream Bioprocessing
		Membrane Characterization
		Membranes for Electrochemical Applications I
Thursday May 21	11:00 AM – 12:40 PM (Parallel Oral Sessions)	Molecular and Process Modeling
		Membrane Synthesis and Casting II
		Membrane Distillation and Pervaporation
		Module Modeling and Design
		Membranes for Electrochemical Applications II
	1:00 PM – 3:00 PM (Parallel Oral Sessions)	Organic Solvent Separations
		Contaminant Removal from Water
		Industrial Applications
		Awards Session
		Membranes for Electrochemical Applications III

***EDT – US Eastern Daylight Savings time, UK +5 hrs, Germany +6 hrs, Saudi Arabia (in Riyadh) +7 hrs, Inda (IST) +9.5 hrs, Singapore +12 hrs, China +12 hrs, South Korea +13 hrs, Japan +13 hrs.**

NAMS 2020 online – SESSION GRID — MONDAY, MAY 18TH

11:00 AM	**Plenary Talk - [Room 1]** **Accelerating Technology Development By Making Data More Reproducible: Lessons Learned From Literature Meta-Analysis** **Prof. David Sholl, School of Chemical & Biomolecular Engineering, Georgia Institute of Technology**				
Parallel Sessions & Session Chairs	**Membrane Fouling I** **Oral 1, [Room 1]** **Steven Weinman** The University of Alabama stweinman@eng.ua.edu **Dan Miller** Lawrence Berkeley National Lab danieljmiller@lbl.gov	**Carbon Capture I** **Oral 2, [Room 2]** **Winston Ho** Ohio State University ho.192@osu.edu **David Hopkinson** National Energy Technology Lab David.Hopkinson@netl.doe.gov **Katherine Hornbostel** University of Pittsburg hornbostel@pitt.edu	**Seawater Desalination** **Oral 3, [Room 3]** **Bill Phillip** wphillip@nd.edu University of Notre Dame	**Emerging Materials for Liquid Separations I** **Oral 4, [Room 4]** **Hee-Jeung Oh** The Pennsylvania State University hjoh@psu.edu **Baoxia Mi** University of California, Berkeley mib@berkeley.edu **Piran Kidambi** Vanderbilt University piran.kidambi@vanderbilt.edu	**Inorganic Materials** **Oral 5, [Room 5]** **Kumar Varoon Agrawal** Ecole polytechnique fédérale de Lausanne kumar.agrawal@epfl.ch **Luis Francisco Villalobos** Ecole polytechnique fédérale de Lausanne francisco.villalobos@epfl.ch
12:00 PM	Oral 1 - Electrochemical Prevention of Mineral Scale on Electrically Conducting Desalination Membranes Bongyeon Jung (University of California, Lo Angeles)	Oral 6 - Novel Facilitated Transport Membrane and Process for Post-Combustion Carbon Capture Yang Han (The Ohio State University)	Oral 11 - New Insights into Solute-Selectivity Relationship of Thin-Film Composite Polyamide Membranes Xi Chen (Columbia University)	Oral 16 - Charge-modified polysulfones as a platform for membrane separations Matthew Green (Arizona State University)	Oral 21 - Synthesis of polycrystalline ZIF-8 membranes in few minutes for CO2/N2 and CO2/CH4 separation Kumar Varoon Agrawal (Ecole polytechnique fédérale de Lausanne)
12:20 PM	Oral 2 - Assessment of Oil Fouling By Oil-Membrane Interaction Energy Henry Tanudjaja (Nanyang Technological University)	Oral 7 - Carbon Molecular Sieves with Superior H2/CO2 Separation Properties at Elevated Temperatures for Pre-combustion CO2 Capture Haiqing Lin (The State University of New York at Buffalo)	Oral 12 - Omniphobic PVDF-HFP nanofibrous membrane with Pine-needle-like hierarchical TiO2 nanostructures: Towards enhanced performance for membrane distillation Weihua Qing (New Jersey Institute of Technology)	Oral 17 - Diffusion-Induced in situ Growth of Covalent Organic Frameworks for Composite Membranes Priyanka Manchanda (King Abdullah University of Science and Technology)	Oral 22 - Zeolite membrane separator for safe Li-ion batteries Jerry Lin (Arizona State University)
12:40 PM	Oral 3 - Role of Instability Phenomena in Mitigating Membrane Fouling Weiyi Li (Southern University of Science and Technology)	Oral 8 - Carbon Capture Performance of Crosslinked Polyphosphazene Blend Membranes Victor A Kusuma (National Energy Technology Laboratory)	Oral 13 - Energy Barriers for Decoupled Cation and Anion Transport in Sub-nanometer Pores Xuechen Zhou (Yale University)	Oral 18 - Lyotropic Liquid Crystals with Water-Continuous Nanostructure as Thin-film Composite Nanofiltration Membranes Yizhou Zhang (University of Pennsylvania)	Oral 23 - Preparation and nanofiltration properties of thin film meso-porous oxide membranes via the sonochemical precipitation method Minghui Qiu (Nanjing Tech University)
1:00 PM	Oral 4 - Do membrane hydrophobicity and hydrodynamic shear affect the initial deposition and pioneering colonization of anaerobes isolated from an anaerobic membrane bioreactor? Yang Yang (Ben-Gurion University of the Negev)	Oral 9 - Design and Techno-Economic Analysis of a Zeolite Membrane Reactor Intensified IGCC Process Jerry Lin (Arizona State University)	Oral 14 - High Free Volume, Ion-containing Pentiptycene-based Polysulfone Membranes for Water Purification Application Tao Wang (University of Notre Dame)	Oral 20 - Two-Dimensional Covalent Organic Frameworks with Various Backbones and Pore Surface Engineering as Selective Layers of Thin-Film Composite Membranes Katie Li-Oakey (University of Wyoming)	Oral 24 - Optimization of the performance of the photocatalytic TiO2 membranes fabricated by suspension plasma spray (SPS) process Elnaz Alebrahim (Concordia University)
1:20 PM	Oral 5 - Nanofiltration of saline oil-water emulsions: Combined effect of salt concentration polarization and fouling by oil on flux performance Vlad Tarabara (Michigan State University)	Oral 10 - Synthesis of Sterically Hindered Polyvinylamine and Its Application in Facilitated Transport Membranes for CO_2 Capture from Flue Gas Ting-Yu Chen (The Ohio State University)	Oral 15 - Zwitterionic Polysulfone Membranes for Pervaporation Tejaswi Nori (Arizona State University)		Oral 25 - Few-nanometer thick organosilica membranes for high-temperature H2/CO2 separation Haiqing Lin (The State University of New York at Buffalo)
1:40 PM	Break				
2:00 PM	Poster Session 1: Materials (link will be online)				

NAMS 2020 online – SESSION GRID — TUESDAY, MAY 19TH

Parallel Sessions & Session Chairs	Membrane Fouling II Oral 6, [Room 1] **Steven Weinman** The University of Alabama stweinman@eng.ua.edu **Dan Miller** Lawrence Berkeley National Lab danieljmiller@lbl.gov	Carbon Capture II Oral 7, [Room 2] **Winston Ho** Ohio State University ho.192@osu.edu **David Hopkinson** National Energy Technology Lab David.Hopkinson@netl.doe.gov **Katherine Hornbostel** University of Pittsburg hornbostel@pitt.edu	Emerging Materials for Liquid Separations II Oral 8, [Room 3] **Hee-Jeung Oh** The Pennsylvania State University hjoh@psu.edu **Baoxia Mi** University of California, Berkeley mib@berkeley.edu **Piran Kidambi** Vanderbilt University piran.kidambi@vanderbilt.edu	Osmotically Driven Processes Oral 9, [Room 4] **Tony Straub** University of Colorado Boulder tonystraub90@gmail.com **Andrea Achilli** The University of Arizona achilli@email.arizona.edu **Milad R. Esfahani** The University of Alabama mesfahani@eng.ua.edu	Water Reuse Oral 10, [Room 5] **Vicky Karanikola** The University of Arizona vkaranik@email.arizona.edu **Jack Gilron** Ben-Gurion University of the Negev jgilron@bgu.ac.il
11:00 AM	Oral 26 - Exploring and modeling the effect of pattern geometry on nanofiltration membrane fouling Anna Malakian (Clemson University)	Oral 31 - High flux CO2-selective single-layer graphene membranes: synthesis and scale-up Kumar Varoon Agrawal (Ecole polytechnique fédérale de Lausanne)	Oral 36 - Ultrathin Two-Dimensional Covalent Organic Framework (COF) Membranes for Molecular Separations Devin Shaffer (University of Houston)	Oral 41 - Reverse-fluxed Draw Solutes Removal and Microalgae Production in an Osmotic Photobioreactor System For Complete Water Recovery from Secondary Effluent Zixuan Wang (Washing University in St. Louis)	Oral 46 - Desalination for a Circular Water Economy Meagan Mauter (Stanford University)
11:20 AM	Oral 27 - Oscillation induced aggregation to reduce colloidal fouling in microfiltration Jack Gilron (Ben Gurion University)	Oral 32 - Tuning the transport properties of CANAL ladder polymers by alkyl substitutions and backbone conformation modification Francesco Maria Benedetti (Massachusetts Institute of Technology)	Oral 37 - Toward Systematic Tuning of Thin-Film Composite Membrane Selectivity Mackenzie Anderson (University of California, Los Angeles)	Oral 42 - Zwitterionic Hydrogels with Polyamide-Skin Layers Mitigating Concentration Polarization for Forward Osmosis Thien N Tran (The State University of New York at Buffalo)	Oral 47 - The impact of pre-chlorination on microfiltration fouling during municipal wastewater effluent filtration for water reuse Kunal Gupta (Texas A&M University)
11:40 AM	Oral 28 - Living Filtration Membranes Demonstrate Anti-Biofouling Properties Katherine Zodrow (Montana Tech)	Oral 33 - Subambient Carbon Capture using Next Generation Hollow Fiber Membrane Modules at the National Carbon Capture Center David Hasse (Air Liquide)	Oral 38 - Dry-Wet Phase Inversion Synthesis Role on Graphene Oxide-Polysulfone Membrane for Membrane Distillation Lucy M Camacho (Texas A&M University Kingsville)	Oral 43 - A Simplified Modeling Framework to Investigate Osmotically Driven Processes at the System-scale Zachary Binger (The University of Arizona)	Oral 48 - Understanding colloidal fouling formation during inline coagulation/ultrafiltration process Bianca M. Souza Chaves (The University of Arizona)
12:00 PM	Oral 29 - Covalently grafting graphene oxide onto UF membranes to improve antifouling properties Xiaoyi Chen (The State University of New York at Buffalo)	Oral 34 - Scalable charge-modified polysulfone IL support membrane morphologies for CO2 capture in space Mani Modayil Korah (Arizona State University)	Oral 39 - Phosphorene- Based Antifouling Membranes: Synthesis, Fabrication and Applications Joyner Eke (University of Kentucky)	Oral 44 - Experimental Demonstration of Power Generation from Fertilizer via Pressure Retarded Osmosis Jonathan Maisonneuve (Oakland University)	Oral 49 - Combination of AnMBR with outdoor microalgae cultivation for industrial wastewater treatment, resource recovery and near-zero waste Roy Bernstein (Ben-Gurion University at Negev)
12:20 PM	Oral 30 - Laser-Induced Graphene for Charged Membrane Applications Enabled by Sequential Infiltration Synthesis David S Bergsman (Massachusetts Institute of Technology)	Oral 35 - Ultra-thin supported graphene oxide membrane for CO2 capture Yi Zhou (Ohio State University)	Oral 40 - Understanding Virus Prefilters in Improving the Filtration Performance of the Planova BioEX Virus Filter for Monoclonal Antibodies Solomon Isu (University of Arkansas)	Oral 45 - Mitigation of Biological Fouling in a Forward Osmosis Membrane Bioreactor Sage Hiibel (University of Nevada, Reno)	Oral 50 - Membrane Surface Characterization Upon Cleaning of a Heavily Fouled Reverse Osmosis Membrane from an Advanced Reclamation Facility Bilal Abada (Texas A&M University)
12:40 PM	Oral 151 – Mineral scale formation and dissolution in RO operation in cyclic modes of operation Anditya Rahardianto (University of California, Los Angeles)				
1:00 PM	Poster Session 2: Processes (link will be online)				
2:00 PM	Poster Session 3: Applications & Poster 4: Undergraduate Competition (link will be online)				

NAMS 2020 online – SESSION GRID

WEDNESDAY - AM, MAY 20TH

Parallel Sessions & Session Chairs	**Polymeric and Mixed-Materials – Gas Separation I** **Oral 11, [Room 1]** **Ben Sundell** Armaco benjamin.sundell@aramcoservices.com **Zach Smith** Massachusetts Institute of Technology zpsmith@mit.edu **Xiaoli Ma** University of Wisconsin-Milwaukee ma26@uwm.edu	**Bioinspired and Biomimetic Materials** **Oral 12, [Room 2]** **Yuexiao Shen** Texas Tech University Yuexiao.Shen@ttu.edu **Patrick Saboe** National Renewable Energy Laboratory psaboe@gmail.com	**Microporous / Catalytic Materials** **Oral 13, [Room 3]** **Ayse Asatekin** Tufts University ayse.asatekin@tufts.edu **Will Tarpeh** Stanford University wtarpeh@stanford.edu **Vlad Tarabara** Michigan State University tarabara@egr.msu.edu	**High Salinity Streams, Brine Minimization and ZLD** **Oral 14, [Room 4]** **Kerri Hickenbottom** The University of Arizona klh15@email.arizona.edu **Jon Brant** University of Wyoming jbrant1@uwyo.edu	**Process Intensification and Integration** **Oral 15, [Room 5]** **Hannah Murnen** Compact Membrane Systems hmurnen@compactmembrane.com **Ed Sanders** Air Liquide ed.sanders@airliquide.com **Albert S Kim** The University of Hawaii at Manoa albertsk@hawaii.edu
11:00 AM	Oral 51 - Effect of thermal treatment on the structure and gas transport properties of a triptycene-based polybenzoxazole exhibiting configurational free volume Michele Galizia (University of Oklahoma)	Oral 56 - Artificial Water Channels-toward biomimetic membranes for desalination Mihail Barboiu (Institut Europeen des Membranes)	Oral 61 - Membranes with Integrated Nanomaterials: Material Science Advances and Applications Dibakar Bhattacharyya (University of Kentucky)	Oral 66 - Minimal and zero liquid discharge with reverse osmosis using low-salt-rejection membranes Zhangxin Wang (Yale University)	Oral 71 - Energy Efficiency of Electro-Driven Brackish Water Desalination: Electrodialysis Significantly Outperforms Membrane Capacitive Deionization Mohan Qin (University of Wisconsin-Madison)
11:20 AM	Oral 52 - Optimizing Plasticization Benefits of Polyimides for Membrane-based Natural Gas Sweetening Yang Liu (Georgia Institute of Technology)	Oral 57 - Nature-inspired ion conducting polymers for energy conversion and storage devices Shudipto K Dishari (University of Nebraska-Lincoln)	Oral 62 - Efficient Ammonia Decomposition in a Catalytic Membrane Reactor to Enable Hydrogen Storage and Utilization Colin Wolden (Colorado School of Mines)	Oral 67 - Evaluation of Integrated Electrocoagulation-Microfiltration and Direct Contact Membrane Distillation Processes for Treating Produced Water Mahmood Jebur (University of Arkansas)	Oral 72 - CO_2-Selective Membrane for Enhancing H2 Utilization in Solid Oxide Fuel Cells Kai Chen (The Ohio State University)
11:40 AM	Oral 53 - New Facilitated Transport Membranes for Hydrogen Purification from Coal-Derived Syngas Yang Han (The Ohio State University)	Oral 58 - Ultra-High Ionic Exclusion Through Carbon Nanomembranes Armin Gölzhäuser (Bielefeld University)	Oral 63 - Tuning Ultrafiltration Membrane Performance via Surface Graft Polymerization of Acrylic Acid Enabled by Air Plasma Activation Yian Chen (University of California, Los Angeles)	Oral 68 - A Novel Membrane Distillation-Crystallization Approach for Inland Brine Treatment Sage Hiibel (University of Nevada, Reno)	Oral 73 - Flow Patterns through densely packed hollow fibers: a CFD study for membrane distillation Albert Kim (University of Hawaii)
12:00 PM	Oral 54 - High Performance Gas Separation Membranes from CANAL Ladder Polymerization Holden Lai (Stanford University)	Oral 59 - High Density Membrane Protein-Polymer Nanosheets-Based Biomimetic Membranes Yu-Ming Tu (The University of Texas at Austin)	Oral 64 - Monitoring protein fouling on patterned membranes by light microscopy and simulation Anna Malakian (Clemson University)	Oral 69 - Critical Analysis of Membrane Distillation at High Salinity Mukta Hardikar (The University of Arizona)	Oral 74 - Analysis of Membrane Processes for CO2 Removal and H2 Reuse for Solid Oxide Fuel Cells Kai Chen (The Ohio State University)
12:20 PM	Oral 55 - Carbon Molecular Sieve Membranes for Petrochemical Gas Separations Surendar R Venna (Dow Inc.)	Oral 60 - Biomimetic carbon nanotube and AAO systems utilizing electro-dynamic interfaces Bruce Hinds (University of Washington)	Oral 65 - Influences of Microwave Irradiation on Performances of Membrane Filtration and Catalytic Degradation of Perfluorooctanoic Acid (PFOA) Fangzhou Liu (New Jersey Institute of Technology)	Oral 70 - Testing mineral scaling propensities of fluorinated and non-fluorinated polyamide thin film composite membranes Sankaranarayanan Ayyakudi Ravichandran (University of Colorado)	Oral 75 - Experimental evaluation of an optimized concentration gradient battery Fei Liu (University of North Carolina at Chapel Hill)
12:40 PM	Break				

Parallel Sessions & Session Chairs	**Polymeric and Mixed-Materials – Gas Separation II** **Oral 16, [Room 1]** **Ben Sundell** Armaco benjamin.sundell@aramcoservices.com **Zach Smith** Massachusetts Institute of Technology zpsmith@mit.edu **Xiaoli Ma** University of Wisconsin-Milwaukee ma26@uwm.edu	**Membrane synthesis and Casting I** **Oral 17, [Room 2]** **Christine Duval** Case Western Reserve University ced84@case.edu	**Downstream Bioprocessing** **Oral 18, [Room 3]** **Prity Bengani-Lutz** Repligen plutz@repligen.com **Onur Kas** UCB Onur.Kas@ucb.com **James McGrath** University of Rochester jmcgrath@bme.rochester.edu	**Membrane Characterization** **Oral 19, [Room 4]** **Santiago Romero** University of Edinburgh Santiago@ed.ac.uk	**Membranes for Electrochemical Applications I** **Oral 20, [Room 5]** **Geoff Geise** University of Virginia gmg9j@virginia.edu **Orlando Coronell** Univ. of North Carolina at Chapel Hill coronell@ad.unc.edu **Chris Arges** Louisiana State University carges@lsu.edu **Shudipto Dishari** University of Nebraska sdishari2@unl.edu
1:00 PM	Oral 76 - Microporous Pentiptycene-based Polymers with Heterocyclic Rings for High Performance Gas Separation Membranes Zihan Huang (University of Notre Dame)	Oral 82 - Polyamide Nanofiltration Membrane with Highly Uniform Sub-nanometre Pores for Sub-1Å Precision Separation Yuanzhe Liang (Vanderbilt University)	Oral 87 - Membrane adsorbers for medical isotope purification Christine E Duval (Case Western Reserve University)	Oral 92 - Tailoring the Structure and Performance of Model Polyamide-based Membranes for Desalination Christopher M Stafford (National Institute of Standards and Technology)	Oral 97 - Modelling and validation of concentration dependence of ion exchange membrane permselectivity: significance of convection and Manning's counter-ion condensation theory Ryan S Kingsbury (University of North Carolina at Chapel Hill)
1:20 PM	Oral 77 - High-Performance Gas Separation Membranes Based on Poly(benzimidazole) Alexander Bridge (The University of Texas at Austin)	Oral 83 - Synthesis of Ultrahigh MW Polyvinylamine for Incorporation in Membranes for CO2 Capture Kai Chen (The Ohio State University)	Oral 88 - New Multimodal Anion-Exchange Membranes for Polishing of Biologics Joshua Osuofa (Clemson University)	Oral 93 - Applying Transition-State Theory to Model Solute Transport in Membranes with Sub-nanometer Pores Razi Epsztein (Technion – Israel Institute of Technology)	Oral 98 - The Influence of Ion Association on Ion Solubility in Ion Exchange Membranes Rahul Sujanani (The University of Texas at Austin)
1:40 PM	Oral 78 - Polybenzimidazole-Derived Carbon Molecular Sieve Hollow Fiber Membranes with Tailored Oxygen Selective Transport Jong Geun Seong (Los Alamos National Laboratory)	Oral 84 - Surface Patterning of Microporous Membranes using Thermally Induced Phase Separation under Confinement Shouhong Fan (University of Colorado at Boulder)	Oral 89 - Fouling and Retention Behavior of BioOptimal MF-SL Microfilter during Cell Culture Harvesting Xianghong Qian (University of Arkansas)	Oral 94 - Unraveling the Clay-Membrane Interactions during An Osmotically Driven Process via Optical Coherence Tomography Xin Liu (Southern University of Science and Technology)	Oral 99 - Functionalized Ceramic Ion Exchange Membranes for Electrodialysis Gregory M Newbloom (Membrion, Inc.)
2:00 PM	Oral 79 - Tailoring structural and functional features of high-performance ionenes and ionic composites designed for membrane-based gas separations Kathryn E O'Harra (University of Alabama)	Oral 85 - Membrane with cross-linked zwitterionic nanopores achieves sub-nanometer separations Samuel J Lounder (Tufts University)	Oral 90 - Development of nanopocket membranes for tangential flow analyte capture (TFAC) of extracellular vesicles Thomas Gaborski (Rochester Institute of Technology)	Oral 95 - Local density and free volume inhomogeneities govern transport properties in reverse osmosis membranes Michael Geitner (The Pennsylvania State University)	Oral 100 - Tradeoff in membranes for artificial photosynthesis: Ion transport and product crossover Sarah M Dischinger (Lawrence Berkeley National Laboratory)
2:20 PM	Oral 80 - Azo-UiO-66 MOF for low-energy CO2 capture adsorbents and mixed matrix membranes for CO2/N2 separation Bradley P. Ladewig (Karlsruhe Institute of Technology)	Oral 86 - Highly Permeable and Selective Crosslinked Polymer Membranes for Energy Efficient Gas Separation Si Li (University of Notre Dame)	Oral 91 - Catch and Display Diagnostics: Capturing Non-protein Biologics on Ultrathin Silicon Membranes Using Size and Affinity-based Separations Michael Klaczko (University of Rochester)	Oral 96 - Probing the Structure and Dynamics of Membranes under Realistic Operating Conditions, using Operando Spectroscopy Casey O'Brien (University of Notre Dame)	Oral 101 - Highly Selective Ion Separations with Pressure-driven Flow through Charged Nanoporous Membranes Chao Tang (University of Notre Dame)
2:40 PM	Oral 81 - Physical aging of sub-micron PBI membranes at elevated temperatures Melanie M Merrick (The University of Texas at Austin)	Oral 152 – Lyotropic liquid crystal templating for making ultrafiltration membranes Reza Foudazi (New Mexico State University)			

NAMS 2020 online – SESSION GRID

THURSDAY - AM, MAY 21ST

Parallel Sessions & Session Chairs	**Molecular and Process Modeling** **Oral 21, [Room 1]** **David Warsinger** Purdue University david.warsinger@gmail.com **Shihong Lin** Vanderbilt University Shihong.lin@vanderbilt.edu	**Membrane Synthesis and Casting II** **Oral 22, [Room 2]** **Christine Duval** Case Western Reserve University ced84@case.edu	**Membrane Distillation and Pervaporation** **Oral 23, [Room 3]** **Lee Vane** United States Environmental Protection Agency Vane.Leland@epa.gov	**Module Modeling and Design** **Oral 24, [Room 4]** **David Ladner** Clemson University ladner@clemson.edu **Nils Tilton** Colorado School of Mines ntilton@mines.edu **Grigorios Panagakos** National Energy Technology Lab Grigorios.Panagakos@netl.doe.gov	**Membranes for Electrochemical Applications II** **Oral 25, [Room 5]** **Geoff Geise** University of Virginia gmg9j@virginia.edu **Orlando Coronell** Univ. of North Carolina at Chapel Hill coronell@ad.unc.edu **Chris Arges** Louisiana State University carges@lsu.edu **Shudipto Dishari** University of Nebraska sdishari2@unl.edu
11:00 AM	Oral 102 - Effect of packing nonuniformity at the fiber bundle-case interface on flow distribution and separation performance for hollow fiber membrane modules Glenn Lipscomb (University of Toledo)	Oral 107 - Atomically Thin Graphene Membranes for Desalination and Molecular Separations Piran Kidambi (Vanderbilt University)	Oral 112 - Elucidating the Trade-off between Membrane Wetting Resistance and Water Vapor Flux in Membrane Distillation Jongho Lee (University of British Columbia)	Oral 117 - Rough or wiggly? Membrane topology and morphology for fouling control Bowen Ling (Stanford University)	Oral 122 - Understanding water-splitting thermodynamics and kinetics in micropatterned bipolar membranes Subarna Kole (Louisiana State University)
11:20 AM	Oral 103 - Estimating fluid pressure gradients within crosslinked aromatic polyamide using molecular dynamics Riley Vickers (University of North Carolina at Chapel Hill)	Oral 108 - A Comparison of PolarClean, Gamma-Valerolactone and their Mixture as Bio-derived solvents for Polysulfone Membrane Fabrication Xiaobo Dong (University of Kentucky)	Oral 113 - Cost optimization of gap membrane distillation Timothy V Bartholomew (National Energy Technology Laboratory)	Oral 118 - Characterization of laminar, transitional, and turbulent flow regime in membrane modules Alexander V Dudchenko (Stanford University)	Oral 123 - Co-permeation behavior of methanol and acetate in polyether-based cation exchange membranes Jung Min Kim (Auburn University)
11:40 AM	Oral 104 - High-throughput computational prediction of the cost of carbon capture using mixed matrix membranes Janice A. Steckel (U.S. Department of Energy)	Oral 109 - Electrospun Pretreatment Membranes Husain Mithaiwala (Arizona State University)	Oral 114 - Modeling Pilot-Scale Air Gap Membrane Distillation as a Special Case of Permeate Gap Membrane Distillation Mukta Hardikar (The University of Arizona)	Oral 119 - 3D Numerical simulations of temperature polarization in vacuum membrane distillation systems with active membrane heating Mark A Dudley (Colorado School of Mines)	Oral 124 - Carbon nanotube thin film deposition on hollow fiber membranes: addressing the scalability challenges of conductive membranes Melissa J Larocque (McMaster University)
12:00 PM	Oral 105 - Opportunities for high productivity and selectivity desalination via osmotic distillation with improved membrane design Sangsuk Lee (University of Colorado Boulder)	Oral 110 - Aqueous phase separation of responsive copolymers for sustainable and mechanically stable membranes Wouter Nielen (University of Twente)	Oral 115 - Economic Evaluation of Membrane Distillation as a Competitive Brackish Water Desalination Technology Haamid S Usman (Concordia University)	Oral 120 - Direct numerical simulations of unsteady mixing in direct contact membrane distillation systems with different spacer blockages and vertical offsets Nils Tilton (Colorado School of Mines)	Oral 125 - Selective ion conducting membranes for non-aqueous redox flow battery applications Geoffrey M Geise (University of Virginia)
12:20 PM	Oral 106 - Ion association as key element in RO and NF modeling Viatcheslav Freger (Technion-Israel Institute of Technology)	Oral 111 - Stepwise synthesis of oligoamide film on porous supports for preparing new types of membranes Ron Kasher (Ben Gurion University of the Negev)	Oral 116 - Conducting thermal energy to the membrane/water interface for the enhanced desalination of hypersaline brines using membrane distillation Jingbo Wang (University of California, Los Angeles)	Oral 121 - Hydrodynamic load exerted on a moving bundle of hollow fibers using constraint dissipative hydrodynamics: hydrorattle simulation Albert S. Kim (University of Hawaii at Moana)	Oral 126 - Graphene/Nafion Based Membrane Structure for Crossover Mitigation in Energy Storage and Conversion Systems Saheed A Bukola (National Renewable Energy Laboratory)
12:40 PM	Break				

NAMS 2020 online – SESSION GRID

THURSDAY - PM, MAY 21ST

Parallel Sessions & Session Chairs	**Organic Solvent Separations** **Oral 26, [Room 1]** **Ryan Lively** Georgia Institute of Technology ryan.lively@chbe.gatech.edu **Steve White** Membrane Technology and Research steve.white@mtrinc.com **Michele Galizia** University of Oklahoma mgalizia@ou.edu **Neel Rangnekar** ExxonMobil neel.d.rangnekar@exxonmobil.com	**Contaminant Removal from Water** **Oral 27, [Room 2]** **Prakhar Prakash** Chevron Prakhar.Prakash@chevron.com	**Industrial Applications** **Oral 28, [Room 3]** **Dibakar Bhattacharyya** University of Kentucky DB@uky.edu **CJ Kurth** Solecta Membranes cj.kurth@solectamembranes.com	**Awards Session** **Oral 29, [Room 4]** **Caleb Funk** DuPont caleb.funk@dupont.com **Bruce Hinds** University of Washington bjhinds@uw.edu	**Membranes for Electrochemical Applications III** **Oral 30, [Room 5]** **Geoff Geise** University of Virginia gmg9j@virginia.edu **Orlando Coronell** Univ. of North Carolina at Chapel Hill coronell@ad.unc.edu **Chris Arges** Louisiana State University carges@lsu.edu **Shudipto Dishari** University of Nebraska sdishari2@unl.edu
1:00 PM	Oral 127 - Molecular origin of flux non-linearity in Organic Solvent Nanofiltration Michele Galizia (University of Oklahoma)	Oral 132 - Adsorption of organic micropollutants to polymer surfaces probed by second harmonic scattering laser spectroscopy Dan Miller (Lawrence Berkeley National Laboratory)	Oral 137 - Produced water desalination using high temperature membranes Xiaofei Huang (Hydranautics), Cheng Chen (Chevron)	Oral 141 - 3D Printed Adsorber for Capturing Chemotherapy Drugs before They are Released in the Body Hee Jeung Oh (Pennsylvania State University)	Oral 147 - Tailoring Non-Aqueous Electrolyte Concentration enables Optimum Ion Transport in Pentablock Terpolymer Cation Exchange Membranes Michelle Lehmann (University of Tennessee)
1:20 PM	Oral 128 - Thin-film nanocomposite membranes with enhanced permeability and selectivity for organic solvent forward osmosis Bofan Li (National University of Singapore)	Oral 133 - Laser-induced graphene polymer composite membranes as electrically active filters for contaminant removal Christopher J. Arnusch (Ben Gurion University of the Negev)	Oral 138 - Membrane based upgrading of Biogas: Using Realtime Controls and Selective Distribution of Membrane Types to Effectively Respond to Varying Gas Flow and Gas Quality. Steven K Pedersen (Evonik)	Oral 142 - Mechanical properties of thin-film composite membranes and the roles they play on transport in osmotic processes Jaime Idarraga-Mora (Clemson University)	Oral 148 - Powering up Biomedical Devices by Harnessing Energy from Reverse Electrodialysis using Sodium Concentrations from the Flow of Human Blood Efecan Pakkaner (University of Arkansas)
1:40 PM	Oral 129 - Studies in separation of organic solvent mixtures during reverse osmosis and nanofiltration by a perfluorodioxole copolymer membrane Kamalesh K Sirkar (New Jersey Institute of Technology)	Oral 134 - Printing zwitterionic copolymer thin film composite (TFC) membranes: Enabling Tunability of Thickness and Remarkable Permeances for Nanofiltration Applications Xin Qian (University of Connecticut)	Oral 139 - Industrial Applications of Membranes in the Membrane Science, Engineering and Technology (MAST) Center at the University of Arkansas Ranil Wickramasinghe (University of Arkansas)	Oral 143 - Engineering Selective Desalination Membranes by Controlling Functional Group Configuration Hongxi Luo (University of Virginia)	Oral 149 – Ionic conductivity of ion exchange membranes: Measurement techniques and salt concentration dependence Jovan Kamcev (University of Michigan, Ann Arbor)
2:00 PM	Oral 130 - Complex aromatic hydrocarbon mixture OSRO separation using defect-engineered Torlon® hollow fiber membranes Hye Youn Y Jang (Georgia Institute of technology)	Oral 135 - Enhancing antimicrobial properties of Chitosan/ Graphene Oxide membrane for waste water treatment application Raheleh Daneshpour (University of Arkansas)	Oral 140 - High Throughput Development of Carbon Molecular Sieve Membranes and Adsorbents for Olefin/Paraffin Separations Jay (Junqiang) Liu (The Dow Chemical Company)	Oral 144 - More than wetting resistance: How membrane surface wettability regulates water vapor permeability and mineral scaling in membrane distillation Tiezheng Tong (Colorado State University)	
2:20 PM	Oral 131 - Organic solvent nanofiltration (OSN) membranes for separation of close-boiling mixtures and high-boilers Petrus Cuperus (SolSep BV)	Oral 136 - Early-stage membrane fouling detection for improving membrane cleaning efficiency Thomas Schäfer (Polymat, University of the Basque Country)		Oral 145 - Comprehensive study of dissolved methane harvesting using omniphobic membrane contactor Abhishek Dutta (University of British Columbia)	
2:40 PM	Oral 150 - Polymer Membrane-based Liquid Hydrocarbon Fractionation Ronita Mathias (Georgia Institute of technology)			Oral 146 - Quantification of the CO_2-induced plasticization of glassy polymer and CMS film membranes from the behavior of CH_4 diffusivity, and how observations of multicomponent sorption and permeation rectified recurring misconceptions Giuseppe Genduso (KAUST)	

Welcome to NAMS 2020

We hope that you and loved ones are healthy and well. Thank you for joining us for NAMS 2020 ONLINE, in response to COVID-19 making a physical meeting impossible. Though there are many advantages to in-person meetings, we hope that this online experiment brings both personal connections and great technical content. In particular we are excited the on-line poster format that is extremely interactive in content. We've kept the meeting format essentially intact, though shortened presentation times to give more manageable 4hr periods over 4 days. We will be working hard to have networking options such as chat hallways. Regrettably, this may be the only membrane-focused conference for the year, so we hope the experiment serves our membership well. Special thanks to co-chairs Manish Kumar and Mary Laura Lind for going far above the normal significant service to community.

Importantly this is an annual gathering of a vibrant community that wants to tell the world about these important technologies. The society is working hard bring together industry and academia and mentor a new generation of membraneologist. Important are our awards efforts to recognize career-long success and enable young scientist to be part of the community. Congratulations to our new NAMS fellows Jerry Lin, Ingo Pinnau, and Ed Sanders for decades of impressive contributions the membrane community. Also congratulations to Andrew Zydney recipient of our most prestigious triennial Micheals Award for Membrane Innovation. Additionally, we have 2 new Educational Innovation Fellows team awards to tangibly promote membrane education. Also be aware of our young scientist award technical session that represent the best works across diverse membrane fields. Though the annual meeting is the pinnacle of society activity, throughout the year we work hard at getting the word out as with our Membrane Quarterly newsletter and personalized mentorship program with sustained interactions. Behind the scenes we have enacted and Industrial Advisory Committee as well as a formal worldwide body to coordinate events. We look forward to new educational initiatives to make NAMS the 'go to place' for membrane science and engineering and are excited to see our 'Members Only' section going live. We look forward to "seeing" you at the meeting and above all, please enjoy this dynamic and engaging community.

Bruce Hinds, NAMS President and Christina Carbrello, NAMS Vice-President

Message from the 2020 NAMS Conference Chairs

It is our pleasure to welcome you to the 29th Annual Meeting of the North American Membrane Society, and the first ever NAMS meeting online. It has been our honor to serve as co-chairs for this year's NAMS meeting given the challenges of the times. We felt that it was even more necessary for our community to get together and "see" each other in this socially distant and isolating environment. This was the motivation for us to do what we could to put this conference together despite some daunting logistical challenges and a steep learning curve.

We are expecting ~500 conference attendees at this conference which will be run completely online. We will have approximately 150 posters presented and the same number of talks. These contributions are of the highest quality (as seen from the abstracts), with the posters already published so far being particularly innovative. Please check them out!

A very warm thank you to our dedicated oral and poster session chairs that have stuck with us through the challenging planning process. Also special thanks to the NAMS 2020 Conference Advisory Committee (Glenn Lipscomb, Isabel Escobar) and NAMS board members and leadership (especially Jamie Hestekin, Christina Carbrello, and Bruce Hinds) for all the input and support. We would also like to acknowledge our Fundraising chairs Keith Murphy and JR Johnson as well as our open-minded sponsors (NSF, Air Products, Air Liquide, MTR, ExxonMobil, Hiden Isochema, Evonik, Sterilitech, W. L. Gore & Associates, ISSEC (GeorgiaTech), and MDPI publications) who helped us raise substantial funds to support the conference and conference attendees.

We are truly excited to "see" you at the conference. Hopefully you are excited to attend NAMS online too!

Manish Kumar and Mary Laura Lind
NAMS 2020 Conference Co-Chairs

Alan S. Michaels Award for Innovation in Membrane Science and Technology

Dr. Alan Michaels was one of the true innovators and pioneers in membrane science and technology. His academic and industrial work on ultrafiltration, membrane-based drug delivery systems, and new membrane processes for the biopharmaceutical industry all represented breakthroughs that helped redefine the membrane field.

This award consists of a $10,000 prize, and is given by the North American Membrane Society to honor the late Dr. Michaels and to recognize individuals who have made outstanding innovations and/or exceptional lifetime contributions to membrane science and technology.

2020 Alan S. Michaels Award

Dr. Andrew Zydney

Dr. Andrew Zydney, Bayard D. Kunkle Chair and Professor of Chemical Engineering, Penn State University received the Alan S. Michaels Award for his work in microfiltration, membrane fouling, and applications of membranes in bioprocessing, a field that was also at the core of much of Alan Michaels' pioneering efforts. Zydney's work has not only provided important fundamental understanding of the behavior of membrane separations, it has also had a major impact on the design and development of important commercial membrane processes while helping to shape the overall direction of membrane-related research in bioseparations.

In addition to his research, Zydney has been a tireless contributor to the broader membrane community. As Editor-in-Chief he has the led the Journal of Membrane Science for the past 10+ years through a period of rapid growth and dramatic increase in impact factor. He previously served as Editor of Separation and Purification Reviews and Associate Editor of Separation Science and Technology. Zydney has been an active member of the Board of Directors of the North American Membrane Society, first from 1998-2006 and then from 2016-2019. He served as President of NAMS from 2002-2003 and has co-organized two highly successful NAMS meetings: Baltimore in 1997 and Pittsburgh in 2019. Zydney is also an active consultant, assisting numerous companies in the design, development, and successful commercial implementation of a wide range of membrane processes.

Professor Zydney is an outstanding researcher who has successfully addressed a diverse range of significant problems. He has pioneered the development of membrane systems in bioprocessing, including membranes for the purification of monoclonal antibodies, vaccines, and DNA. He has educated literally thousands of students in membrane separations, and his former PhD students have positions at leading bioprocessing, chemical, and membrane companies, as well as several academic institutions, around the world. His outstanding innovations and exceptional lifetime contributions to membrane science and technology are highly deserving of recognition with the Alan S. Michaels Award.

NAMS Fellows are recognized members of NAMS who have demonstrated ongoing excellence in membrane engineering and practice. Election as Fellow is in recognition of both "service to NAMS" and "highly significant professional accomplishment in the membrane field."

2020 Fellow

Dr. Jerry Y.S. Lin

Dr. Jerry Y.S. Lin, Regents' Professor of Chemical Engineering at Arizona State University, is an internationally recognized leader in inorganic membrane science. He has made numerous original contributions developing new materials and synthesis methods and understanding fundamental separation and transport properties of mesoporous, microporous and dense inorganic membranes for filtration and molecular separation. Dr. Lin has three decades of fundamental and applied research which has contributed to advancing inorganic membrane science from its infancy to a major subdivision of membrane science. His technical contributions to membrane science are documented in his 350 papers and 9 issued patents. He has been an editor of Journal of Membrane Science since 2008 and recently become Editor-in-Chief of the journal. Dr. Lin also co-chaired two highly successful NAMS Annual Meetings (1998 and 2013).

2020 Fellow

Dr. Ingo Pinnau

Ingo Pinnau is founding Director of the Advanced Membranes and Porous Materials Center and Professor in Chemical Engineering at King Abdullah University of Science and Technology (KAUST). His career in membrane science and technology spans over 35 years with internationally recognized contributions in the design of high-performance materials and membranes for energy-intensive separations. His research has resulted in several commercial membrane products for large-scale use in gas- and vapor separations. He has authored or co-authored 170 research papers and is named as inventor on 45 granted U.S. patents. He joined NAMS in 1987, served twice as NAMS President and was co-organizer of Annual NAMS meetings in 2004 and 2018 and ICOM in 2008.

NAMS Fellows are recognized members of NAMS who have demonstrated ongoing excellence in membrane engineering and practice. Election as Fellow is in recognition of both "service to NAMS" and "highly significant professional accomplishment in the membrane field."

2020 Fellow

Dr. Edgar (Ed) S. Sanders, Jr.

Dr. Edgar (Ed) S. Sanders, Jr. is Chief Technology Officer and an International Fellow for Air Liquide. He has made critical contributions to understanding small molecule transport in polymeric materials and commercialization of membrane gas separation technology over the past four decades. From his pioneering work on mixed gas sorption and permeation in glassy polymers to commercialization of state-of-the-art oxygen and carbon dioxide selective membranes, modules, and processes, Dr. Sanders has provided technical leadership to the membrane community. These efforts are documented in approximately 40 patents and a nearly equal number of technical publications. Dr. Sanders' specific expertise in membrane separations and broad expertise in chemical engineering product/process conception, development, and design led to his stewardship of Air Liquide's advanced separation technology portfolio and signoff responsibility for separation process design. He has received multiple awards from The Dow Chemical Company and Air Liquide. He also is highly sought plenary speaker for membrane meetings. Dr. Sanders served NAMS as a Board of Directors Member for nine years, chair of the Workshop committee for six years, Vice-President, President, and co-chair of the 2011 Annual Meeting in Las Vegas. He holds BS and PhD degrees in chemical engineering from North Carolina State University.

NAMS Education Innovation Fellow

The NAMS Board of Directors would like to congratulate the inaugural NAMS Education Innovation Fellows. They were selected for their creative proposals to advance membrane science education through engaging, easily disseminated means, as well as a proven track record of educational innovation. As part of this honor, $7,500 will be awarded to fund their respective projects, and additional funds will allow them to present the results of their work at NAMS 2021.

2020 Fellows

Dr. David Latulippe

Dr. David Latulippe, Associate Professor of Chemical Engineering at McMaster University, is recognized as a 2020-2022 Education Innovation Fellow for his proposal to develop a coursework package based on membrane process simulation software. The objectives of the course are to introduce students to the capabilities of ultrafiltration, nanofiltration, reverse osmosis, and ion exchange processes and to teach the factors required to design and operate real-world membrane systems. Dr. Latulippe's work will combine highly engaging design project case studies with an ability to be disseminated to institutions around the world with ease. Such a result exemplifies the objectives of the NAMS Education Innovation Fellowship program and the society looks forward to Dr. Latulippe sharing the final product with the rest of the membrane community.

Dr. Daniel Anastasio   Dr. Marissa Tousley

Dr. Daniel Anastasio and Dr. Marissa Tousley, both Assistant Professors of Chemical Engineering at Rose-Hulman Institute of Technology, are recognized as 2020-2022 Education Innovation Fellows for their proposal to develop low-cost, customizable, 3-D printed plate-and-frame membrane systems. The modules, which can be used to demonstrate microfiltration, ultrafiltration, nanofiltration, reverse osmosis, and electrodialysis, will be developed to connect to inexpensive auxiliary equipment and will be made broadly available with CAD files, microcontroller code, assembly instructions, and educational materials. The intended combination of flexibility for education and ease of dissemination perfectly embodies the intent of the Education Innovation Fellowship program. NAMS anxiously awaits the completion of Dr. Anastasio and Dr. Tousley's work to see how the benefits spread throughout the membrane community beyond the Rose-Hulman campus.

NAMS 2020 AWARDS

Awards Session - Oral 29 [Room 4], 1:00 pm - 3:00 pm (EDT), Thursday, May 21ST

Four award recipients* will be featured and give their oral presentations
at 12TH International Congress on Membranes and Membrane Processes (ICOM) in London, UK.
Visit the 2020 ICOM website for more information: http://www.icom2020.co.uk

Student Fellowship Award

The NAMS Student Fellowhsip Awards are presented annually to outstanding graduate students in the area of membrane science and technology.

Woochul Song
University of Texas at Austin
USA, (ICOM 2020*)

Holden Lai
Stanford University
USA, (ICOM 2020*)

Jamie Idarrage-Mora
Clemson University
USA

Abhishek Dutta
University of British Columbia
Canada

Hongxi Luo
University of Virginia
USA

Young Membrane Scientist Award

The NAMS Young Membrane Scientist Awards are awarded to those individuals who are within five years of completing their Ph.D. degree and have already proven to be outstanding membrane scientists in the area of membrane science and technology.

David Warsinger
Purdue University
USA, (ICOM 2020*)

Xiaoli Ma
University of
Wisconsin-Milwaukee
USA, (ICOM 2020*)

Hee Jeung Oh
Pennsylvania State
University
USA

Giuseppe Genduso
KAUST
Saudi Arabia

Tiezheng Tong
Colorado State University
USA

[Room 1], 11:00 AM (EDT), Monday, May 18TH

Accelerating Technology Development By Making Data More Reproducible: Lessons Learned From Literature Meta-Analysis

Plenary Speaker

Dr. David Sholl

School of Chemical & Biomolecular Engineering
Georgia Institute of Technology

Dr. David Sholl is the John F. Brock III School Chair of Chemical & Biomolecular Engineering at Georgia Tech. David's research uses computational materials modeling to accelerate development of new materials for energy-related applications, including generation and storage of gaseous and liquid fuels and carbon dioxide mitigation. His group has published over 350 papers. He has also written a textbook on Density Functional Theory, a quantum chemistry method that is widely applied through the physical sciences and engineering, and a novel, *Polyphony*. David was instrumental in the development of RAPID, a $70M DOE-funded Manufacturing Institute focused on process intensification run by AIChE. He is on the Board of Directors of AIChE and in January 2020 chaired the inaugural Gordon Research Conference on Chemical Separations. More information on David's research group is available from sholl.chbe.gatech.edu

Abstract

Nearly every researcher has had the frustrating experience of not being able to reproduce data previously reported in the literature. This experience has important implications for individual research productivity, the success of "big data" techniques and the ability of sub-fields to advance new technologies. But what can be done to improve this situation? I will describe a variety of quantitative conclusions that were drawn from systematic literature meta-analysis of thousands of papers describing synthesis of porous materials and molecular adsorption in these materials. These conclusions motivate steps that individuals, research groups and research communities can take to make data more reliable.

NAMS

The North American Membrane Society

2021

Convergence
Connecting Membrane Science with Societal Challenges

3 ½ days of presentations including 2 poster sessions and Tuesday afternoon free for networking time

Chairs:

Uwe Beuscher, W.L. Gore & Assocs.
Yifu Ding, University of Colorado - Boulder
John Pellegrino, University of Colorado - Boulder

NAMS 2020 ONLINE - SPONSORS

EMERGING MEMBRANE MATERIALS, PROCESSES, AND APPLICATIONS

DIAMOND LEVEL

National Science Foundation

GOLD LEVEL

Air Liquide

Air Products

SILVER LEVEL

Membrane Technology & Research

ExxonMobil

BRONZE LEVEL

Hiden Isochema Ltd

Evonik

MDPI AG - Membranes, Polymers (Journals)

Separation Science Center, Chemical and Biomolecular Engineering, Georgia Tech

Sterlitech

W.L. Gore & Associates, Inc.

National Science Foundation

Summary of ALaS for ICOM

Air Liquide

Air Liquide provides industry-leading, advanced separation technology for use in Aerospace, Clean Energy, GHG capture, Biogas, Natural Gas, Chemical and Inerting markets. Committed to innovation in membrane and complementary separation technology for more than 30 years, Air Liquide is the preferred technology partner for customers worldwide.

Air Liquide Advanced Separations (ALaS) has the most diversified gas membrane and separation product portfolio, including yet not limited to, cryogenic, adsorption and absorption technologies. ALaS specializes in the separation of air gases for inerting and aerospace, recovery/purification of hydrogen, methane and helium, and upgrading of subquality natural gas (H_2S, CO_2 and H_2O removal). Aligned with the corporate climate commitment to reduce GHG emissions and improve air quality, ALaS offers a full line of biogas recovery products for digesters and landfills with an option for Green H_2 production.

Air Liquide's breadth and experience in membranes and complementary separation technology provides an optimized membrane (or hybrid) solution, while contributing to a new low carbon society.

We Twist High-Performance Polymers to Highly Efficient Separation Membranes
Evonik offers innovative, hollow-fiber membranes of the SEPURAN® brand for efficient gas separation and the unique PuraMem® and DuraMem® flat sheet membranes for filtration of organic solvents.
www.evonik.com/sepuran
www.evonik.com/puramem
EVONIK
Leading Beyond Chemistry

Hiden Isochema is a world leader in the design and manufacture of sorption instruments for research, development and production applications in materials science and related fields.
HIDEN ISOCHEMA
We offer a range of fully automated gravimetric and manometric instruments. Our full product range also includes dedicated breakthrough analyzers and unique climate control systems.
GAS AND VAPOR SOLUBILITY
PERMEATION AND DIFFUSION
SORPTION ISOTHERMS AND KINETICS
hidenisochema.com
+44 (0)1925 244678 | info@hidenisochema.com

GORE

Together, improving life

STERLITECH

Corporation

NAMS 2020 SPONSORS - click the logos to visit our sponsors!

DIAMOND SPONSOR

National Science Foundation

GOLD SPONSORS

AIR PRODUCTS

Air Liquide

SILVER SPONSORS

ExxonMobil

MTR MEMBRANE TECHNOLOGY & RESEARCH

BRONZE SPONSORS

HIDEN ISOCHEMA

STERLITECH Corporation

EVONIK Leading Beyond Chemistry

GORE

Georgia Tech ISSEC

polymers an Open Access Journal by MDPI

membranes an Open Access Journal by MDPI

Abstracts – Oral Presentations, Monday

Presenting authors are indicated by an asterisk*. See the most up-to-date version of the program book with full abstracts at **www.membranes.org/nams-2020/**

Oral 1 - MEMBRANE FOULING I

12:00 pm (Room 1, Monday)

Oral 1 - Electrochemical Prevention of Mineral Scale on Electrically Conducting Desalination Membranes

David Jassby (University of California, Los Angeles), Unnati Rao (University of California, Los Angeles), Bongyeon Jung* (University of California, Los Angeles).

A major challenge faced by membrane-based desalination processes is the growth of mineral crystals on the membrane surface (scaling). Scaling restricts the flow of water through the membrane and can cause physical destruction of the membrane material. Scaling occurs when super saturation conditions develop along the membrane surface, a process known as concentration polarization. The depletion of water along the membrane due to its passage to the permeate side leads to the formation of a concentration polarization layer. To mitigate scaling, concentration polarization must be minimized by limiting water recovery, encouraging turbulent conditions and the use of anti-scalants. Here, we report on an energy efficient technique that uses electrokinetic mixing to nearly eliminate calcium sulfate and silicate scaling in membrane distillation and nanofiltration separation processes. The application of low potentials (2Vpp cell potential) to the membrane had a dramatic impact on scaling, with the impact being dependent on the frequency of the potential applied. Theoretical calculations reveal that the concentrations of ions along the membrane surface respond rapidly to the applied electrical potentials. We hypothesize that frequent shifts in membrane polarization induce electrokinetic mixing in the stagnant boundary layer and reduce available time for crystal nucleation, shifting the system from rapid heterogeneous nucleation on the membrane surface to slow homogeneous nucleation in the bulk fluid, which nearly eliminated membrane scaling.

12:20 pm (Room 1, Monday)

Oral 2 - Assessment of Oil Fouling By Oil-Membrane Interaction Energy

Henry Tanudjaja* (Nanyang Technological University), Jia Wei Chew (Nanyang Technological University).

Membrane-filtration is promising for treating the voluminous oily wastewater, especially when the oil emulsions are smaller than 20 μm. However, studies on the inevitable membrane fouling phenomenon by oil are rather scarce. In particular, a question that remained to be addressed was whether the DLVO or XDLVO model provides better predictions of the oil-membrane interfacial interactions and thereby the extent of fouling. Accordingly, this study investigated four oil types (namely, hexadecane, soybean oil, fish oil and crude oil) that were stabilized by the same non-ionic surfactant (namely, Tween 20) and had similar mean droplet diameters. The direct observation through the membrane (DOTM) technique was used to quantify the critical flux of the different emulsions, and both the XDLVO and DLVO models were used to quantify the foulant-membrane and foulant-foulant interactions. DOTM results indicated that the critical flux values were similar for all oils under the conditions tested, except crude oil. Although the XDLVO model appears to be more comprehensive than the DLVO in terms of accounting for the additional Lewis acid-base polar (AB) interaction that is acknowledged to be important in membrane-filtration, results indicate that the dominance of the AB component drowns out the other interactions like that of electrostatics (EL) in this case, which impedes accurate prediction of the different fouling tendencies by the different oil types.

12:40 pm (Room 1, Monday)

Oral 3 - Role of Instability Phenomena in Mitigating Membrane Fouling

Weiyi Li* (Southern University of Science and Technology), Xin Liu (Southern University of Science and Technology).

It has been recognized that hydrodynamic instabilities could play a critical role in generating different secondary flows in a fluid channel; more complex hydrodynamic interactions would be involved in a membrane process, where the membrane can be viewed as a permeable wall and membrane spacers with various structures are usually employed in the channel. This study exploited optical coherence tomography (OCT) to in-situ characterize the membrane fouling while highlighting the instability phenomena induced by the wall permeation and the centrifugal acceleration. The evolution of the cake layer on the membrane surface was visualized and quantified to infer the interplay between the foulant particles and the secondary flows associated with the instability phenomena. The characterization results indicate that the centrifugally induced instability could result in an extremely uneven distribution of the deposits and thereby compromise the filtration performance enhanced by the additional shear stresses arising from the secondary flows. This study would shed light on the development of design heuristics for membrane spacers or modules with novel structures.

1:00 pm (Room 1, Monday)

Oral 4 - Do membrane hydrophobicity and hydrodynamic shear affect the initial deposition and pioneering colonization of anaerobes isolated from an anaerobic membrane bioreactor?

Yang yang* (Ben-Gurion University of the Negev), Gideon Oron (Ben-Gurion University of the Negev), Moshe Herzberg (Ben-Gurion University of the Negev), Roy Bernstein (Ben-Gurion University of the Negev).

Introduction

This study investigated the initial deposition and pioneer colonization of anaerobes, isolated from an AnMBR, on membranes with increased hydrophobicity.

Methods

Initial deposition experiments were conducted in an anaerobic chamber under low shear rates (0.45 s-1), high shear rates (3.09 s-1) without filtration and under filtration conditions (59 s-1, 17 LHM), with PAN, PES, PVDF UF membranes.The deposited cell density was quantified by CLSM for cell transfer coefficient (Kd) calculation. Cell imaging was done by SEM and deposited pioneering bacteria and archaea was identified by DNA sequencing.

Results and Discussions

Under non-filtration conditions, the hydrodynamic shear had no significant effect on the Kd of the cells deposited on hydrophobic PVDF. In contrast, Kd was elevated on the hydrophilic PAN under low shear rates. Under filtration conditions, the Kd of the cells deposited on PVDF was significantly higher than the case using PAN. These results were consistent with the results under high shear rates and reversed to the results under low shear rates without filtration. Notably, the Kd values under the filtration conditions were two orders of magnitude higher than the ones without filtration, suggesting that both conditions can be described by the convection-diffusion-migration model. However, under filtration conditions, both elevated cross-flow rate and the convection of cells to the membrane, which provides drag force, result in a higher cell deposition rate. The sequencing results indicate that the main selective force exerted on dissimilarity of pioneer bacterial diversity was hydrodynamic shear, whereas the membrane type was the major selective force on the dissimilarity of archaeal diversity. This study demonstrates that the hydrophobic membranes get fouled by anaerobes much easier and faster than the hydrophilic membranes in AnMBR. Physical cleaning for maintaining membrane performance should be applied for hydrophilic membranes.

1:20 pm (Room 1, Monday)

Oral 5 - Nanofiltration of saline oil-water emulsions: Combined effect of salt concentration polarization and fouling by oil on flux performance

Charifa Hejase (Michigan State University), Vlad Tarabara* (Michigan State University).

The exploration and production of oil and gas generates large volumes of oily wastewater known as produced water. Produced water composition varies significantly depending on the type of hydrocarbon product generated, the geographic location of the field, and the amount of time the water remains in contact with the geological formation.

However, the main constituent is suspended oil (100 to 5000 mg/L) often accompanied by elevated levels of salt. The high salinity in produced water is primarily attributed to the dissolved sodium and chloride ions where the concentration of total dissolved solids can reach ~ 300 g/L. If not properly treated, discharged produced water can pose significant environmental risks.
The present study focused on the separation of saline oil-in-water emulsions by nanofiltration membranes. We show that the interaction of emulsified oil and rejected dissolved species (NaCl and surfactant) in the vicinity of the membrane surface affects permeate flux and selectivity of separation. Concentration dependencies of both the salt permeability coefficient and the reflection coefficient are determined and used to evaluate relative contributions of osmotic pressure and oil fouling to the overall flux decline. The decrease in the reflection coefficient with solute concentration minimizes the effect of concentration polarization and makes oil layer formation on the membrane surface the dominant fouling mechanism. We conclude that the overall performance of a nanofiltration membrane treating saline oil-water emulsion is determined by emulsion stability and is relatively insensitive to feed salinity. Understanding the behavior of oil within the mass transfer boundary layer with high salt concentration should help devise better strategies for minimizing membrane fouling.

Oral 2 – CARBON CAPTURE I

12:00 pm (Room 2, Monday)

Oral 6 - Novel Facilitated Transport Membrane and Process for Post-Combustion Carbon Capture

Yang Han* (The Ohio State University), Kai Chen (The Ohio State University), Witopo Salim (The Ohio State University), Dongzhu Wu (The Ohio State University), Winston Ho (The Ohio State University).

Large-scale application of membrane in post-combustion carbon capture has been limited by the trade-off between CO2 permeance and CO2/N2 selectivity of most polymeric membrane materials. In order to overcome this limitation, research efforts on facilitated transport membrane (FTM) have been devised with the objectives of (1) developing carriers with high CO2 loading capacity and reactive diffusivity, and (2) designing membrane processes that can capitalize on the outstanding selectivity shown by FTM. In this presentation, a novel FTM was synthesized in a composite membrane configuration with a 170-nm selective layer coated on a nanoporous support. In the selective layer, polyvinylamine was used as fixed-site carrier and an amino acid salt, synthesized by deprotonating sarcosine with 2-(1-piperazinyl)ethylamine, was blended as mobile carrier. The membrane was used to fabricate 1.4-m2 spiral-wound modules, which exhibited a CO2 permeance of 1450 GPU and a CO2/N2 selectivity of 185 at 67°C with actual flue gas at the National Carbon Capture Center in Wilsonville, AL, USA.
In addition, for the range of CO2 partial pressure relevant to carbon capture from coal-derived flue gas, significant increase in CO2 permeance can be achieved upon the bulk removal of CO2. For instance, at 67°C, the CO2 permeance can increase from 1464 to 1918 GPU when the CO2 partial pressure reduces from 74.1 to 3.9 kPa. Such a carrier saturation phenomenon has been modeled and incorporated into a two-stage membrane process featuring partial retentate recycle. It has been demonstrated that the bulk CO2 removal reduces the CO2 partial pressure gradually in the membrane module. This feature mitigates the carrier saturation and results in an uprising CO2 permeance upon the CO2 removal. For the membrane performance at 67°C, an attractive capture cost of $41.75/tonne can be achieved for 90% CO2 capture in addition to offering the reduced system footprint.

12:20 pm (Room 2, Monday)

Oral 7 - Carbon Molecular Sieves with Superior H_2/CO_2 Separation Properties at Elevated Temperatures for Pre-combustion CO_2 Capture

Haiqing Lin* (The State University of New York at Buffalo), Hien Nguyen (The State University of New York at Buffalo), Maryam Omidvar (The State University of New York at Buffalo).

Membrane materials with high H2 permeability and H_2/CO_2 selectivity at 200–300 °C hold promise for pre-combustion CO_2 capture. However, polymers with high permeability tend to exhibit weak size-sieving abilities and thus low selectivity. Herein we demonstrate that carbonization of suitable polymers to form carbon molecular sieves (CMS) can break the tradeoff. Polybenzimidazole (PBI) is one of the leading polymers for this separation with H_2 permeability of 27 Barrer and H_2/CO_2 selectivity of 14 at 150 °C. Its carbonization at 600 oC and 900 oC leads to pure-gas H_2 permeability of 370 Barrer and 54 Barrer, and H_2/CO_2 selectivity of 8.9 and 80 at 150 oC, respectively. These results are consistent with the decrease of the free volume element size from 5.08 Å to 4.90 Å (determined using PALS). The pyrolysis forms micro-cavities leading to high gas permeability and ultra-microporous channels leading to strong size-sieving ability. The PBI pyrolyzed at 900 oC shows mixed-gas H2 permeability of 39 Barrer and H_2/CO_2 selectivity of 53 at 150 °C, and good stability for 40 h in the presence of 0.31 mol% water vapor. Moreover, PBI can be doped with phosphoric acid (H_3PO_4) before the pyrolysis. The H_3PO_4 crosslinks the PBI chains (increasing the size-sieving ability), while it is thermally lable (lowering the carbonization temperatures needed). For instance, PBI with a doping level of 0.11 (defined as the molar ratio of H3PO4 to PBI repeating units) was carbonized at 600 °C and exhibits mixed-gas H_2 permeability of 110 Barrer and H_2/CO_2 selectivity of 48 at 200 °C. When the doping level increases to 0.23, the mixed-gas H_2 permeability decreases to 46 while the H_2/CO_2 selectivity increases to 80 at 200 °C, above Robeson's 2008 upper bound and demonstrating their robustness against physical aging and CO2 plasticization.

12:40 pm (Room 2, Monday)

Oral 8 - Carbon Capture Performance of Crosslinked Polyphosphazene Blend Membranes

Victor A Kusuma* (National Energy Technology Laboratory), Zi Tong (National Energy Technology Laboratory), Lingxiang Zhu (National Energy Technology Laboratory), Joshua McNally (Idaho National Laboratory), James Baker (National Energy Technology Laboratory), Christopher Orme (Idaho National Laboratory), David Hopkinson (National Energy Technology Laboratory), Frederick Stewart (National Energy Technology Laboratory).

Thiol-ene photopolymerization was a highly effective technique to cross-link gummy polyphosphazene elastomers containing methoxyethoxyethoxy (MEE) side groups. The solid material has a pure gas CO_2 permeability of 610 barrer and CO_2/N_2 selectivity of 35 at 40°C. The immediate benefit of this facile cross-linking technique was enabling the material to be applied as a solid selective coating on a porous membrane support. In this talk, we will demonstrate recent exciting advances we have achieved with this membrane. The material was successfully tested under coal-derived flue gas at the National Carbon Capture Center, demonstrating stable performance over 500 hours of continuous testing. The material performance has been characterized with mixed gases containing various levels of humidity: gas permeability decreased with increasing humidity, and was fully reversible, with selectivity remained barely changed. Finally, we successfully fabricated a thin film composite membrane with CO_2 permeance of 1200 gpu with CO_2/N_2 selectivity of 31. The membrane performance, monitored over 2000 hours, did not exhibit reduction due to physical aging because of the rubbery nature of the coating material.

1:00 pm (Room 2, Monday)

Oral 9 - Design and Techno-Economic Analysis of a Zeolite Membrane Reactor Intensified IGCC Process

Lie Meng (Arizona State University), Haoren Lu (Nexant, Inc.), Gerald Choi (Nexant, Inc.), Jerry Lin* (Arizona State University).

The application of membrane reactors into an integrated gasification combined cycle (IGCC) plant for electrical power generation offers an attractive route to intensify the water-gas shift (WGS) reaction with CO2 capture. Palladium-based membranes are only permeable to H2, however, their practical applications are limited by their high cost and scarcity on the chemical stability towards H2S. Recently, we have developed a pore-modified MFI zeolite membrane with good H2 perm-selectivity, excellent chemical stability, and relatively low manufacturing costs. The use of MFI zeolite membrane reactor (ZMR) in high-temperature (>400oC) WGS confirms a significant enhancement in the CO conversion. In this work, using an integrative approach that includes experimental investigation on membrane property, mathematic modeling of ZMR, and techno-economic analysis (TEA) for a cost-benefit comparison, we uncovers features that determine the power output, capital expenditure (CAPEX), operating expenditure (OPEX), cost of electricity (COE) and cost of CO2 capture in a 550-MW ZMR-IGCC process. We optimize the performance of single-stage ZMRs with currently available membrane property for process design and achieve CO conversion >98%, CO2 capture ratio >93%, together with more than 96% H2 is recovered. The TEA on the ZMR-based IGCC reveals a decrease in the COE from 138.6 mills/kWh (reference Case 2 from DOE/NETL 2010/1397) to 136.3 mills/kWh. Other benefits of the ZMR-IGCC includes more power is generated from the steam turbines due to less steam consumption in the Selexol unit, and lower power consumption in CO2 compression because of a high-pressure retentate stream. It is also found that an increase in H2 permeance from 10-7 to 10-6 mol s-1 m-2 Pa-1 could offer an at-least 3-time reduction in the total membrane area requirement. This work is financially supported by the United States Department of Energy and the National Energy Technology Laboratory (DE-FE0026435).

1:20 pm (Room 2, Monday)

Oral 10 - Synthesis of Sterically Hindered Polyvinylamine and Its Application in Facilitated Transport Membranes for CO2 Capture from Flue Gas

Ting-Yu Chen* (The Ohio State University), Xuepeng Deng (The Ohio State University), Li-Chiang Lin (The Ohio State University), Winston Ho (The Ohio State University).

Membrane separation technology requires highly CO2-selective and permeable membranes for a cost-effective post-combustion CO2 capture from flue gas. Amine-containing facilitated transport membranes are able to realize an increased permeability due to the reversible reaction between CO2 and amino groups. Here, we have developed an improved method for the synthesis of a high-molecular-weight sterically hindered polyvinylamine (SH-PVAm) as the new fixed-site carrier in our ultra-thin CO2 selective membranes. Commercial polyvinylamine (PVAm) was mono-methylated into SH-PVAm using the stepwise reductive amination with a highly polar fluorinated alcohol as solvent to enhance the equilibrium shift to the imine formation. The method was found to prevent the over-alkylation, resulting in the increased yield of the target product. The synthesized SH-PVAm was characterized and confirmed by FTIR and NMR. The SH-PVAm solution retained a sufficiently high viscosity after incorporating the aminoacid salt, 2-(1-piperazinyl)ethylamine sarcosinate, as the mobile carrier for the membrane coating without any penetration issue. The SH-PVAm exhibited a significantly enhanced performance in the blended membrane of about 170 nm thickness over the unmodified PVAm for CO2/N2 separation at 57oC. The former membrane demonstrated a CO2 permeance of 1071 GPU and a CO2/N2 selectivity of 183 whereas the latter showed 975 GPU CO2 permeance and 163 selectivity. The improved membrane performance corresponded well with our computational results on the reaction chemistry of CO2 between the sterically hindered vinylamine monomer (SH-VAm) and the pristine vinylamine monomer (VAm) by density functional theory (DFT). SH-VAm showed a preference for the more efficient bicarbonate formation pathway and contributed to a higher CO2 loading capacity. Therefore, the steric hindrance effect of SH-PVAm enhanced the solubility of CO2 in the polymer matrix and resulted in the higher CO2 permeance.

Oral 3 – SEAWATER DESALINATION

12:00 pm (Room 3, Monday)

Oral 11 - New Insights into Solute-Selectivity Relationship of Thin-Film Composite Polyamide Membranes

Xi Chen* (Columbia University), Chanhee Boo (Columbia University), Ngai Yin Yip (Columbia University).

Thin-film composite polyamide (TFC-PA) membranes are widely used in separation technologies. The solution-diffusion (S-D) mechanism is the most widely accepted theory for transport in osmotically-driven membrane processes. Importantly, the model predicts a tradeoff relationship between permeability and selectivity. However, fundamental understanding of the first principles governing the relationship in aqueous-separation membranes is still lacking, and there is no rigorous theoretical framework relating transport phenomena in TFC-PA membranes to permeant properties. In this study, we advance a framework for better understanding the permeability-selectivity tradeoff relationship in TFC-PA membranes. Water and solute permeabilities of TFC-PA membranes of different selectivities are examined using neutral organic solutes with various sizes and dimensions. A second transport regime in addition to S-D is identified for the first time. A set of characteristic elements is proposed for specific descriptions of the extended permeability-selectivity framework, and underlying factors governing transport are analyzed through relating the transport elements to solute properties. Our study shows that solute size is the principal factor governing the S-D transport. The transport elements can be predicted based on first principles using the classic S-D framework adapted from gas-separation membranes. Influences of solute size on transport in the additional regime are studied. Transport elements are shown interrelated. Lastly, we reconcile the empirical evidence with theoretical principles to propose a fundamentals-based approach for a priori predicting S-D transport behavior in TFC-PA membranes, using only intrinsic physical property of the permeant. Findings of this study provide significant new insights for understanding transport mechanisms in all osmotically-driven membrane processes and establishes the foundation for rationally-informed design of next-generation membranes.

12:20 pm (Room 3, Monday)

Oral 12 - Omniphobic PVDF-HFP nanofibrous membrane with Pine-needle-like hierarchical TiO2 nanostructures: Towards enhanced performance for membrane distillation

Weihua Qing* (New Jersey Institute of Technology), Xianhui Li (The University of Hong Kong), Chuyang Tang (University of Hong Kong / University of New South Wales).

Wetting and fouling phenomena are main concerns for membrane distillation (MD) in treating high salinity industrial wastewater. This work developed an omniphobic membrane by growing TiO2 nanorods on electrospun polyvinylidene fluoride-co-hexafluoropropylene (PVDF-HFP) nanofibers using hydrothermal technique. The TiO2 nanorods form a uniform pine-needle-like hierarchical nanostructure on PVDF-HFP fibers. A further fluorination treatment provides the membrane with a robust omniphobic surface, displaying contact angles of 168o and 153o for water and mineral oil, respectively. Direct contact MD experiments demonstrated that the resulting membrane shows a high and stable salt rejection of >99.9%, while the pristine PVDF-HFP membrane suffers a rejection decline caused by intense oil fouling and pore wetting in the desalination process in presence of surfactant and mineral oil. The superior anti-wetting and anti-fouling behaviors were ascribed to a non-wetting Cassie-Baxter state established by the accumulating of a large amount of air in the hydrophobized hierarchical structures. The development of omniphobic membranes with pine-needle-like hierarchical nanostructures provides an approach to mitigate membrane fouling and wetting in the MD process for the water reclamation from industrial wastewater.

12:40 pm (Room 3, Monday)

Oral 13 - Energy Barriers for Decoupled Cation and Anion Transport in Sub-nanometer Pores

Xuechen Zhou* (Yale University), Razi Epsztein (Technion), Jaehong Kim (Yale University), Menachem Elimilech (Yale University).

In previous studies, energy barriers for coupled cation and anion transport through polyamide thin-film composite (TFC) membranes have been measured with concentration gradient as driving force, and mechanisms underlying individual ion transport (i.e., cation or anion transport) have been only speculated through the comparison between these obtained energy barriers. In this work, to elucidate transport mechanisms of the individual ions in sub-nanometer pores, we employed electric field as driving force to decouple the cation and anion transport and, for the first time, measured the energy barriers for individual cation (i.e., sodium) and anion (i.e., fluoride, chloride, bromide, and iodide) permeation through negatively charged TFC membranes. Notably, we found that the counterion (sodium) experiences higher energy barrier for membrane transport than the co-ions, even when the co-ion has a higher hydration energy (i.e., fluoride). This finding highlights the contribution of pore-ion interactions during diffusion inside the membrane (e.g., electrostatic attraction), in addition to the ion dehydration at the pore mouth, to the overall energy barrier. Further comparison between the energy barriers for individual cation and anion transport and those for their coupled transport validates the hypothesis of previous experimental studies that the energy barrier for coupled cation and anion transport is the addition of those for each of them. Taken together, our study shows this developed electrochemical method is a useful tool for future investigation on ion transport mechanisms.

1:00 pm (Room 3, Monday)

Oral 14 - High Free Volume, Ion-containing Pentiptycene-based Polysulfone Membranes for Water Purification Application

Tao Wang* (University of Notre Dame), Ruilan Guo (University of Notre Dame), Feng Gao (University of Notre Dame), Bill Phillip (University of Notre Dame).

Sulfonated polysulfones have shown promising performance for membrane desalination applications. However, the water permeance and water/salt selectivity still have a lot of room for improvement to be of commercial and practical interests. Previous study showed that incorporation of hierarchical bulky structure units into the backbone of sulfonated polysulfones might be an effective way to enhance desalination performance. Herein, we report a new class of high-free-volume, sulfonated polysulfone copolymers containing hierarchical pentiptycene-based structure as desalination membranes. A series of pentiptycene-containing sulfonated polysulfone random copolymers was prepared with systematically varied pentiptycene content and sulfonation degree and was examined in terms of fundamental water and salt transport properties. The incorporation of space-occupying pentiptycene moiety generated additional free volume and increases the tortuosity of transport pathways, thus enhancing the water permeance and water/salt selectivity simultaneously. For example, a membrane with 33 mol % pentiptycene units and 40 mol % sulfonation degree shows a decent salt rejection of 93% in the dead-end cell filtration experiments, surpassing the BPSH35 (~86%) polysulfone without pentiptycene units at comparable water permeance. To examine the membrane morphological effect, a series of pentiptycene-containing polysulfone multi-block copolymers (PENT0-BPS100) with alternating hydrophilic-hydrophobic blocks (5k-15k) were prepared, with pentiptycene units in the hydrophobic blocks. As a result, the nanophase separated morphology in multi-block copolymers could further improve the water permeance and water/salt selectivity, approaching the empirical upper bound. In this talk, synthesis and characterization of these new polymers will be presented. Discussions will focus on the fundamental relationships between water/salt transport properties and microscopic structures within each polymer series.

1:20 pm (Room 3, Monday)

Oral 15 - Zwitterionic Polysulfone Membranes for Pervaporation

Tejaswi Nori* (Arizona State University).

One of the resources without which human beings cannot continue is water, especially potable water. As the human population rapidly increases, it consumes water at a rapid rate due to which the total freshwater sources available on Earth are starting to decrease. The decrease in the available sources creates a need to find alternative sources of potable water. Desalination of seawater is one of the options available to us, which is mostly done through Reverse Osmosis (RO). RO produces concentrated brine solutions which are difficult to manage; pervaporation uses vacuum on the permeate side to create a concentration gradient and is less sensitive to the salt concentration relative to RO. However, very few investigations have probed the relationship between material properties and pervaporation desalination performance. This work investigates the effect of sulfobetaine zwitterion-modified poly(arylene ether sulfone). PSf membranes were synthesized by step growth polymerization with 2,2–diallyl bisphenol A (DABA) as one of the monomers. The zwitterionic charges were introduced by the addition of 2-dimethyl ethanethiol and 1,3-propane sultone. The charge percent in the polymer was controlled by controlling the amount of DABA added. Membranes with 25 wt% and 75 wt% were prepared. The membranes were fabricated using the non-solvent induced phase separation method. Membrane composition was determined using Fourier-transform infrared spectroscopy and the membrane thickness and morphology were determined using scanning electron microscopy. The separation performance was evaluated using a dead-end pervaporation cell. The flux, pure water permeance and salt rejection were obtained, and fouling performance was evaluated.

Oral 4 – EMERGING MATERIALS FOR LIQUID SEPARATIONS I

12:00 pm (Room 4, Monday)

Oral 16 - Charge-modified polysulfones as a platform for membrane separations

Matthew Green* (Arizona State University), Yi Yang (Arizona State University), Mani Modayil Korah (Arizona State University), Hoda Shokrollahzadeh Behbahani (Arizona State University), Tejaswi Nori (Arizona State University).

Polysulfones have been used for membrane-based separations because of their processability, oxidative stability, and their robust thermomechanical properties. Recently, we have identified a synthetic pathway to introduce a variety of functional groups to the polysulfone backbone, which can tune the transport characteristics as well as introduce fouling resistance. This talk will highlight the adaptable nature of the synthetic pathway and the impact the added functionality has on surface and cross-section characteristics, processability, and thermomechanical performance. In addition, the membrane performance, both the energy efficiency and permselectivity, for the treatment of saline waters with foulants will be discussed. To date, a variety of source waters with a range of salinities and compositions have been tested, which speaks to the modularity of these promising membranes. Finally, the talk will highlight the modularity of the post-polymerization functionalization route employed, which enables the membrane performance characteristics to be tailored for alternative applications, such as CO2 capture. In summary, this talk will highlight recent efforts to develop a platform membrane technology suitable to remove a range of contaminants at variable concentrations in order to deliver potable water.

12:20 pm (Room 4, Monday)

Oral 17 - Diffusion-Induced in situ Growth of Covalent Organic Frameworks for Composite Membranes

Priyanka Manchanda* (King Abdullah University of Science and Technology), Suzana Nunes (King Abdullah University of Science and Technology).

Covalent organic frameworks (COFs) represent an emerging class of crystalline porous polymers. The well-ordered structures with opportunities for diverse structural design and functionality have offered the COFs superior potential in various applications, such as molecular separation, catalysis, gas adsorption. The fully ordered microcrystalline structure of COF is however frequently reached with long reaction times and at high reaction temperatures thereby limiting their applicability to membrane fabrication.

In the present work, we discuss a new methodology for in situ developing thin layers of a covalent organic framework (COF) on a flexible polymeric support induced by diffusion under ambient conditions with reaction times as short as 3 hours. The method is simple and the size of the membrane is limited only by the size of the diffusion cell, which can be scaled up. We demonstrate it, by using 1,3,5-tri-(4-aminophenyl) benzene (TPB) and 2,5-dimethoxybenzene-1,4-dicarboxaldehyde (DMTP) as reactants placed in one side of the diffusion cell and scandium triflate as catalyst on the other chamber, separated by a crosslinked porous polyacrylonitrile membrane. We followed the formation of an ordered COF layer on the substrate by X-ray diffraction and scanning electron microscopy. The diffusion method is advantageous, when compared to previously reported ones for COF sheets formation, as there is no need for transferring a pre-formed layer, which would be a sensitive step and a frequent cause of defects. Investigation of the growth process via analysis of crystallinity and morphology of the COF composite membrane by varying parameters is discussed. The resultant composite membranes have been evaluated for water vapor permeance and exhibit ultrahigh water vapor permeance of 2973 GPU, which is more than 10 times higher than that of composite membranes reported for Pebax®, which is considered promising for dehumidification.

12:40 pm (Room 4, Monday)

Oral 18 - Lyotropic Liquid Crystals with Water-Continuous Nanostructure as Thin-film Composite Nanofiltration Membranes

Yizhou Zhang* (University of Pennsylvania), Chinedum Osuji (University of Pennsylvania).

Manufacturing precisely engineered thin-film composite (TFC) nanofiltration membranes, is of critical importance for meeting the increasing societal demands for wastewater remediation and freshwater purification. The self-assembly of liquid crystal materials provides a potentially attractive approach to realize nanostructured materials with low size dispersity of transport-regulating features in the 0.5 to 2 nm range. To date, however, the realization of practical NF membranes based on liquid crystal materials has been stymied variously by the difficulty associated with aligning nanostructures in films, the poor retention of structure during processing to create stable solid films, and the challenge of fabricating sufficiently thin films (with desired nanostructure orientation and structure preservation) for membrane operation at reasonable pressure drops. Here, we demonstrate a system that enables these critical challenges to be surmounted. Highly selective NF membranes are realized through a cross-linked lyotropic liquid crystal mesophase based on a polymerizable cationic surfactant. The composition of the system permits the deposition of thin films, down to ~ 100 nm, by solution coating. The system assembles to form hexagonally packed cylinders with a center-center distance of roughly 4 nm. Solution deposition of sub-micron films results in the planar arrangement of the cylinders with respect to the film surface, with the ~ 1 nm separation between cylinder surfaces providing the transport regulating feature. Membranes were successfully fabricated on microporous supports using a sacrificial layer strategy. The resulting membranes demonstrate hydraulic permeance over 10 L m-2 h-1 bar-1 with a molecular cut-off at ~ 250 Da, and robust rejection of divalent salts. The ease of fabrication, self-assembly of nm-scale features and competitive nanofiltration performance suggest that these materials may be attractive for next-generation nanofiltration membranes.

1:00 pm (Room 4, Monday)

Oral 20 - Two-Dimensional Covalent Organic Frameworks with Various Backbones and Pore Surface Engineering as Selective Layers of Thin-Film Composite Membranes

Phuoc Duong (University of Wyoming), Valerie Kuehl (University of Wyoming), Veronica Spaulding (University of Wyoming), Jiashi Yin (University of Wyoming), John Hoberg (University of Wyoming), Bruce Parkinson (University of Wyoming), Katie Li-Oakey* (University of Wyoming).

Two-dimensional (2D) materials providing high permeance and selectivity have rapidly emerged as the advanced membranes materials. The inherent features of 2D materials offer significant advantages in water treatment and reuse, including high separation efficiency, low energy consumption, and low cost. Graphene oxide (GO) membranes have been one of the most studied 2D materials with the great potential to achieve both high flux and selectivity in water treatment and reuse. However, the well-known tortuous water molecules transfer path through interstitial GO flakes significantly limits water flux. A new generation of 2D membrane materials with high density tunable nanopores, often referred to as 2D covalent organic frameworks (COFs), enables ultra-fast transportation of water molecules through the nanopores. In this talk, various COF thin-film composite membranes with nanopores of approximately 1-3 nm will be presented and discussed, with a focus on how inner pore functional groups affect water flux and solute rejection. Specifically, the COF membranes in this talk exhibit the increased water flux by orders of magnitude compared to the GO membranes. Additionally, high ion selectivity was observed, demonstrating highly desirable molecular separation for filtration processes. These experimental observations of high selectivity and high permeance support the hypothesis that COF pores may be well aligned throughout COF membranes. This hypothesis is also consistent with cross sectional scanning electron microscopic (SEM) images of COF membranes. In contrast to the distinct laminar structure of GO membranes, the COF membranes appear to be isotropic. Beside the advantage of achieving high selectivity and permeance, the tunable nature of COF pores resulting from the functional groups lining the pores, provides a great platform to probe the separation mechanism of different feed conditions through the membranes, including the effects of pH, charges, size and shape of solutes.

Oral 5 – INORGANIC MATERIALS

12:00 pm (Room 5, Monday)

Oral 21 - Synthesis of polycrystalline ZIF-8 membranes in few minutes for CO2/N2 and CO2/CH4 separation

Kumar Varoon Agrawal* (Ecole polytechnique fédérale de Lausanne), Jian Hao (Ecole polytechnique fédérale de Lausanne), Deepu Babu (Ecole polytechnique fédérale de Lausanne).

The superior performance of ZIF membranes in gas separation has propelled intensive efforts to synthesize polycrystalline submicron-thick ZIF films in a scalable way [1]. The traditional solvothermal crystallization route is disadvantageous compared to recently reported vapor-phase crystallization routes [2,3] as the synthesis time for membranes in the former case is often several hours/days. This is mainly because the crystallization in the solvothermal route (dipping a substrate in a growth precursor solution) is difficult to control with the precursor concentration dropping as a function of time.
In this presentation, I will discuss novel crystallization using sustained precursors (CUSP) route that maintains a high precursor concentration in the growth step, hindering the undesired Ostwald ripening observed in the late stage of growth [4-5]. As a result, well-intergrown polycrystalline MOF films hosting a uniform grain size and a thickness of a few hundred nanometers could be obtained at room temperature in just 8 minutes, yielding attractive H2/C3H8 selectivity (2433) and C3H6/C3H8 (30) selectivities.
Further, I will discuss post-synthetic treatment, which shrinks the ZIF-8 lattice by ca. 1%, drastically improves the molecular-sieving performance of these films [6]. Lattice stiffening is confirmed by the appearance of a temperature-activated transport, attributed to a stronger interaction of gas molecules with the pore aperture, with activation energy increasing with the molecular size (CH4 > CO2 > H2). High CO2/CH4, CO2/N2, and H2/CH4 selectivities of up to 32.5, 29.3, and 175, respectively, and complete blockage of C3H6 is achieved.

12:20 pm (Room 5, Monday)

Oral 22 - Zeolite membrane separator for safe Li-ion batteries

Kishen Rafiz (Arizona State University), Jerry Lin* (Arizona State University).

Lithium-ion batteries have emerged as the most popular energy storage system, but they pose safety issues under extreme temperatures or in the event of a thermal runaway due to flammable polymer membrane separators used to separate electrodes in the batteries and organic solvent used for electrolyte. The recent development of salt-concentrated electrolytes with fire-retarding solvents and ceramic membrane separators offer promise to solving the problems related to the safety concerns of Li-ion batteries. However, these salt-concentrated electrolytes have an inherently high viscosity and low wettability on commercially dominant polymeric separators, which results in a bottleneck for commercialization in terms of a long wetting-treatment for separator and cathode materials. Here we report electrode (LiNi1/3Mn1/3Co1/3O2 - NMC) cathode)-coated zeolite membrane separators synthesized using a scalable blade-coating process. The zeolite membrane separator shows a much higher wettability towards these electrolytes due to their higher surface energy and intra-particle pores. These coated zeolite membrane separators when used in full cells with salt-concentrated electrolytes, demonstrate higher charge retention on long-term cycling, lower charge transfer, and SEI resistance compared to polymeric separators. The cells with the silicalite separator and the salt-concentrated electrolyte also demonstrate stable performance at temperatures up to 65 °C. This presents avenues to utilize this electrolyte and separator configuration for cells requiring stable performance under conditions of thermal stress. Due to their higher wettability, these membrane separators would require a lower processing time to wet the cathode, anode and separator materials, and thus significantly increase the production rate of batteries if this salt-concentrated electrolyte were to be used in a commercial process.

12:40 pm (Room 5, Monday)

Oral 23 - Preparation and nanofiltration properties of thin film meso-porous oxide membranes via the sonochemical precipitation method

Minghui Qiu* (Nanjing Tech University), Zhihao Jin (Nanjing Tech University), Hendrik Verweij (The Ohio State University), Yiqun Fan (Nanjing Tech University).

Ceramic nanofiltration (NF) membranes attract more attention due to their good thermal and chemical stability, narrow pore size distribution, and high permeance [1]. Here, we propose a synthesis of colloidal sol via the sonochemical precipitation method for the preparation of thin film NF membranes. Ceria nanoparticle dispersions with a narrow size distribution around 4.2 nm was obtained via sonochemical precipitation in salt solutions, followed by controlled growth with the addition of TMAOH and N,N-bis(2-hydroxyethyl)glycine (bicine). With the processes of ceria precursor, twice "coating-drying" cycle and sintering at 500ºC, the achieved ceria membranes on the flat supports displayed the only cubic structure, the defect-free layer with thickness of ~200nm and the pore size around 2-3 nm. They exhibited stable water permeance of 43 L/(m2·h·bar) and salt rejection of 73% for 1 mol/m3 NaCl at pH=4 and 25C.

Then the similar sonochemical precipitation method has been applied to synthesize homogeneous zirconia colloidal sol from the precursor of zirconyl chloride octahydrate solution. The particle size of the resulting sols can be controlled as small as 1.5 nm, and coated on the tubular asymmetric alumina supports to obtain the zirconia NF membranes. The as-prepared zircoina NF membranes displayed a MWCO and water permeance of 500 Da and 20 L/(m2·h·bar) respectively, and showed high performance in purifying ordinary oligofructose.

Acknowledgements

This study was financially supported by the National Natural Science Foundation of China (91534108), the National High Technology Research and Development Program of China (2012AA03A606) and the Water Resources Research Institute Program under grant number 2013OH300B.

References

[1] Lu Y., Chen T., Chen X., et al., Fabrication of TiO2-doped ZrO2 nanofiltration membranes by using a modified colloidal sol-gel process and its application in simulative radioactive effluent. Journal of Membrane Science, 2016. 514: p. 476-486.

1:00 pm (Room 5, Monday)

Oral 24 - Optimization of the performance of the photocatalytic TiO2 membranes fabricated by suspension plasma spray (SPS) process

Elnaz Alebrahim* (Concordia University), Md. Saifur Rahaman (Concordia University), Christian Moreau (Concordia University).

Porous ceramic filtration membranes with anisotropic microstructures are considered to be effective and eco-friendly tools for treating diverse industrial wastewaters. Depending on the contaminants of concern, submicron to a few micron-sized pores are essential for these applications. Suspension plasma spray (SPS) is an emerging thermal spray technique that enables the deposition of a few microns to nano-sized particles in a suspension to generate coatings with the desired microstructure. The potential of the SPS process for fabricating hybrid titanium dioxide-based photocatalytic ceramic membranes, where pores are generated as a result of the presence of un-melted particles has previously been demonstrated by the authors. This study aims to optimize the porosity and the resulting flux of the membranes by investigating the influence of the various SPS process parameters namely the suspension flow rate, standoff distance, and suspension solid content. Moreover, the influence of the particle size of the feedstock on the pore size of the membranes is studied. Two aqueous suspensions of titanium dioxide powder with the mean particle sizes of 200 nm and 50 nm were used to prepare the suspensions. The influence of spray parameters on the characteristics of the membranes was investigated. SEM and X-ray diffraction were used to characterize the microstructure and the crystalline composition of the membranes, respectively. The preliminary results of porosity evaluation using mercury intrusion porosimetry showed an increase in porosity up to 50 percent. The coated membranes showed photocatalytic activity under both UVA and visible lights. Water flux and rejection performance of the membranes will be investigated shortly.

1:20 pm (Room 5, Monday)

Oral 25 - Few-nanometer thick organosilica membranes for high-temperature H2/CO2 separation

Haiqing Lin* (The State University of New York at Buffalo), Lingxiang Zhu (Energy National Energy and Technology Laboratory), Liang Huang (The State University of New York at Buffalo).

Traditional silica membranes exhibit excellent H2/CO2 separation properties for H2 purification and CO2 capture, but they are brittle and faced with challenges for low-cost and large-scale production. Herein, we demonstrate a facile fabrication of ultrathin silica layer on polymeric membranes via plasma treatment to achieve superior H2/CO2 separation properties and excellent scalability of membrane production. Specifically, organosilica membranes were produced by applying oxygen plasma on polydimethylsiloxane (PDMS) thin films supported by microporous polymeric supports. The plasma treatment increases the O/Si ratio to 2 on the top of the PDMS, determined by X-ray photoelectron spectroscopy (XPS), and the ratio at different depths confirms the formation of ~10 nm thick SiO2. With a plasma exposure time of 120 seconds, the membrane comprising 1.5 μm PDMS exhibits mixed-gas H2 permeance of 370 GPU (1 GPU = 10-6 (STP) cm-3 cmHg-1) and H2/CO2 selectivity of 51 at 200 oC, which surpasses the Robeson's upper bound. The membrane shows stable mixed-gas separation performance in the presence of water vapor at 200 oC for 47 h. When the PDMS layer thickness was decreased to 200 nm, the 120-s plasma treatment increases the mixed-gas H2 permeance to 930 GPU while the mixed-gas H2/CO2 selectivity decreases to 32, which is still above the upper bound. The robust H2/CO2 separation properties coupled with the facile fabrication demonstrate the potential of these organosilica membranes for practical H2 purification and CO2 capture.

Abstracts – Oral Presentations, Tuesday

Presenting authors are indicated by an asterisk*. See the most up-to-date version of the program book with full abstracts at **www.membranes.org/nams-2020/**

Oral 6 - MEMBRANE FOULING II

11:00 am (Room 1, Tuesday)

Oral 26 - Exploring and modeling the effect of pattern geometry on nanofiltration membrane fouling

Anna Malakian* (Clemson University), Lucas Messick (Clemson University), Tara Spitzer (Clemson University), Scott Husson (Clemson University).

This presentation will report findings from a systematic study to understand the roles of colloidal chemistry and membrane surface properties on membrane fouling using constant flux filtration. Commercial polyamide nanofiltration membranes were modified with several line-and-groove patterns differing in period dimensions using nanoimprint lithography. Threshold flux measurements were made for as-received and patterned membranes by the flux-stepping measurement method using solutions of silica nanoparticles with different surface chemistry as model foulants. A combined intermediate pore blocking and cake filtration model was applied to the experimental data to determine threshold flux values. Model fits were in excellent agreement with experimental data, indicating that it is an effective tool for determining threshold flux with a sparse data set. Patterned membranes generally exhibited 20-25% higher threshold flux than as-received membranes. Differences in Coulombic interactions and hydrophilicity between the foulants and membrane surface influenced fouling rates. Nevertheless, patterning influenced the threshold flux more significantly than differences in the surface chemistry of foulant particles. This result appears to be based on selective deposition caused by regions of low shear within the pattern grooves. Based on this postulate, the model was altered by incorporating selective deposition into the cake filtration constant. Model validation was performed by threshold flux measurements using patterns with different geometries.

11:20 am (Room 1, Tuesday)

Oral 27 - Oscillation induced aggregation to reduce colloidal fouling in microfiltration

Hamdy Abo Zaid (Ben-Gurion University of the Negev), Levi Gheber (Ben-Gurion University of the Negev), Jack Gilron* (Ben-Gurion University of the Negev).

Oscillation induced aggregation can be an alternative non-chemical method for coagulating particles prior to membrane filtration. This has been directly observed microscopically with 4.5 micron silica. However particles that are in the size range 0.1-0.3 microns tend to be much more of a fouling hazard in membrane processes. Oscillation induced aggregation of fluorescent 0.1 micron silica particles was observed at various oscillatory frequencies as the formation of fluorescent bands. Deposition of fluorescent nanoparticles was then studied in a microscopic flow cell lined with a 0.2 micron microfiltration membrane. Operating the membranes in semi-deadend mode on a suspension of 1500 mg/L nanoparticles at a flux of 40 L/m2-h resulted in deposition that caused transmembrane pressure to increase to 350 mbar. The developed transmembrane pressure was only 100 mbar when the suspension was first subjected to oscillation at 14 Hz. The impact of oscillation on cake formation and removal will be further presented.

11:40 am (Room 1, Tuesday)

Oral 28 - Living Filtration Membranes Demonstrate Anti-Biofouling Properties

Carson Bechtel (Montana Technological University), Katherine Zodrow* (Montana Technological University).

Membrane fouling, particularly biofouling, is considered the 'kryptonite' of membrane processes. Fouling can result in reduced efficiency, altered selectivity, and more frequent membrane replacement. Biofouling is especially critical because microbes in the fouling layer secrete extracellular polymeric substances (EPS), making the layer difficult to remove. Living Filtration Membranes (LFM's) are a sustainable membrane technology composed of bacterial cellulose and native microorganisms. In this study, water from Butte, Montana's three drinking water sources (Basin Creek Reservoir, Moulton Reservoir, and the Big Hole River) during Winter and Spring was pre-treated and used in 7-hour bench scale dead-end filtration tests. After filtration, the initial stages of biofilm formation on the two types of membranes were compared: LFM's vs. commercially produced nitrocellulose. Both types of membranes had a similar molecular weight cutoff (MWCO) of 50kDa. The starting fluxes for both membranes were also similar for each trial, and ranged from 227 to 186 (L m -2 h-1). The starting fluxes for the nitrocellulose membranes ranged from 227 to 187 (L m -2 h-1), and decreased to 5 (L m -2 h-1) for both trials after 400 minutes. The starting fluxes for the LFMs ranged from 205 to 186 (L m -2 h-1), and decreased to the range of 111 to 79 (L m -2 h-1) after 400 minutes. To understand these differences in fouling, a confocal microscope using epifluorescence and the software COMSTAT2 was used to quantify the biofilms. The surface roughness, charge, and contact angles of the membranes were determined using atomic force microscopy (AFM), a SURPASS3 streaming potential analyzer, and a Biolin Scientific Tensiometer. We predict that characterizing the fouling layers and physical properties of living filtration and nitrocellulose membranes will lead to more effective control of biofouling for membrane technologies in the future.

12:00 pm (Room 1, Tuesday)

Oral 29 - Covalently grafting graphene oxide onto UF membranes to improve antifouling properties

Xiaoyi Chen* (State University of New York at Buffalo), Erda Deng (State University of New York at Buffalo), Dongwon Park (State University of New York at Buffalo), Blaine Pfeifer (State University of New York at Buffalo), Haiqing Lin (State University of New York at Buffalo).

Graphene oxide (GO) is negatively charged and has good antifouling properties. However, its good dispersibility in water makes it challenging for membrane surface modification to improve antifouling properties. Herein, we demonstrate a versatile approach to covalently graft GO onto UF membrane surfaces with different hydrophilicity in aqueous solutions at room temperature. Specifically, we first functionalize the PSf membrane surface with polydopamine (PDA), a bio-adhesive that can deposit on membranes with a variety of surface characteristics. Second, the GO with carboxyl acid groups on the edges can be activated by 1-ethyl-3-(3-dimethylaminopropyl) carbodiimide (EDC) and N-hydroxysuccinimide (NHS). Finally, the activated GO can be covalently grafted onto the PDA layer through the reaction of the carboxyl groups and amines in the PDA. The GO was prepared using the modified Hummers' method and then etched using H2O2 to reduce the size and maximize the grafting efficiency. The GO and modified membranes were thoroughly characterized using SEM, AFM, Raman spectroscopy, and zeta-potential analyzer. GO grafting improves the surface hydrophilicity without decreasing the water permeance and decreases the bacteria adhesion by 82%. The GO-modified membranes exhibit 63% less fouling rate and 20% greater critical flux than the pristine one using 1 g/L sodium alginate in a constant-flux crossflow system. The modified membranes demonstrate stability for continuous 48-h operation and interval cleanings using NaOH solutions. Given the simplicity of the GO grafting and the versatility of the PDA for various surfaces, our approach can also be used as post-modification of modules for industrial use.

12:20 pm (Room 1, Tuesday)

Oral 30 - Laser-Induced Graphene for Charged Membrane Applications Enabled by Sequential Infiltration Synthesis

David Bergsman* (Massachusetts Institute of Technology), Beza Getachew (Massachusetts Institute of Technology), Jeffrey Grossman (Massachusetts Institute of Technology).

Membrane-based processes are becoming increasingly popular for water treatment due to their relatively high energy efficiency and low cost. However, these advantages are mitigated by the need for additional pre-treatment steps that are required to maintain effective membrane operation. The treatment and prevention of membrane fouling constitutes a large fraction of typical operational costs. One potential approach to combat fouling is to design conductive membrane coatings that can prevent the attachment and growth of biofoulants, both electrostatically and via electrochemical generation of reactive oxygen species. Despite their potential, these coatings are often expensive, requiring additional chemicals and non-scalable methods to produce, e.g. carbon nanotube mats or other graphitic coatings deposited by vacuum filtration. In this work, we explore the use of laser-induced graphene (LIG) for the creation of conductive ultrafiltration membranes. Porous polyethersulfone (PES) membranes are first treated with alumina using sequential infiltration synthesis (SIS) before being irradiated with an infrared laser. We show that this alumina, which can be scalably produced using spatial SIS, can localize LIG formation to the surface of the membrane, preventing the buried, un-lased areas of PES from melting and losing their porosity during the lasing process. This allows the top-most layer of the PES to be a conductive coating that can be used to charge the membrane surface and used to improve membrane performance (e.g. fouling mitigation). The formation of LIG is verified by scanning electron microscopy and Raman spectroscopy. The conductive layer is also shown to possesses relatively high conductivity, which is important for reducing power consumption in devices. Insight into the mechanism behind the improved stability to melting provided by SIS is provided by thermogravimetric analysis, differential scanning calorimetry, and Fourier-transform infrared spectroscopy.

12:20 pm (Room 1, Tuesday)

Oral 151 - Mineral scale formation and dissolution in RO operation in cyclic modes of operation

Yoram Cohen (University of California, Los Angeles), Anditya Rahardianto* (University of California, Los Angeles), Yeunha Kim (University of California, Los Angeles), Tae Lee (University of California, Los Angeles), Muhammad Bilal (University of California, Los Angeles).

Remote and disadvantaged communities in various regions of the United States and across the globe are confronted with impaired local potable groundwater supplies. In this regard, desalination and water purification of impaired groundwater via reverse osmosis (RO) membranes has been promoted as a potential approach for wellhead water treatment in remote communities. Although RO technology is mature and in use in various sectors, mineral scaling (by sparingly soluble mineral salts) of reverse osmosis (RO) membranes remains a serious impediment that places a limit on the attainable upper limit of product water recovery. The use of antiscalants to mitigate mineral scaling is effective to some degree as is pH adjustment (e.g., for calcium carbonate scaling control), but both increase operating cost and the attainable upper recovery is constrained even when using the above measures. In the present work, alternate approaches were explored to mitigate RO membrane mineral scaling including (a) periodic freshwater or undersaturated feedwater flush for enhancing RO recovery in RO operations in the modes of cyclic semi-batch and steady-state with partial recycle, and (b) RO operation in the mode of cyclic feed flow reversal that also incorporates feed flushing. Studies with RO pilot systems utilizing spiral-wound RO elements were carried out with direct membrane surface scale observations to quantify the kinetics of mineral scaling and mineral crystal dissolution. Flushing efficiency was shown to be paramount to retarding the progression of minerals scaling. Irrespective of the RO operational mode, when freshwater or feedwater (provided it is undersaturated with respect to the problematic mineral scalants) flushing is optimally triggered one can achieve improved scale mitigation and enhanced recovery.

Oral 7 – CARBON CAPTURE II

11:00 am (Room 2, Tuesday)

Oral 31 - High flux CO2-selective single-layer graphene membranes: synthesis and scale-up

Kumar Varoon Agrawal* (Ecole polytechnique fédérale de Lausanne), Shiqi Huang (Ecole polytechnique fédérale de Lausanne).

Nanoporous single-layer graphene (N-SLG), prepared by incorporating subnanometer vacancy defects in the graphene lattice, is highly promising for high flux gas separation because the resistance to diffuse is controlled by a single transition state at the nanopore [1–3]. Molecular sieving resolution (MSR), defined as the difference in the kinetic diameters of molecules to be separated, of 0.3-0.5 Å is needed for CO_2/N_2 and CO_2/CH_4 separation, respectively [4-5].
In this presentation, I will share recent developments in our laboratory on advances on N-SLG membranes including their scale-up to practical dimensions [6-8]. Strategies including controlled etching of graphene lattice, functionalization of graphene lattice as well as hybrid graphene-polymer membranes will be discussed. Specifically, I will discuss a novel etching chemistry, allowing controlled nucleation and expansion of vacancy defects, and yielding MSR of 0.2 Å. We show that molecular cutoff can be adjusted by 0.1 Å by a slow expansion of nanopores, making O_2/N_2 separation possible from atom-thick membranes. Large CO_2 permeances (above 10000 GPU) combined with attractive CO_2/N_2 separation factor (above 20), makes N-SLG membranes highly promising for postcombustion carbon capture. We show that molecular sieving (selectivities > 20) can be obtained from large graphene coupons reaching 1 cm^2 in size.

References

[1] Jiang et al., Nano Lett. 9, 4019–24 (2009).
[2] Celebi et al,. Science. 344, 289–92 (2014).
[3] Li et al., Science. 342, 95–98 (2013).
[4] Yuan et al., ACS Nano. 11, 7974–7987 (2017).
[5] Koenig et al., Nat. Nanotechnol. 7, 728–732 (2012).
[6] He et al., Energy Environ. Sci. 12, 3305–3312 (2019).
[7] Zhao et al., Sci. Adv. 5, eaav1851 (2019).
[8] Huang et al., Nat. Commun. 9, 2632 (2018).

11:20 am (Room 2, Tuesday)

Oral 32 - Tuning the transport properties of CANAL ladder polymers by alkyl substitutions and backbone conformation modification

Francesco Maria Benedetti* (Massachusetts Institute of Technology), Holden Lai (Stanford University), Jun Myun Ahn (Stanford University), Zexin Jin (Stanford University), Albert Wu (Massachusetts Institute of Technology), Maria Grazia De Angelis (University of Bologna), Yan Xia (Stanford University), Zachary Smith (Massachusetts Institute of Technology).

In this work, a new family of polymers based on Catalytic Arene-Norbornene Annulation (CANAL) polymerization is discussed for its application as gas separation membranes. CANALs are rigid ladder polymers similar in structure to PIMs, but they feature norbornyl benzocyclobutene (NBC) backbones, thereby introducing another high-performance ladder architecture to the gas separation field, since PIMs contain only Tröger's base- or dioxane-based connectivity.
We have investigated how subtle changes in short alkyl substituents have an impact on transport properties. The unique purely hydrocarbon and nonpolar rigid backbone of CANALs make them an exceptional model system for fundamental investigations of the structure-property relationships of ladder polymers. An analysis of the energetic contributions to sorption and diffusion interestingly showed that CANALs present features of both dense and microporous polymers.

Recent developments allowed the formation of a new high molecular weight CANAL polymer featuring five-membered rings in the polymer backbone. By making this modification on the backbone conformation, we have been able to achieve remarkable combinations of high selectivity and permeability. In particular, H2/CH4, H2/N2, and O2/N2 separations showed performance beyond the 2015 upper bounds, as well as 2008 H2/CO2 and CO2/CH4 upper bounds. Mixed-gas data were also performed for a broad range of compositions and pressures to assess the potential of this new material in more realistic conditions, and compared with pure- and mixed-gas upper bounds, showing very promising results. Sorption and diffusion experiments, as well as X-ray scattering, were analyzed to provide a fundamental understanding of the exceptional performance reported.
The structural diversity explored in this work will serve to rationally design new microporous polymers with enhanced transport properties, addressing fundamental limitations in current design strategies for polymer membranes.

11:40 am (Room 2, Tuesday)

Oral 33 - Subambient Carbon Capture using Next Generation Hollow Fiber Membrane Modules at the National Carbon Capture Center

David Hasse* (Air Liquide), Shilu Fu (Air Liquide), Shdhir Kulkarni (Air Liquide), Alex Augustine (Air Liquide), Trapti Chaubey (Air Liquide), Dean Kratzer (Air Liquide), Noemi Collado (Air Liquide).

Introduction

Air Liquide is developing a hybrid CO2 capture process based on sub-ambient temperature hollow-fiber membrane operation in combination with cryogenic distillation (Ref 1, 2, 3). Prior work has shown extremely attractive membrane performance. The current work seeks to demonstrate the performance of the novel membrane fiber, in commercial form, with flue gas from a coal fired power plant.

Methods

Bench scale testing at 1,000 NM3/hr scale has started at the National Carbon Capture Center (NCCC, Wilsonville AL) in a field-test unit incorporating pre-treatment, drying, two commercial membrane bundles, and a slip stream test for novel materials. The membranes can be tested at pressures up to 16 bar with temperatures as low as -52° C.

Results

The novel membrane showed significant performance improvements: up to 10-times the CO2 permeance and similar CO2/N2 selectivity (corresponding to approximately 4-times potential productivity per bundle) as compared to baseline commercial fiber. Testing with flue gas started in December 2017 at the NCCC. Testing was concluded in September 2019.
Technoeconomic analyses of this process will be presented, showing state of the art capture costs

Acknowledgement:

Current development work is supported by the Department of Energy, National Energy Technology Laboratory under Award Number DE-FE0026422 (J. Figueroa, Program Manager). Prior work presented here was done under Award Number DE-FE0013163.

References:

1. https://www.netl.doe.gov/File%20Library/Events/2016/c02%20cap%20review/3-Wednesday/T-Chaubey-AirLiquide-Cold-Membrane-Operation.pdf
https://www.netl.doe.gov/File%20Library/Events/2016/c02%20cap%20review/4-Thursday/A-Augustine-AirLiquide-Hollow-Fiber-Membrane-Modules.pdf
Energy Procedia
Volume 63, Pages 186-193 (2014) http://www.sciencedirect.com/science/article/pii/S1876610214018347

12:00 pm (Room 2, Tuesday)

Oral 34 - Scalable charge-modified polysulfone IL support membrane morphologies for CO2 capture in space

Mani Modayil Korah* (Arizona State University), Matthew Green (Arizona State University), Yi Yang (Arizona State University).

Membrane-based gas separations offer cost-effective, efficient strategies for purifying valuable resource streams or for removing unwanted components and by-products. This work highlights the use of a membrane-based CO2 capture technology for space applications. The carbon cycle in space requires careful attention, as the build-up of CO2 has a detrimental impact on the productivity and health of astronauts. The current state-of-the-art materials are packed beds containing pellets of silica and/or zeolite. While effective, the maintenance and reliability of these assemblies are poor. Zeolite dusting can also be detrimental to astronaut health. Thus, supported ionic liquid membranes have been proposed for CO2 capture.

The tunability of chemical structure, to enable high CO2 solubility and selectivity makes Ionic liquids (IL) an ideal choice for this application. Additionally, ILs are nonvolatile and do not possess thermal transitions over a wide range of temperatures. However, ILs must be supported within a polymer matrix to enable free-standing modules that can be deployed in microgravity environments. With the purpose of providing an efficient polymer matrix for ILs, copolymers containing a poly(arylene ether sulfone) (PAES) backbone and charged side chains were synthesized by step growth polymerization and post-polymerization modifications. The charge groups can be modified based on the IL used.

Further, film, capsule and fiber morphologies of different size ranges were developed. A new electrospray encapsulation method and an emulsion phase inversion method was used to make small (1-10 μm) and large (>100 μm) capsules, respectively. The capsules with IL will be embedded in charged polysulfone nanofiber mats to improve membrane performance. This presentation will review progress in the synthesis, characterization, and processing of novel polymers into the various morphologies. The ability to load with variable IL content and perform CO2 removal will be discussed.

1:20 pm (Room 2, Tuesday)

Oral 35 - Ultra-thin supported graphene oxide membrane for CO2 capture

Yi Zhou* (The Ohio State University)

Large scale CO2 capture through membrane separation is the only viable method for cost-effective emission reduction. However, the fabrication of CO2 separation membranes with sufficient flux and selectivity in the presence of water vapor and an acceptable cost price still presents an almost insurmountable challenge. In this work, it is shown that ultra-thin sup-ported graphene oxide (GO) membranes have the potential to meet this challenge. 15 nm thick GO membranes can be prepared through deposition of a dispersion by spin-coating, drying and rapid thermal processing (RTP). The RTP treatment results in the formation of selective CO2 transport pathways. The effect of any remaining non-selective defects is minimized by an application of a thin (<50 nm) layer of Polydimethylsiloxane (PDMS). The optimized membranes showed CO2 permeance (flux/pressure difference) of and CO2/N2 selectivity of 259 at 57°C in a wet mixture of 20% CO2 and 80% N2.

Oral 8 – EMERGING MATERIALS FOR LIQUID SEPARATIONS II

11:00 am (Room 3, Tuesday)

Oral 36 - Ultrathin Two-Dimensional Covalent Organic Framework (COF) Membranes for Molecular Separations

David Shaffer* (University of Houston), Rahul Shevate (University of Houston).

Covalent organic frameworks (COFs) are an emerging class of separation materials formed from porous crystalline polymers. COFs exhibit high throughput and size-based selectivity due to their high density of uniform pores. The pore size and molecular selectivity of COFs may be controlled by the selection of precursor monomers, and a variety of three-dimensional COF chemistries have been reported. The application of COFs for liquid separations remains limited because of the challenges of synthesizing two-dimensional COF materials and translating them into defect-free, large-area membranes. This presentation describes the development of ultrathin two-dimensional COF membranes using a support-free interfacial polymerization technique. Membranes as thin as 24 nm are fabricated using imine-based COF chemistry. The COF membranes demonstrate pore flow behavior and high permeances for both water and organic solvents. Dye rejection tests reveal a sharp molecular weight cutoff of approximately 800 Daltons. The membranes are mechanically robust and chemically stable, with permeance remaining constant after extreme mechanical and chemical stresses like bending tests, acid treatment, alkali treatment, and sonication. The potential applications of these COF membranes for liquid separations are demonstrated through challenge tests, including removing toxins during water purification and size-based separation of biologically-relevant macromolecules from complex organic mixtures.

11:20 am (Room 3, Tuesday)

Oral 37 - Toward Systematic Tuning of Thin-Film Composite Membrane Selectivity

Mackenzie Anderson* (University of California, Los Angeles).

Efforts to develop new reverse osmosis membrane materials can often be hampered by the difficulty of predicting a material's potential selectivity: a tangled web of structure, properties, and processing. To strengthen the basis for using new classes of polymers for membranes, this work explores the relationship between the polymer chemistry and membrane performance of Polybenzimidazole(PBI), PBI-blends, and sulfonated PBI. Our primary objective is to set the foundations for design parameters for achieving a specific selectivity, that are derived from observed relationships between experimental data, polymer chemistry, and solution diffusion-based predictions. PBIs have become popular in the field of membrane technology because of their chemical tolerance, but have versatile chemical properties, including pH dependent surface charge and several crystalline phases that influence their performance as a separation material. Blends and sulfonated derivatives of PBI are prepared, producing amphiphilic systems and made into practical membranes using thin-film lift off (T-FLO). Using T-FLO, the properties of a pre-made, dense, polymer film can be maintained when the support is cured on the film to form a covalently-bound, thin film composite membrane. The interfacial surface tensions and differences in NaCl permeability, as related to feed stream pH, are measured. We have begun the organization of relationships among Lewis acid/base and Van der Waals surface energies, crystallinity and separation performance. These data inform our understanding of how we can, using T-FLO, take advantage of dense thin-films for highly selective, aqueous separations of trace organic contaminants from drinking water.

11:40 am (Room 3, Tuesday)

Oral 38 - Dry-Wet Phase Inversion Synthesis Role on Graphene Oxide-Polysulfone Membrane for Membrane Distillation

Lucy M Camacho* (Texas A&M University-Kingsville), Samuel Olatunji (Texas A&M University-Kingsville).

Graphene oxide-Polysulfone (GO-PSF) membranes for direct contact membrane distillation (DCMD) were successfully synthesized using a dry-wet phase inversion method. This study demonstrated that both the surface and cross-sectional morphology and DCMD desalination performance of the membranes improved by the acidic pH of the GO sheets. A decrease in the pH of the GO sheets led to membranes with highly porous skin surface, low roughness, high contact angle, and high permeate flux in DCMD. The best GO-PSF membrane fabricated in this study had a roughness of 31.75 nm, contact angle of 90.5º, and porosity of 35%. During the DCMD process with a 25,000 mg/L NaCl solution, the most effective membrane produced a permeate flux of 76.9 L/m2·h and salt rejection of 99.97% at a hot-feed and cold-permeate side temperature of 90ºC and 20ºC, respectively. The most effective membrane produced a permeate flux of 62 L/m^2·h and low permeate conductivity when tested with RO concentrate, and rejected arsenic and other ions much better than a commercial polypropylene membrane tested for comparison purposes. These findings are promising for the application of GO-PSF membranes as MD-dedicated membranes.

12:00 pm (Room 3, Tuesday)

Oral 39 - Phosphorene- Based Antifouling Membranes: Synthesis, Fabrication and Applications

Joyner Eke* (University of Kentucky), Isabel Escobar (University of Kentucky).

During nanofiltration, rejected materials can accumulate on the membrane surface and/or pores to form a fouling layer. Phosphorene is a two-dimensional (2D) exfoliated form of black phosphorus (BP). Phosphorene exhibits useful features, including its anisotropic electric conductance and optical responses, which distinguish it from other isotropic 2D materials such as graphene. Specifically, relevant to the field of membrane science, phosphorene possesses photocatalytic properties, which could be explored in making reactive membranes that can self-clean. Furthermore, phosphorene's ambipolar transport characteristics provide it with anti-microbial capabilities since applying electrical potential on conductive surfaces can hinder the development of biofilm. The overarching goals of this project are to prove that phosphorene's electrical conductivity is anti-microbial, then establish that its photocatalytic properties can destroy organic compounds; hence, develop novel nanocomposite membranes holistically from the initial investigation of phosphorene to testing. To this end, for the first time, membranes have been embedded with phosphorene. After modification, phosphorene-modified membranes were used to filter methylene blue (MB). Phosphorene-modified and unmodified membranes displayed similar rejection of MB; however, after reverse-flow filtration was performed to mimic pure water cleaning, the average recovered flux of phosphorene-modified membranes was four times higher than that of unmodified membranes. Furthermore, coverage of MB on phosphorene membranes after reverse-flow filtration was four times lower than that of unmodified membranes, which supports the hypothesis that phosphorene membranes can become self-cleaning. The antimicrobial potential of phosphorene will be tested using Serratia marcescens. This study will investigate the specific role of electric current in bacterial detachment and inactivation when a constant current is applied through phosphorene.

12:20 pm (Room 3, Tuesday)

Oral 40 - Understanding Virus Prefilters in Improving the Filtration Performance of the Planova BioEX Virus Filter for Monoclonal Antibodies

Solomon Isu* (University of Arkansas), AAA ZZZ (Pennsylvania State University), Xianghong Qian (University of Arkansas), Andrew Zydney (Pennsylvania State University), Ranili Wickramasinghe (University of Arkansas).

Virus clearance is an essential component of the downstream purification of all mammalian cell-derived biopharmaceutical products like monoclonal antibodies. Virus filtration flux and throughput are found to be protein specific as well as filter specific. Moreover, product properties such as hydrophobicity, charge and conformational stability play a major role in virus filtration performance. Use of a prefilter is often critical when developing a practical virus filtration step. The effectiveness of a virus prefilter in reducing the fouling during virus filtration is strongly affected by the properties of the protein molecule and the type of impurities present in the feed. Here the effectiveness of three different prefilters fir the removal of foulants based on size exclusion, ion-exchange and hydrophobic interaction mechanisms were investigated using virus filtration performance of a monoclonal antibody through the Planova BioEX hollow fiber virus filter. Size-based prefiltration was largely ineffective, suggesting that the key foulants are not protein aggregates. Depending on the pH and conductivity of the feed streams, removal of foulants during prefiltration is dominated by either hydrophobic interactions or ion-exchange mechanisms. Both hydrophobic interaction and ion-exchange prefilters are found to be effective in improving the filtration performance of BioEX filters. In addition, proteomic analysis using combined SDS PAGE and LC-MS was successfully performed to identify dominant foulants including product variants and specific host-cell proteins. These results provide important guidance for the design of more effective pre-filtration in the context of virus filtration.

Oral 9 – OSMOTICALLY DRIVEN PROCESSES

11:00 am (Room 4, Tuesday)

Oral 41 - Reverse-fluxed Draw Solutes Removal and Microalgae Production in an Osmotic Photobioreactor System For Complete Water Recovery from Secondary Effluent

Zixuan Wang* (Washington University in St. Louis), Zhen He (Washington University in St. Louis), Yi-Ying Lee (University of Maryland Center for Environmental Science and University of Maryland Baltimore Country), David Scherr (Virginia Polytechnic Institute and State University), Yantao Li (University of Maryland Center for Environmental Science and University of Maryland Baltimore County).

Forward osmosis (FO) is an energy-efficient membrane treatment process. The two major bottlenecks of FO are the draw solutes (DS) regeneration and reversed salts flux (RSF) accumulation that increase the operational expense. An osmotic photobioreactor (OsPBR) process was designed to address these two challenge: 1. using concentrated fertilizer as draw solutes so that its diluted final solution could be directly applied in fertilizer irrigation and thus bypass DS regeneration step; 2. coupling microalgae Chlorella vulgaris UTEX 395 and Scenedesmus sp. HTB1 into FO feed solution side to uptake the accumulated nutrient ions (NH4+ and PO43-) during the water recovery from secondary effluent wastewater. Results showed that microalgae did not exert negative impact on water production and 100% water recovery from secondary effluent was achieved with stable averaged water flux of 0.63 ± 0.07 L m-2 h-1 over 16 days of operation. The high RSF of NH4+ induced uncommonly high forward cations exchange (Na+ and K+). Additional chemical precipitation (Ca2+, Mg2+) during FO process further depleted nutrients level and inhibited algal bioactivity. Intermittent small amount of chemicals supplement was necessary for complete NH4+ and PO43- removal in the OsPBR and microalgae density could be concentrated to 3.63 times as initial (0.24 to 0.88 g L-1). With hydraulic retention time of 3.33 d, OsPBR successfully controlled nutrients in the reactor at low level (NH4+-N < 3 mg L-1, PO43--P < 1 mg L-1, and COD < 30 mg L-1). Solids retention time of 17.82 d was demonstrated to be best for maximized microalgae products yield (9.07 g lipids m-3 d-1) and healthy FO operation. This study is the first of its kind to proactively utilize FO reversed ions to stimulate high-density microalgae growth while reusing agricultural water in an energy-efficient manner.

11:20 am (Room 4, Tuesday)

Oral 42 - Zwitterionic Hydrogels with Polyamide-Skin Layers Mitigating Concentration Polarization for Forward Osmosis

Thien Tran* (University at Buffalo, The State University of New York), Shiwei Pan (Wanhua Chemical Group Co., Ltd), Xiaoyi Chen (The State University of New York at Buffalo), Adrienne Blevins (University of Colorado Boulder), Yifu Ding (University of Colorado Boulder), Haiqing Lin (The State University of New York at Buffalo).

Forward osmosis (FO) membranes are often subjected to internal concentration polarization, drastically lowing water flux. Herein, we demonstrate novel impregnated membrane (IMs) of zwitterionic hydrogels in strong porous supports with a skin layer of highly cross-linked polyamides prepared by gel-liquid interfacial polymerization (GLIP). We first impregnated monomers of sulfobetaine methacrylate (SB) and poly(ethylene glycol) diacrylate in a PTFE support. After photopolymerization, the hydrogels were swollen by piperazine solutions and then exposed to trimesoyl chloride to form polyamide skin layers. Such membrane configuration eliminates the pores and thus internal concentration polarization, and the polyamide layer provides high salt rejection. For example, NF90 and SW30-XLE membranes exhibit water permeance (Aw) of 0.10 and 0.046 LMH/bar in FO mode, respectively, while an IM-PA shows an Aw value of 0.25 LMH/bar and Na2SO4 rejection of 90% or higher. The water permeance of the IM-PA is comparable with the commercial HTI-CTA membrane. The effect of hydrogel compositions and the IP conditions for polyamide formation on the water/salt separation properties is thoroughly investigated. The morphologies and nano-mechanical properties of the PA skin layer are investigated with SEM and AFM. The effect of foulants (such as sodium alginate) on the long-term water flux will also be presented.

11:40 am (Room 4, Tuesday)

Oral 43 - A Simplified Modeling Framework to Investigate Osmotically Driven Processes at the System-scale

Zachary Binger* (The University of Arizona), Andrea Achilli (The University of Arizona).

Osmotically driven membrane processes, like forward osmosis (FO) and pressure retarded osmosis (PRO), may hold key advantages when integrated with reverse osmosis for seawater desalination. Research tools developed to investigate these processes have grown in complexity and rigor but can be both time intensive and size limited. For a more simplified estimation of membrane performance, 2-D finite area models have been used to simulate osmotically driven processes at the bench scale with both accuracy and speed. However, previous 2-D models typically rely on plate-and-frame geometries that don't appropriately represent the flow paths of the two streams in spiral-wound membranes (SWM). SWMs present limitations inherent to their geometry and configuration. Most notably, the different rate of pressure losses in each flow channel require the feed and draw streams to be operated at different flow rates to maintain proper operating pressure. This alters the effective osmotic pressure of the system and affects water and energy recovery. Any error in predicting water recovery in module-scale modeling will extend to system level evaluations and should drive model development. In this work, a model was created to rigorously represent SWMs by creating an iteration scheme, that when applied to the differential areas in a finite area model, simulates the crossflow orientation and interaction of the two process streams present in a spiral-wound module. The module-scale model was then expanded to a system-scale model to evaluate the performance of a system when taking into account operational limitations unique to SWMs. The results suggest that multi-stage configurations with periodic recharge of process streams may hold key advantages for increasing the feed water recovery of FO, allowing up to 90% recovery without recirculation. Recharge is advantageous for PRO because it allows for more feed water to be available for recovery, resulting in increased specific energy recovery.

12:00 pm (Room 4, Tuesday)

Oral 44 - Experimental Demonstration of Power Generation from Fertilizer via Pressure Retarded Osmosis

Pouyan Pourmovahed (Oakland University), Jonathan Maisonneuve* (Oakland University).

Large amounts of energy are currently wasted in hydroponic agriculture when concentrated fertilizers are diluted in water prior to plant delivery. In this study we consider the potential of using of a pressure retarded osmosis (PRO) process to convert the potential energy available in fertilizer gradients, to useful mechanical work. Experimental and numerical analysis are used to demonstrate that high power densities of up to 15 W/m2 are obtained when ~7 bar pressure is applied to single fertilizer solutes such as potassium chloride and calcium nitrate, and common fertilizer blends such as Hoagland solution. In addition, we investigate the practical limits of energy recovery that are imposed by the need to maintain reasonable power throughout the fertilizer dilution batch process. Up to 2 Wh per kg of fertilizer solute is recovered while maintaining at least 5 W/m2 power density with commercially available membranes. The possibility of losing valuable fertilizer towards the feed stream is discussed, and is evaluated as less than 1 % of the initial fertilizer mass over the whole fertilizer dilution process. Implications at the food, energy, and water nexus are discussed.

12:20 pm (Room 4, Tuesday)

Oral 45 - Mitigation of Biological Fouling in a Forward Osmosis Membrane Bioreactor

Derrick Satterfield (University of Nevada, Reno), Jack Griffin (University of Nevada, Reno), Terin George (University of Nevada, Reno), Sage Hiibel* (University of Nevada, Reno).

T Thermal osmotic backwashing and mechanical air scouring were evaluated for mitigation of biological fouling in a forward osmosis membrane bioreactor. Elevated backwash temperatures (60 and 80 °C) had improved short-term flux recovery relative to ambient backwashes; however, energy requirements for heating the backwash water may be cost prohibitive at large scale. Biological growth was also enhanced on the feed side of the membrane surface at with 60 °C backwash, leading to decreased long-term flux. Continuous air scouring had the lowest specific energy requirements in the bench-scale setup, but is not economically feasible at large scale. Thus, a combination of reduced duration (2.5 min) and intermittent frequency (every 12 hr) backwashing was found to be the most favorable cleaning method to maximize water production and minimize energy inputs.

Oral 10 – WATER REUSE

11:00 am (Room 5, Tuesday)

Oral 46 - Desalination for a Circular Water Economy

Megan Mauter* (Stanford University).

Today's water systems are enabled by ample fresh water sources, low-cost centralized treatment, and facile wastewater disposal. Climatic change, aging infrastructure, and source water contamination have exposed the vulnerabilities of this linear water paradigm. While seawater desalination enables coastal communities to augment their supply, more broadly securing water systems will require new technologies for distributed desalination and fit-for-purpose reuse of non-traditional waters. Technology innovations coupled with policy changes will enable the evolution of a linear water economy into a resilient circular water economy where water is continuously used and reused and "contaminants" become the feedstocks for other economically valuable processes. This talk will highlight roadmapping and R&D investments by the National Alliance for Water Innovation (NAWI) that will support the development of a circular water economy built on autonomous, precise, resilient, intensified, modular, and electrified desalination systems. This talk will also highlight opportunities for accelerating R&D progress in water reuse through enhanced federal coordination, information gathering efforts to direct technology development, and an expanded role for consulting engineers in knowledge dissemination.

11:20 am (Room 5, Tuesday)

Oral 47 - The impact of pre-chlorination on microfiltration fouling during municipal wastewater effluent filtration for water reuse

Kunal Gupta* (Texas A&M University), Shankar Chellam (Texas A&M University).

Microfiltration (MF) is an integral step of the water reclamation process; however, it's fouling continues to be an impediment. In this study, we looked at chlorination, a commonly employed step before MF in water reuse treatment trains, as a pretreatment step. That is pre-chlorination to control fouling during MF of municipal wastewater effluent. We evaluated the effect of NaOCl dosage (0 to 15 ppm, 90% effective chlorine) on the fouling of 0.1 μm hollow fiber polyvinylidene fluoride (PVDF) membranes. The experiments were performed with disinfected effluent from a local wastewater treatment plant on a bench scale constant flux system for five filtration-backwashing cycles. Fouling after pre-chlorination exhibited a dose-dependent behavior. Higher doses (7.5-15 ppm) employed in this study effectively reduced fouling; however, lower doses (1.5-5 ppm) resulted in increased fouling.
The dose-dependent alteration in fouling was analyzed in terms of interactions of effluent organic matter (EfOM) with the membrane. We investigated (i) sorption tendency of raw and chlorinated EfOM with colloidal PVDF particles (DOC removal during batch adsorption) and membranes (streaming potential variation) and (ii) free energy of adhesion between EfOM and membrane using extended Derjaguin-Landau-Verwey-Overbeek theory. Both sorption tendency and free energy of adhesion varied as a function of chlorine dose and showed a strong correlation to the observed dose-dependent fouling behavior. Also, pre-chlorination effectively changed the dominant fouling mechanism from cake filtration during raw effluent filtration to standard and/or intermediate blocking as revealed by constant flux blocking laws analysis. This change in the fouling mechanism was ascribed to the reduction in particle size with chlorination, which was evinced by particle size distribution measurements. The results from this research suggest the importance of careful consideration while choosing chlorine dosage before MF.

11:40 am (Room 5, Tuesday)

Oral 48 - Understanding colloidal fouling formation during inline coagulation/ultrafiltration process

Bianca Souza Chaves* (The University of Arizona), Andrea Achilli (The University of Arizona).

The inline coagulation/ultrafiltration (IC/UF) process has become a pretreatment strategy for reverse osmosis (RO) fouling control. Integrating coagulation with membrane filtration may enhance the removal of organic matter in the UF, reduce RO membrane fouling, and reduce the treatment time and footprint compared with conventional processes. Our previous results from an engineering-scale UF/RO system using treating tertiary effluent suggest that RO fouling may be associated to colloidal dissolved organic carbon (DOC) that is not retained by the UF membrane. IC/UF may be an alternative pretreatment to reduce RO fouling by aggregation of DOC and retention on the UF membrane surface. This work evaluates the membrane-DOC-coagulant interactions in the IC/UF process for the treatment of tertiary treated wastewater effluent. Aluminum chlorohydrate (ACH) and polyaluminum chloride were tested at different doses (1-10 mgAl3+/L) in a UF bench-scale unit using membrane discs (12 cm2) to compare the mechanisms of hydrolysis and precipitation behaviors. DOC removal, turbidity, excitation-emission matrix fluorescence, filterability, and fouling resistance were utilized to determine colloidal fouling formation, composition, and foulant-surface interactions. Results showed that the lowest filterability (higher permeation resistance) was achieved using 1 mgAl3+/L of ACH, which resulted in a net 20% increase in DOC removal compared with the UF filtrate without coagulant. Scanning electron micrograph images were used to fundamentally understand the formation mechanism of the fouling layer. Furthermore, experiments were performed using a UF unit with a larger membrane surface area (140 cm2) and backwash experiments were assessed to evaluate the resistance-in-series model. Results suggested that residual foulants might interact with the membrane (irreversible fouling) but hydrolysis products/precipitates formed during coagulation lead to mostly reversible accumulation on the membrane.

12:00 pm (Room 5, Tuesday)

Oral 49 - Combination of AnMBR with outdoor microalgae cultivation for industrial wastewater treatment, resource recovery and near-zero waste

Roy Bernstein* (Ben-Gurion University at Negev).

Over the last two decades, anaerobic membrane bioreactor (AnMBR) research gained much interest due to its main advantages throughout wastewater treatment: high effluent quality, nutrient retention, low produced sludge and biogas production. In this study, the treatment of food industry wastewater using AnMBR coupled with cultivation of microalgae using AnMBR effluent was tested under outdoors conditions. The wastewater contained average values of 1.5 gr/L total organic carbon, 160 mg/L total nitrogen, and 9 mg/L total phosphate. The AnMBR process removed 97% of the wastewater TOC retaining TN and TP in the effluent. UF membranes were operated under constant sub-critical flux (4-8 L/m2·h) and the trans-membrane pressure was also relatively stable and low during the whole experiment (15-40 mbar). The membranes required chemical cleaning only after every 3-4 months. The AnMBR effluent was used as a growth medium for the cultivation of three microalgae species (Chlorella, Scenedesmus and Coelastrella) under outdoors experiments during all four seasons in a desert climate. All microalgae species perform efficiently with TN and TP mean uptakes of 25-55 mg/L·day and 1-5 mg/L·day, respectively. Uptakes changed seasonally in accordance with solar radiation. Microalgae biomass productivity were 0.3-0.7 gr/L·day and the required standards of TN and TP for discharge were reached within 2-5 days – in accordance with season. Monod specific growth regression corroborate with the ability to efficiently reach N and P standards with sufficient specific growth rates (0.3-0.8 day-1). Moreover, membrane harvesting feasibility was initially assessed by measuring critical fluxes with correspond to algae concentrations using a UF harvester. These results indicate that an integrated process can be highly efficient in treating high-strength industrial food wastewater in sun-rich dryland areas and could have positive energy, water, and nutrient balances together with low GHG emissions.

12:20 pm (Room 5, Tuesday)

Oral 50 - Membrane Surface Characterization Upon Cleaning of a Heavily Fouled Reverse Osmosis Membrane from an Advanced Reclamation Facility

Bilal Abada* (Texas A&M University), Shankar Chellam (Texas A&M University).

Reverse osmosis (RO) membrane from the third stage of a three RO-pass train in an advanced water purification facility (Orange County, CA) was thoroughly examined for foulants identification and chemical cleaning assessment. Silicon-based foulants were the most dominant foulants on the membrane surface and existed in different forms like silica, silicate, and organosilicons. Biological fouling was observed to a lesser extent as compared to silicon foulants. Minor contributions of aluminum and calcium scales accompanied the dominant combined silicon-biological foulants. Membranes were cleaned with either NaOH or a mixture of sodium tripolyphosphate and sodium dodecylbenzene sulfonate.

Neither of the chosen agents successfully recovered the declined flux with a single cleaning cycle. Due to silicon-based fouling dominance, multiple NaOH cleaning cycles were attempted. Ten cleaning cycles effectively restored the flux, with no improvement in salt rejection. Fourier transform infrared (FTIR) spectroscopy showed evidence of near-complete removal of biological fouling. X-ray photoelectron spectroscopy (XPS) scans indicated substantial removal of Ca scales but only negligible removal of Si and Al from the membrane surface. Therefore, this partial removal of foulants did not explain complete flux restoration.

Interestingly, virgin membrane cleaning within the same cleaning protocols also showed a flux increase. Both XPS and FTIR spectra did not reveal chemical alteration to the membrane active layer. Thus, flux increase could be linked to membrane swelling, rather than a deterioration of the polyamide layer. Hence, we propose that the efficient flux restoration in fouled membranes was due to a combination of partial removal of foulants and swelling of the membranes. These results underline the importance of conducting an in-depth analysis of cleaned membranes that is beyond only flux and salt rejection measurements to assess foulants removal and polyamide layer integrity.

Abstracts – Oral Presentations, Wednesday

Presenting authors are indicated by an asterisk*. See the most up-to-date version of the program book with full abstracts at **www.membranes.org/nams-2020/**

Oral 11 – POLYMERIC AND MIXED-MATERIALS – GAS SEPARATION I

11:00 am (Room 1, Wednesday)

Oral 51 - Effect of thermal treatment on the structure and gas transport properties of a triptycene-based polybenzoxazole exhibiting configurational free volume

Ryan D. Crist (University of Oklahoma), Zihan Huang (University of Notre Dame), Ruilan Guo (University of Notre Dame), Michele Galizia* (University of Oklahoma).

Polymer membranes exhibiting superior stability at temperatures higher than 200°C would improve the thermal and economic efficiency of hydrogen separation from pre-combustion syngas in the Integrated Gasification Combustion Cycle (IGGC), as well as of H2/CH4 separation from refinery processes.
The effect of thermal treatment at 220°C for 10 days on the structure and transport properties of 30 □m thick films of a triptycene-based polybenzoxazole (TPBO) was investigated experimentally and theoretically. In sharp contrast with the behavior of other rigid glassy polymers, such as PIMs and Matrimid polyimide, which lose most of their permeability upon exposure to high temperatures for a few hours, gas and water vapor diffusion and permeability coefficients in TPBO decrease by just 20%, while sorption is virtually unaffected by thermal treatment. Remarkably, the CO2/CH4 selectivity exhibit a negligible change. Fluorescence spectroscopy, WAXD and FTIR analysis indicate that, in contrast with typical behavior of glassy polymers, TPBO does not experience accelerated physical aging, and rule out formation of intermolecular charge transfer complexes upon thermal treatment. According with this physical picture, the diffusion coefficient of penetrant molecules sorbed in the Langmuir's mode, DH, does not change after treatment. Small molecule diffusivity and permeability decline is caused by a decrease in polymer chain mobility, which makes more difficult opening gaps to allow penetrant diffusion jumps. According to this picture, the Henry's mode diffusion coefficient, DD, substantially decreases upon thermal treatment. The higher stability exhibited by TPBO relative to other high Tg glassy polymers is ascribed to the presence of configurational free volume, which is not related to the non-equilibrium transient conformation, but to the molecular configuration and, as such, it is not relaxed upon protracted exposure to high temperatures.

11:20 am (Room 1, Wednesday)

Oral 52 - Optimizing Plasticization Benefits of Polyimides for Membrane-based Natural Gas Sweetening

Yang Liu* (Georgia Institute of Technology), William Koros (Georgia Institute of Technology), Gongping Liu (Nanjing Tech University), Mohamed Eddaudi (King Abdullah University of Science and Technology), Zhijie Chen (King Abdullah University of Science and Technology).

The consumption of natural gas will continue to rise steadily over the next several decades. Methane (CH4) is the major component, but a significant amount of undesirable impurities, especially acidic CO2 and H2S, also present in the raw gas. At present, roughly 30% of proven natural gas resources in the U.S. require treatment. The membrane process offer energy-saving and smaller footprints comparing with commercialized thermally driven gas amine absorption process. An attractive progress for membrane-based natural gas sweetening relies upon the recently revealed "plasticization benefits" of glassy polymer membranes. Unlike traditional observations, plasticization of glassy polymer membranes can promote both H2S permeability and H2S/CH4 selectivity under aggressive conditions, providing opportunities for realistic natural gas sweetening using membranes. Excessive plasticization must be avoided to achieve optimum and stable gas separation performance. In this study, we outline a strategy to achieve plasticization benefits using metal-organic frameworks (MOFs). Based on a proven platform of M-fcu-MOFs and 6FDA-based polyimides, the developed MOF-polyimide hybrid membranes show extraordinary performance for H2S/CH4 separation under realistic gas separation conditions, e.g. > 50 bar, well above the trade-off limit of rubbery polymer membranes. Moreover, the membranes remain high CO2/CH4 separation efficiencies provided by the polyimide. We show that MOF-controlled plasticization benefits were successfully achieved. In addition, the MOFs also tailored the molecular competitions in the gas permeation process, further increasing H2S/CH4 and CO2/CH4 selectivities. Furthermore, we extended our studies to a commercial cellulose triacetate (CTA) hollow fiber membrane to illustrate the controlled-plasticization benefit. Our study establishes an excellent MOF-polyimide formulation with simultaneous advantages on H2S and CO2 removal for future natural gas sweetening applications.

11:40 am (Room 1, Wednesday)

Oral 53 - New Facilitated Transport Membranes for Hydrogen Purification from Coal-Derived Syngas

Yang Han* (The Ohio State University), Xuepeng Deng (The Ohio State University), Li-Chiang Lin (The Ohio State University), Winston Ho (The Ohio State University).

CO2-selective, amine-containing facilitated transport membranes (FTMs) are of great interest for syngas purification since high-pressure H2 can be retained upon CO2 removal. Various FTMs have shown decent chemical and thermal stability at aggressive conditions, but their CO2/H2 separation properties are largely limited by the severe carrier saturation at high syngas pressure. Herein, we report a new approach to enhance the CO2 permeability by manipulating the steric hindrance of the amine carrier. A series of α-aminoacids with different alkyl or hydroxyethyl substituents were deprotonated by 2-(1-piperazinyl)ethylamine, resulting in nonvolatile amine carriers with different degrees of steric hindrance. For hosting the low MW amine carriers, a water-swellable polymer network was synthesized from poly(vinyl alcohol) crosslinked by a bidentate tertiary aminosilane. In the presence of moisture, a bulkier alkyl substituent increased the steric hindrance and hence destabilized the carbamate adduct to afford bicarbonate through hydrolysis. Thus, this drastically increased the chemisorption of CO2. Further, density functional theory (DFT) calculations were conducted to study the function of the hydroxyethyl substituent, which indicated that the hydroxyl group stabilized the bicarbonate through strong hydrogen bonding, thus further improving the CO2 sorption. The enhanced CO2 solubility significantly mitigated the carrier saturation, and an unprecedented CO2/H2 selectivity greater than 130 was demonstrated at 107°C and 12.5 atm of CO2 partial pressure. As the CO2 partial pressure reduced to 0.4 atm, a less hindered amine yielded a higher reactive diffusivity of CO2, resulting in a CO2 permeance of 435 GPU with a selectivity greater than 500. The CO2/H2 separation performance of these reaction-mediated polymeric membranes is well above the theoretical upper bound, and they open up a new avenue for designing a highly selective membrane process for syngas purification.

12:00 pm (Room 1, Wednesday)

Oral 54 - High Performance Gas Separation Membranes from CANAL Ladder Polymerization

Yan Xia (Stanford University), Holden Lai* (Stanford University).

Microporous polymers, best known as polymers of intrinsic microporosity (PIMs), have emerged as a new generation of membrane materials for gas separation applications, defining the most recent upper bounds for many industrially relevant gas pairs. PIMs usually exhibit very high gas permeability but moderate selectivity, with pure-gas ideal permselectivity of <50 for H2/CH4 and <40 for CO2/CH4. We recently developed an efficient annulation polymerization (termed CANAL) to synthesize microporous ladder polymers with rigid norbornyl benzocyclobutene backbones from readily available norbornadiene and bromoarenes. A diverse family of microporous polymers have been developed using CANAL chemistry, including PIMs containing various contorted linkages. We discovered that the gas separation performance and mechanical properties of membranes from these polymers strongly depend on the molecular designs, such as substitutions, backbone conformations, and types of linkages. With optimal design, we have obtained mechanically robust membranes with an unprecedented combination of high selectivity and ideal permselectivity, with performance beyond the 2015 $O2/N2$, $H2/CH4$, $H2/N2$ upper bounds and 2008 $H2/CO2$ and $CO2/CH4$ upper bounds.

12:20 pm (Room 1, Wednesday)

Oral 55 - Carbon Molecular Sieve Membranes for Petrochemical Gas Separations

Surendar Venna* (Dow Inc.), Thomas Fitzgibbons (Dow Inc.), Derrick Flick (Dow Inc.), James Heard (Dow Inc.), Jay (Junqiang) Liu (The Dow Chemical Company), Hali McCurry (Dow Inc.), Nikki Montanez (Dow Inc.), Gerard Rogers (Dow Inc.), Li Tang (Dow Inc.), Cole Witham (Dow Inc.), Abhishek Roy (Dow Inc.).

The Energy-efficient olefins/paraffins separation is important for the petrochemical industry. Membrane technology is emerging as an energy and capital efficient petrochemical separation process alternative to or in conjunction with the conventional thermally driven, energy-intensive cryogenic distillation. The application opportunities of membranes in the petrochemical plant along with the performance requirements and limitations will be discussed. Carbon molecular sieves (CMS) are an interesting class of porous membranes that display exceptional gas separation properties for petrochemical separations along with good plasticization resistance compared to the precursor polymer and polymeric membranes. Polyimide was selected as a promising precursor for CMS membranes based on its high rigidity and good processability. This study considers the differences in the gas separation performance of the flat sheet and hollow fiber derived CMS membranes for further development of these membranes. This talk will also discuss the structure-property relationships of the polyimide-derived CMS membranes. Preliminary observations on the effect of the petrochemical gas composition on the gas separation performance of these CMS membranes and the challenges of these separations along with the gaps in this technology development will be presented.

Oral 12 – BIOINSPIRED AND BIOMIMETIC MATERIALS

11:00 am (Room 2, Wednesday)

Oral 56 - Artificial Water Channels- toward biomimetic membranes for desalination

Mihail Barboiu* (Institut Europeen des Membranes).

This lecture discusses the incipient development of the first artificial water channels systems. We include only systems that integrate synthetic elements in their water selective translocation unit. Therefore, we exclude peptide channels because their sequences derive from the proteins in natural channels. We review many of the natural systems involved in water and related proton transport processes. We describe how these systems can fit within our primary goal of maintaining natural function within bio-assisted artificial systems. In the last part, we present several inspiring breakthroughs from the last decade in the field of biomimetic artificial water channels. All these examples demonstrate how the novel interactive water-channels can parallel biomolecular systems. At the same time these simpler artificial water channels offer a means of understanding water structures useful to understanding many biological scenarios. Moreover they can be used for the preparation of highly selective membranes for desalination.

[1] M. Barboiu and A. Gilles, Acc. Chem. Res. 2013, 46, 2814–2823.
[2] M. Barboiu, Angew. Chem. Int. Ed. 2012, 51, 11674-11676.
[3] Y. Le Duc, et al., Angew. Chem. Int. Ed. 2011, 50(48), 11366-11372.
[4] E. Licsandru, et al., J. Am. Chem. Soc., 2016, 138, 5403-5409.
[5] M. Barboiu, Chem. Commun., 2016, 52, 5657- 5665.
[6] I. Kocsis, et al. Science Adv. 2018, 4, eaao5603

11:20 am (Room 2, Wednesday)

Oral 57 - Nature-inspired ion conducting polymers for energy conversion and storage devices

Shudipto K Dishari* (University of Nebraska-Lincoln).

Ion conducting polymers (ionomers) are an integral part of energy conversion and storage device. In a hydrogen fuel cell or redox flow battery system, ionomers conduct protons through a bulk membrane from anode to cathode and through a sub-micron thick layer at the catalyst interface for oxygen reduction reaction (ORR) to occur. However, due to the ion transport limitations at the catalyst interfaces, ORR becomes sluggish. While the current state-of-the-art fluorocarbon and hydrocarbon-based ionomers conduct protons in bulk membrane efficiently, proton conduction through the thin films of the same ionomers is very poor. To address this issue, we have recently synthesized a novel class of ionomers inspired by biological ion channels which leverages water wire effect to facilitate ion conduction under confinement. The ion conduction, mechanical and morphological properties of these nature-mimicking ionomers were explored in sub-micron thick films and bulk membranes. Through this work, we were able to unravel the ion conduction mechanism within these ionomers and identify the critical parameters controlling the ion conduction under thin-film confinement.

11:40 am (Room 2, Wednesday)

Oral 58 - Ultra-High Ionic Exclusion Through Carbon Nanomembranes

Yang Yang (Bielefeld University), Roland Hillmann (Bielefeld University), Yubo Qi (Bielefeld University), Riko Korzetz (Bielefeld University), Niklas Biere (Bielefeld University), Daniel Emmrich (Bielefeld University), Michael Westphal (Bielefeld University), Björn Büker (Bielefeld University), Andreas Hütten (Bielefeld University), André Beyer (Bielefeld University), Dario Anselmetti (Bielefeld University), Armin Gölzhäuser* (Bielefeld University).

The collective "single-file" motion of water molecules through natural nanoconduits, such as aquaporins, inspires the development of high-performance membranes for water separation. Two-dimensional (2D) materials are particularly promising, however, a membrane that combines rapid water flow with superior ion rejection is still highly desireable. In this presentation, it will be demonstrated that 1.2 nm thick carbon nanomembranes (CNMs) made by cross-linking of terphenylthiol (TPT) self-assembled monolayers possess an extremely high pore density of one sub-nm channel per square nanometer [1]. Conductivity measurements in salt solutions show that TPT CNMs efficiently hinder the translocation of ions, including protons, while water molecules pass through exceptionally high permeance (~700 LMH) [2]. The membrane resistance reaches 10^4 Ω·cm2 in 1 M Cl− solutions, comparable to lipid bilayers. Consequently, a single CNM channel yields a ~10^8 higher resistance than pores in lipid membrane channels and carbon nanotubes. The ultra-high ionic exclusion by CNMs is likely dominated by steric hindrance mechanisms, coupled with electrostatic repulsion, surface functional groups and entrance effects. We demonstrate the operation of TPT CNMs as semipermeable membranes in forward osmosis, and discuss the potential of CNMs for applications like cold concentration and membrane distillation.

[1] Y. Yang, P. Dementyev, N. Biere, D. Emmrich, P. Stohmann, R. Korzetz, X. Zhang, A. Beyer, S. Koch, D. Anselmetti, A. Gölzhäuser, ACS Nano 2018, 12, 4695.

[2] Y. Yang, R. Hillmann, Y. Qi, R. Korzetz, N. Biere, D. Emmrich, M. Westphal, B. Büker, A. Hütten, A. Beyer, D. Anselmetti, A. Gölzhäuser, Adv. Mater., 2020, 1907850.

12:00 pm (Room 2, Wednesday)

Oral 59 - High Density Membrane Protein-Polymer Nanosheets-Based Biomimetic Membranes

Yu-Ming Tu* (The University of Texas at Austin), Woochul Song (The University of Texas at Austin), Benny Freeman (The University of Texas at Austin), Manish Kumar (The University of Texas at Austin).

Biomimetic and bioinspired membranes containing membrane protein (MP) channels and MP-mimic functionalized materials have emerged as significant platforms to develop filtration membranes with specific solute selectivity. These membranes typically consist of highly permeable biological or bioinspired pore structures or channels with well-defined pore geometries embedded in a relatively impermeable synthetic membrane matrix. Nevertheless, membranes integrating MP-based membranes into current manufactured technology for separations has remained challenges. These challenges are possibly due to the amount of time needed to create these membranes, the use of vesicular morphologies and the low density of MPs in resulting membranes, which have resulted in much lower than anticipated improvements. In this presentation, we will describe a comprehensive approach to fabricate high-performance biomimetic filtration membranes that are based on densely packed MP channel in block copolymers (BCPs) to create robust and flat nanosheets. These highly packed crystalline structures and well-ordered nanosheets were constructed by a two-hour organic solvent evaporation method and further deposited on a porous substrate as high-throughput selective layers to form composite membranes. Three different types of β-barrel MP channels with pore sizes of 0.8 nm, 1.3 nm, and 1.5 nm were utilized in this work. The resulting three different MP-BCP nanosheet membranes created demonstrated high water permeability of ~300 - 1,000 (L m^{-2} h^{-1} bar^{-1}) and still maintained a sharp and tunable selectivity with three different pore sizes of MP channels. The approaches described could be adapted with the insertion of MP channels of different pore sizes or chemistries into polymer matrices to conduct specifically designed separations.

12:20 pm (Room 2, Wednesday)

Oral 60 - Biomimetic carbon nanotube and AAO systems utilizing electro-dynamic interfaces

Bruce Hinds* (University of Washington).

An important challenge for the membrane community is to mimic natural protein channels that outperform, by orders of magnitude, man-made systems based on pore size and coarse chemical selectivity. To mimic protein channel pumping on a robust engineering membrane platforms, applied bias can be used to actuate charged gatekeepers and induce ionic pumping. Carbon nanotubes have three key attributes that make them of great interest for novel membrane applications 1) atomically flat graphite surface allows for ideal fluid slip boundary conditions and extremely fast flow rates 2) the cutting process to open CNTs inherently places functional chemistry at CNT core entrance for chemical selectivity and 3) CNT are electrically conductive allowing for electrochemical reactions and application of electric fields gradients at CNT tips. Use of the electro-osmotic phenomenon for responsive/programmed transdermal drug delivery devices for nicotine addiction. Delivery rates are switched between therapeutic levels of background transdermal patch and gum with watch-sized battery life of weeks. Another approach is to mimic natural protein channel transport cycles with binding/transport/release/reset events. Porous alumina (AAO) membranes have top and bottom electrodes coated with thin Au layers with pore dimension tuned to match protein dimensions. At this thin layer at pore entrances, Ni-NTA is able to bind to hys-tag residue on target protein, as is commonly employed in chromatography. A binding voltage pulse attracts anionic target protein to top electrode and blocking the pore, while repelling the cationic imidazole release agent. The second voltage cycle attracts cationic release agent to top of membrane while pumping anionic target protein to bottom permeate and resetting the pumping cycle, with non-optomized throughput comparable to chromatographic methods.

Oral 13 – MICRPOROUS / CATALYTIC MATERIALS

11:00 am (Room 3, Wednesday)

Oral 61 - Membranes with Integrated Nanomaterials: Material Science Advances and Applications

Dibakar Bhattacharyya* (University of Kentucky), Hongyi (Derek) Wan (University of Kentucky), Anthony Saad (University of Kentucky), Ronald Vogler (University of Kentucky), Trisha Nickerson (University of Kentucky), Ashish Aher (University of Kentucky), Saiful Islam (University of Kentucky).

Membrane pores functionalized with responsive polymers and/or nano-catalyst provide applications ranging from water treatment and reuse to environmental catalysis to advanced affinity-based separations. The use of macromolecules in microfiltration membrane pores, such as, poly-acrylic acid (PAA), poly-glutamic acid (PLGA) provides tunable water permeability with pH response. The integration of pollutant degrading enzymes by layered assembly or metal/metal oxide (iron, or bimetallic) nano-particles in polymeric and GO membrane domain allow catalytic detoxification (for ex, PCBs, TCE) of water through both reductive and oxidative pathways. In addition, we demonstrate that temperature responsive membranes/gels (p-NIPAM) can be used to adsorb (such as PFOA, PFOS) and desorb organic pollutants through hydrophobicity/hydrophilicity transition near LCST. Our approach also includes direct polymerization of monomers (such as, acrylic acid) in robust microfiltration membranes pores for high capacity capture of toxic and in-situ synthesis of metal nanoparticles for chloro-organic degradation. For organic degradation, a common metal or metal oxide functionalized platform allow both reductive and oxidative approaches to detoxify organics, such as, TCE, PCE, and PCB's. The presentation will include: (1) polymer membrane pore and GO functionalization approaches and results of pollutant degradation rates, (2) use of advanced material characterization techniques, such as FIB to understand nanosized catalyst distribution, (3) applications for pH and temperature responsive membranes. This research is supported by NIH-NIEHS-SRP, and by industries.

11:20 am (Room 3, Wednesday)

Oral 62 - Efficient Ammonia Decomposition in a Catalytic Membrane Reactor to Enable Hydrogen Storage and Utilization

Zhenyu Zhang (CSM), Javishk Shah (CSM), J. Douglas Way (CSM), Colin Wolden* (Colorado School of Mines).

The deployment of fuel cell electric vehicles is constrained by the paucity of hydrogen fueling stations and price, which is dominated by the costs of hydrogen storage and transportation. With more hydrogen per volume than liquid H2 and an extensive distribution infrastructure in place, ammonia is a promising vector for efficient hydrogen distribution. In this talk, we will describe the development of catalytic membrane reactor (CMR) technology for the delivery of high purity H2 from ammonia cracking. The CMR integrates state-of-the-art catalysts with thin metal membranes. In our design a dense palladium/palladium alloy membrane a few microns in thickness is formed by electroless deposition on an asymmetric ceramic support. A delivery system providing 30-bar ammonia is built to increase the transmembrane pressure driving force. The catalyst is both impregnated within the membrane support and packed in the lumen of a CMR, and excellent dispersion maximizes its activity. The use of alkali promoters is found to further enhance the kinetics. Ammonia decomposition rate is inhibited by both H2 adsorption and reverse reaction, but the efficient removal of H2 in a CMR enables temperature reductions by up to 200 ºC and decreases catalyst loading requirements by a factor of 2. Similar improvements in H2 flux productivity over leading reports are obtained, further reducing Pd CapEx. The high purity H2 permeate stream is obtained for a stable operation over 680 hours. It is further purified using an adsorption bed with commercial zeolite adsorbents. The final ammonia impurity is less than 0.5 ppm. We have developed a detailed axisymmetric model that captures the reactor performance with high fidelity, and this is used for the analysis and design of these systems.

11:40 am (Room 3, Wednesday)

Oral 63 - Tuning Ultrafiltration Membrane Performance via Surface Graft Polymerization of Acrylic Acid Enabled by Air Plasma Activation

Yian Chen* (University of California, Los Angeles), Soomin Kim (University of California, Los Angeles), Yoram Cohen (University of California, Los Angeles).

Polysulfone (PSf) based ultrafiltration (UF) membranes have gained popularity in filtration/treatment of aqueous streams in various industries. However, there are operational constraints due to membrane fouling. Accordingly, the present work explored reduction of PSf membrane fouling propensity and performance tuning via membrane surface modification by exposure to Air, He, or He/O2 atmospheric pressure plasma (APP) for surface activation, followed by acrylic acid graft polymerization. The APP type (i.e., source gas) can have a significant influence on membrane surface treatment and thus impact membrane structure and chemistry and also affect the subsequent graft polymerization. For example, membrane surface treatment or activation by Air APP has been generally avoided since surface etching has been often encountered. However, given the low cost of Air APP surface treatment, the present work explored optimizing APP surface treatment conditions, such as exposure time and plasma source-substrate separation distance, with the aim of minimizing aggressive surface etching by Air APP to avoid compromising the base PSf membrane. Indeed, the present results demonstrated that fine tuning of PSf UF membrane performance, as quantified based on membrane permeability and MWCO, via surface graft polymerization activated by Air APP was as effective as with He or He/O2 APP. Tuning the operating conditions for Air, He, or He/O2 treatment enabled synthesis of SNS-PSf membranes of similar permeability and MWCO ranges of 12.7-24.0 L·m2·h/bar and 5.5-11 kDa, respectively. Fouling stress tests with bovine serum albumin (BSA) as a model foulant demonstrated that SNS-PAA-PSf membranes, modified post Air APP surface activation, had the lowest permeability decline (~9.5%) followed by He (~12.2%) and He/O2 APP (~12.5%). Moreover, the BSA fouled SNS-PSf membranes were easily cleaned via DI water backwash (100% permeability recovery) without traces of irreversible protein fouling.

12:00 pm (Room 3, Wednesday)

Oral 64 - Monitoring protein fouling on patterned membranes by light microscopy and simulation

Anna Malakian* (Clemson University), Bowen Ling (Stanford University), Ilenia Battiato (Stanford University), Scott Husson (Clemson University).

The goal of this project is to develop the basic science needed to design new fouling-resistant membranes. We present findings on the relationship between membrane surface patterning and protein fouling that support two hypotheses (i) geometric patterns will reduce membrane fouling compared to the control, and (ii) computational models will accelerate the discovery of new membrane surface morphologies to improve membrane fouling resistance.

Experimental efforts combined flux decline measurements with confocal laser scanning microscopy (CLSM) to study the fouling of as-received and patterned polyethersulfone membranes during ultrafiltration of bovine serum albumin (BSA, model foulant). Patterned membranes were prepared by embossing ordered micropatterns on the surface. Both membrane types were labelled with a fluorophore, 5-DTAF. BSA labelled with Alexa Fluor 488 was used to visualize the fouling profiles with CLSM, which provided three-dimensional images of membrane surface patterns and co-localized protein foulant. In parallel, numerical simulations were used to compare fouling patterns with visual analysis of the CLSM images and to explore the full potential of the micropatterning. We utilized the customized simulator SUMs (Stanford University Membrane solver) within the OpenFOAM framework. The solver uses a finite volume method to study the dynamic couplings among flow, solute transport, and surface fouling. From the numerical study, we found that the micropatterning acts as a fouling-focusing mechanism. When the flow alternates, the foulant shifts its preferential accumulation zone and self-cleans the previously fouled region. This innovative design provides a possibility to consecutively alternate flow direction during operation to periodically clean zones of the membrane surface. The overall approach is expected to aid in the design of new membranes with tailored surface structures that prevent the irreversible deposition of foulants in prone-to-foul regions.

12:20 pm (Room 3, Wednesday)

Oral 65 - Influences of Microwave Irradiation on Performances of Membrane Filtration and Catalytic Degradation of Perfluorooctanoic Acid (PFOA)

Fangzhou Liu* (New Jersey Institute of Technology).

Perfluorooctanoic Acid (PFOA), one of the common per- and poly fluorinated alkylated substances (PFASs), is increasingly detected in the environment due to the diverse industrial applications and high resistance to degradation processes. This study evaluated degradation of PFOA in microwave-assistant catalytic membrane filtration, a process that integrates microwave catalytic reactions into a ceramic membrane filtration. First, water permeation of the pristine and catalyst-coated membranes were examined under the influence of microwave irradiation to analyse the impacts of the coating layer and water temperature increase on permeate flux, which were well interpreted by the Carman-Kozeny, Hagen-Posieulle and Boussinesq models. Then, the PFOA removal was first assessed in a continuous filtration model with and without microwave irradiation. Our results show that PFOA adsorbed on membrane and catalyst materials and fully penetrated the membrane filter after reaching adsorption equilibrium. Under microwave irradiation (7.2 watt·cm-2), approximate 65.9% of PFOA (25 μg·L-1) in the feed solution was degraded within a hydraulic time of 2 min (at the permeate flow rate of 43 LMH) due to the microwave-Fenton like reactions. In addition, low flow rates and moderate catalyst coating densities are critical for optimizing PFOA removal. Finally, potential degradation mechanisms of PFOA were proposed through the analysis of degradation by-products (e.g., PFPeA). The findings may provide new insight into the development of reactive membrane-enabled systems for destruction of refractory PFAS.

Oral 14 – HIGH SALINITY STREAMS, BRINE MINIMIZATION AND ZLD

11:00 am (Room 4, Wednesday)

Oral 66 - Minimal and zero liquid discharge with reverse osmosis using low-salt-rejection membranes

Zhangxin Wang* (Yale University), Menachem Elimelech (Yale University), Akshay Deshmukh (Yale University), Yuhao Du (Yale University).

Minimal and zero liquid discharge (MLD/ZLD) are wastewater management strategies that are attracting heightened attention worldwide. While conventional reverse osmosis (RO) has been proposed as a promising technology in desalination and MLD/ZLD processes, its application is limited by the maximum hydraulic pressures that current RO membranes and modules can withstand. In this study, we develop low-salt-rejection RO (LSRRO), a novel staged RO process, that employs low-salt-rejection membranes to desalinate or concentrate highly saline feed streams, requiring only moderate hydraulic pressures. Based on process modeling, we demonstrate that LSRRO can overcome the hydraulic pressure limitations of conventional RO, achieving hypersaline brine salinities (>4.0 M NaCl or 234 g/L NaCl) that are required for MLD/ZLD applications, without using excessively high hydraulic pressures (<= 70 bar). In addition, we show that the energy efficiency of LSSRO is substantially higher than traditional thermally-driven phase-change-based technologies, such as mechanical vapor compressor (MVC). For example, to concentrate a saline feed stream from 0.1 to 1.0 M NaCl, the specific energy consumption (SEC) of 4-stage LSRRO ranges from 2.4 to 8.0 kWh of electrical energy per cubic meter of feedwater treated, around four times less than that of MVC, which requires 20–25 kWhe per cubic meter. Furthermore, compared to osmotically mediated RO technologies that require bilateral countercurrent stages to treat hypersaline brines, LSRRO is eminently more practical as it can be readily implemented by using 'loose' RO or nanofiltration membranes in conventional RO. Our study highlights LSRRO's potential for energy efficient brine concentration using moderate hydraulic pressures, which would drastically improve the energetic and economic performance of MLD/ZLD processes.

11:20 am (Room 4, Wednesday)

Oral 67 - Evaluation of Integrated Electrocoagulation-Microfiltration and Direct Contact Membrane Distillation Processes for Treating Produced Water

Mahmood Jebur* (University of Arkansas), Ranil Wickramasinghe (University of Arkansas), Mahdi Malmali (Texas Tech University), Xianghong Qian (University of Arkansas), Yunxia Hu (Tianjin Polytechnic University), Yuhe Cao (University of Arkansas), Yu-Hsuan Chiao (University of Arkansas).

Treating the wastewater (produced water) generated from hydraulic fracturing operations is challenging due to the present of high salinity (total dissolved solids (TDS)), toxic organic & inorganic compounds, oil & grease, and naturally occurring radioactive materials. Membrane distillation (MD) is an emerging technology of interest for the treatment of high salinity brines. In this study, microfiltration (MF) was used to remove the flocs formed during the electrocoagulation (EC) pretreatment step, and the MF permeate was treated further by direct contact membrane distillation (DCMD). Three types of membranes, including commercial hydrophobic PVDF membrane, electrospun PVDF-HFP membrane, and carbon nanotube coated PTFE membrane, were investigated for DCMD. In this work, we aimed to demonstrate the feasibility of an integrated EC-MF-DCMD process for water recovery from high salinity produced water obtained from shale gas extraction facilities. During EC, a high removal efficiency of total organic carbon (48.4%) was obtained by using six aluminum electrodes with bipolar configuration at 9.5 A for 20 min. Keeping the feed tank temperature at 40 oC, while the temperature inside the membrane cell at 60 oC can effectively suppress crystallization on the membrane surface, which can significantly mitigate scaling propensity of the membrane. Up to 50% water recovery was obtained for real produced water with initial TDS about 245268.9 mg/L over two DCMD cycles. However, organic fouling was still a big problem, especially surfactants due to forming micelles and bilayers, which can significantly decrease the flux and eventually block the membrane.

11:40 am (Room 4, Wednesday)

Oral 68 - A Novel Membrane Distillation-Crystallization Approach for Inland Brine Treatment

Evangelos Balis (University of Nevada, Reno), Jack Griffin (University of Nevada, Reno), Sage Hiibel* (University of Nevada, Reno).

Desalination is a potential solution to the increasing global demand for water, however brine disposal comes at a high price for inland operations. Current brine disposal methods, such as deep-well injection or evaporation, can be harmful to the environment, thus zero liquid discharge (ZLD) or near-ZLD technologies are of great interest.
Membrane distillation crystallization (MDC) is a technique that can be used to reduce brine volume and approach ZLD. MD is a thermally-driven separation process that utilizes a hydrophobic, microporous membrane. The driving force of MD is the temperature-induced vapor pressure difference between the brine and distillate. MD is ideal for ZLD applications because it can treat brines at near saturation without requiring significantly higher energy inputs compared to lower salinity feeds. MDC combines the MD process with crystallization to recover solid salts and produce high-quality water.
The objective of this work was to develop an MDC system to treat RO brine from inland desalination. A bench-scale MDC setup was created with a flat-sheet direct contact MD module and a novel crystallizer designed so the brine solution cascades over multiple trays, each containing an extended mesh surface that provides nucleation sites and promote crystal growth out of the liquid phase.
The results compared MDC performance to MD alone under the same MD conditions. Three hyper saline single-salt solutions were prepared: NaCl at 288.4 g/L, KCl at 270 g/L, and NaNO3 at 599.29 g/L. After 8 hr of operation, MDC removed 2.3, 2.1, and 2.6 times more water from the feed solution than MD alone for NaCl, KCl, and NaNO3, respectively. These results demonstrate that the MDC system can effectively be adapted towards ZLD applications for brine management in inland desalination applications. In addition to the reduced brine volumes, the solid salts recovered during MDC can offset operational costs of the process.

12:00 pm (Room 4, Wednesday)

Oral 69 - Critical Analysis of Membrane Distillation at High Salinity

Mukta Hardikar* (The University of Arizona), Andrea Achilli (The University of Arizona).

Membrane distillation (MD) is a desalination technology that has promising applications in the treatment of high salinity brines. If properly designed, these systems could assist in achieving near zero-liquid discharge because of high water recovery potential. Successful deployment of this technology can unleash water reuse and brackish water desalination in inland application where brine discharge and management are the main challenges to widespread use of these alternative water sources. For these reasons it is important to analyze salinity in membrane distillation and elucidate the role of size of the system in the relationship between salinity and water flux. While several engineering-scale experiments treating high salinity (220 g/L) water have reported a significant decline in water-flux with increase in salinity, a much smaller decline is often reported at bench-scale experiments. Increase in salinity results in a reduction of water activity which is directly proportional to water vapor pressure. Thus, a feed with higher salinity results in a smaller vapor pressure difference between feed and distillate side and a lower driving force for water production. While this effect exists even at the bench-scale, it is more prominent in a engineering-scale design where the objective is to be as energy efficient as possible. An engineering-scale operates at low transmembrane temperature differences as all the heat from the feed is recovered by the cold side. This exacerbates the effect of salinity on water flux as the presence of salt further reduces transmembrane temperature difference. In this talk, we show how feed salinity decides the limiting operating conditions in an engineering-scale MD system and the minimum transmembrane temperature difference that needs to be maintained for a given feed salinity to achieve positive water-flux. We also discuss its implications on MD energy requirements as a function of feed properties.

12:20 pm (Room 4, Wednesday)

Oral 70 - Testing mineral scaling propensities of fluorinated and non-fluorinated polyamide thin film composite membranes

Sankaranarayanan Ayyakudi Ravichandran* (University of Colorado Boulder), Josue Velasco (University of Colorado Boulder), Saied Delagah (U.S. Department of Interior), John Pellegrino (University of Colorado).

The rationale behind this work is to develop scaling resistant polyamide thin film composite (TFC) membranes for use in membrane crystallizers to treat brine streams that are supersaturated with sparingly soluble salts. A membrane crystallizer can concentrate supersaturated brine feeds to recover more water and minimize reject waste. Our approach uses a mono-fluorinated derivative of trimesoyl chloride and m-phenylenediamine as monomers to make polyamide-TFCs. The underlying theory is that the fluoro groups on the TFC would retard mineral scales from strongly adhering to the membrane surface. We studied the scaling behavior of this novel fluorinated TFC under cross flow conditions with a batch electrolyte feed which is constantly being concentrated and supersaturated due to membrane selectivity. This type of experiment allows us to study how supersaturation chemistry, concentration polarization, crystallization of salts on the membrane surface and in the bulk influence membrane scaling. To understand the role played by bulk and surface crystallization, we have conducted further experiments wherein we study how the surface chemistry of the membrane (fluorinated polyamide-TFCs vs. conventional polyamide-TFCs) influences surface scaling and bulk crystallization when exposed to supersaturated waters. In these experiments we affix TFC membrane coupons of different chemistries to the lids of glass jars and fill the jars with supersaturated feeds and from these batch crystallization tests we have studied the influence of how the different polyamide-TFCs chemistries affect scaling propensity of the membranes. In this type of experiment, as there is no selective transport, the factor of concentration polarization is removed. Through these two types of experiments we can decouple the effects of heterogeneous crystallization and concentration polarization on the scaling propensities of polyamide-TFCs with different surface chemistries.

Oral 15 – PROCESS INTENSIFICATION AND INTEGRATION

11:00 am (Room 5, Wednesday)

Oral 71 - Energy Efficiency of Electro-Driven Brackish Water Desalination: Electrodialysis Significantly Outperforms Membrane Capacitive Deionization

Mohan Qin* (University of Wisconsin-Madison), Sohum Patel (Yale University), W. Shane Walker (The University of Texas at El Paso), Menachem Elimelech (Yale University).

Electro-driven technologies are viewed as a potential alternative to the current state-of-the-art technology, reverse osmosis, for the desalination of brackish waters. Capacitive deionization (CDI), based on the principle of electrosorption, has been intensively researched under the premise of being energy efficient. However, electrodialysis (ED), despite being a more mature electro-driven technology, has yet to be extensively compared to CDI in terms of energetic performance. In this study, we utilize Nernst-Planck based models for ED and constant-current membrane capacitive deionization (MCDI) to systematically evaluate the energy consumption of the two processes. By ensuring equivalently sized ED and MCDI systems — in addition to using the same feed salinity, salt removal, water recovery, and productivity across the two technologies — energy consumption is appropriately compared. We find that ED consumes less energy (has higher energy efficiency) than MCDI for all investigated conditions. Notably, our results indicate that the performance gap between ED and MCDI is substantial for typical brackish water desalination conditions (e.g., 3 g L-1 feed salinity, 0.5 g L-1 product water, 80% water recovery, and 15 L m-2 h-1 productivity), with the energy efficiency of ED often exceeding 30% and being nearly an order of magnitude greater than MCDI. We provide further insights into the inherent limitations of each technology by comparing their respective components of energy consumption, and explain why MCDI is unable to attain the performance of ED, even with ideal and optimized operation.

11:20 am (Room 5, Wednesday)

Oral 72 – CO2-Selective Membrane for Enhancing H2 Utilization in Solid Oxide Fuel Cells

Kai Chen* (The Ohio State University), Witopo Salim (Membrane Technology & Research Inc.), Yang Han (The Ohio State University), Mike Gasda (Bloom Energy Corporation), Winston Ho (The Ohio State University).

In order to utilize the H2 remaining in the anode exhaust of a solid oxide fuel cell (SOFC), a CO2-selective membrane was developed to remove CO2 from the anode exhaust and recycle the concentrated H2 back to the SOFC. As the cooled exhaust still has an average temperature of 120ºC, it is challenging for polymeric membranes to maintain both high CO2/H2 selectivity and CO2 permeance. Previously, facilitated transport membranes were developed using tetramethylammonium hydroxide (TMAOH) as the mobile carrier to selectively remove CO2. However, TMAOH has been reported to degrade via a nucleophilic substitution mechanism, and a possible higher temperature of the anode exhaust in practice may exacerbate the degradation.

Herein, more stable fluoride- and hydroxide-containing species were investigated for their potentials of replacing TMAOH. The thermal stabilities of the quaternaryammonium-containing compounds were characterized by TGA. Moreover, quantitative results from NMR showed that tetramethylammonium fluoride (TMAF) was 14 times more stable than TMAOH at 130ºC. Next, membranes with the optimal composition containing TMAF were synthesized, and their transport performances at different temperatures and feed CO2 partial pressures were measured, respectively. The membrane showed a CO2 permeance of 108 GPU and a CO2/H2 selectivity of 106 at 120ºC, and it was stable for 120 h. The TMAF-containing membrane also displayed an improved stability of about 4 times better than the TMAOH-containing membrane at 130ºC. To date, scale-up of 21"-wide membranes with a total length of 720 ft has been demonstrated, and the membranes exhibited consistent transport results as the lab-scale samples. Additionally, a techno-economic analysis showed that, when the H2 recovery of the membrane process was set at 99%, the CO2 removal cost was calculated to be $62.9/tonne. Lastly, a CO2 removal of 76.9% could be obtained when the air sweep was replaced by vacuum on the permeate side.

11:40 am (Room 5, Wednesday)

Oral 73 - Flow Patterns through densely packed hollow fibers: a CFD study for membrane distillation

Albert Kim* (University of Hawaii), Hyeon-Ju Kim (Korea Research Institute of Ships and Ocean Engineering), Deok-Soo Moon (Korea Research Institute of Ships and Ocean Engineering).

Modeling research on membrane distillation (MD) requires simulations of coupled momentum, mass, and heat transfer phenomena. Hollow fiber (HF) modules are preferred in industrial applications due to a high packing ratio, which results in the number of fibers on the order of 100 to 1,000, packed in a vessel. In HF-MD processes, computational fluid dynamics (CFD) simulations of multi-physics require high-quality meshes for accurate and efficient calculations. Mesh interfaces between two different phase regions should conform to satisfy continuity conditions of mass and heat transfer. Due to the distinct characteristics of HF packing structures, a scalable meshing method is of great necessity , which has not been actively researched to the best of our knowledge. The current study developed a numerical method to generate hexagonally packed structures of many HFs using a standard unit-cell approach. An hexagonal unit cell is made to hold three mesh regions of lumen, membrane, and shell. These individual cells are packed to form a globally hexagonal packing structure of many HFs used for CFD simulations, which investigate specific flow fields and pressure distributions. Several flow regimes and module structures are investigated for a good optimization. The algorithm developed in this study can contribute to the improved scalability of CFD simulations from lab-scale modules to pilot-scale systems.

12:00 pm (Room 5, Wednesday)

Oral 74 - Analysis of Membrane Processes for CO2 Removal and H2 Reuse for Solid Oxide Fuel Cells

Kai Chen* (The Ohio State University), Yang Han (The Ohio State University), Mike Gasda (Bloom Energy Corporation), Winston Ho (The Ohio State University).

In order to enhance fuel utilization in solid oxide fuel cells (SOFCs), part of the anode exhaust may be recycled to join the anode inlet stream. However, the CO2 present in the exhaust would dilute the fuel concentration and lower the efficiency of the SOFCs. Few papers have considered removing CO2 from the recycle stream, and none have reported on using a membrane process for this task. A CO2-selective facilitated transport membrane was developed with fluoride- and hydroxide-containing species for the CO2/H2 separation, and the membrane exhibited a good stability, showing a CO2 permeance of 108 GPU and a CO2/H2 selectivity of 106 at 120°C. Moreover, the membrane was scaled up to 21" in width, which had consistent transport results as the lab-scale samples.
Herein, a process analysis was done on using the developed membrane for an SOFC system. First, due to its wide availability, air was used as the sweep gas. Furthermore, a vacuum was proposed for the permeate side to further enhance the driving force and allow for CO2 capture. Also, as oxygen is absent, amine-containing membranes with better transport performances could be used for the vacuum process at 120°C. At 99% H2 recovery, the air sweep process was estimated to remove 44.7% of CO2 from the recycled anode exhaust, whereas the vacuum process could boost the removal to 76.9%. A less stringent H2 recovery requirement could yield a higher CO2 removal. For instance, a CO2 removal of over 90% could be achieved when the H2 recovery was relaxed to 96.7% for the vacuum operation. In addition, a high-level techno-economic analysis was conducted, without including the H2 reused value. The CO2 removal costs corresponding to the air sweep and vacuum processes were $62.9/tonne and $53.3/tonne, respectively, even with the added cost of the vacuum pump for the amine membrane process. The cost is even lower at $35.4/tonne if the captured CO2 is utilized for enhanced oil recovery.

12:20 pm (Room 5, Wednesday)

Oral 75 - Experimental evaluation of an optimized concentration gradient battery

Fei Liu* (University of North Carolina at Chapel Hill), Ryan Kingsbury (University of North Carolina at Chapel Hill), Mikayla Armstrong (University of North Carolina at Chapel Hill, Orlando Coronell (University of North Carolina at Chapel Hill).

A concentration gradient battery (CGB) is an energy storage system comprised of a series of concentrated and dilute salt solution compartments, separated by ion exchange membranes. The battery is charged by electrodialysis (ED), which increases the concentration gradient between these solutions, and discharged by reverse electrodialysis (RED), which allows these solutions to mix. In both ED and RED, water moves by osmosis from dilute to concentrated compartments, reducing the CGB energy efficiency. In a previous study we used an osmotic ballast in the dilute solution to mitigate this non-ideal water transport. Although the osmotic ballast significantly improved energy efficiency, we lack (i) knowledge on how ballast properties impact energy efficiency and (ii) performance data of a CGB operated with optimized components, operational parameters and ballast. Accordingly, the objective of this study is to: a) investigate the impact of ballast properties (i.e., size, orientation, functional group) on CGB energy efficiency; and b) collect performance data (i.e., energy efficiency, power density) of a CGB operated with components (i.e., membranes, spacer thickness, electrolyte) and operational parameters (i.e., current density, flow rate, salt gradient) identified as optimum by the literature. Our initial results show that ballasts with high molecular weight (MW) generally resulted in better efficiency and lower water transport compared to low MW ballasts. Data also indicates that ballasts without methyl groups resulted in lower water transport than those with methyl groups. Of all ballasts tested, sucrose performed best in terms of reducing non-ideal water transport (by 109% compared to the non-ballasted CGB) and enhancing both Faradaic efficiency (from 47% to 78%) and energy efficiency (from 26% to 38%). Our work contributes to the fundamental understanding of the impact on CGB performance of ballast properties and the interworking of CGB components and operation.

Oral 16 – POLYMERIC AND MIXED-MATERIALS – GAS SEPARATION II

1:00 pm (Room 1, Wednesday)

Oral 76 - Microporous Pentiptycene-based Polymers with Heterocyclic Rings for High Performance Gas Separation Membranes

Zihan Huang* (University of Notre Dame), Claire Yin (University of Notre Dame), Gregory Kline (University of Notre Dame), Ruilan Guo (University of Notre Dame).

Highly rigid polymers with heterocyclic rings, as represented by polybenzoxazoles (PBOs) derived from thermal rearrangement (TR) of ortho-functional polyimides, are one of the high performing gas separation membrane materials. However, most TR membranes possess less attractive selectivity comparing to their ultrahigh permeability due to lack of fine tuning in microporosity. As first demonstrated in our previous work, triptycene-based PBOs (TPBOs) are among the most selective microporous polymers, in which hierarchical triptycene units enabled superior size sieving through their internal free volume. In this study, to further enhance the ultrahigh permeability of PBO-based polymers while maintaining the ultrahigh selectivity by TPBOs, pentiptycene, a bulkier iptycene unit, was integrated in PBOs to form corresponding PPBO membranes. The study has optimized and eventually maximized the membrane performances by TR protocols. Specifically, via adjusting the heating rate and intermediate isothermal temperature, most PPBO films far exceeded the 2015 upper bounds of several gas pairs. For example, PPBO membranes obtained from a slow heating rate showed ultrahigh H2 permeability of 1283 Barrer and H2/CH4 selectivity of 247, which are 60% and 23% higher than those of TPBO counterparts. Moreover, PPBOs showed strong size sieving ability by exceeding the 2015 O2/N2 upper bound for the first time. PPBO further showed excellent plasticization resistance in CO2/CH4 mixed-gas permeation tests by displaying separation performance far above the 2008 mixed-gas upper bound. Noteworthily, a 5-month aged PPBO film showed attractive ageing resistance by performing further above the 2015 H2/CH4 upper bound than its fresh state through 70% selectivity enhancement. In this talk, preparation and characterization of PPBO membranes will be presented. Discussion will focus on elucidating the fundamental relationship between microstructures and gas transport properties of the PPBO membranes.

1:20 pm (Room 1, Wednesday)

Oral 77 - High-Performance Gas Separation Membranes Based on Poly(benzimidazole)

Alexander Bridge* (The University of Texas at Austin), Joshua Moon (The University of California Santa Barbara), Joan Brennecke (The University of Texas at Austin), Benny Freeman (The University of Texas at Austin).

Polybenzimidazole (PBI), commercially produced as Celazole®, is a glassy polymer with high potential to improve and expand applications for commercial gas separation membranes. Unlike current commercial membrane materials which are limited to gas separations below 100°C, PBI exhibits thermal and chemical stability at temperatures up to 400°C. PBI also exhibits size sieving properties that are attractive for dehydrogenation and pre-/post-combustion carbon capture. Challenges preventing integration of PBI membranes into commercial separations include addressing PBI's low gas permeability/permeance and a lack of progress in developing asymmetric membranes formed via phase inversion. In this project, composite membranes were developed by blending PBI with highly permeable thermally rearrangeable polyimides (PI). Phase separated PBI/PI blends exhibited enhanced H2 permeability at 35°C (7.4 Barrer at 10 atmospheres) while maintaining the native H2/CO2 selectivity of PBI (~15). In parallel with this work, concentrated solutions of PBI dissolved in solvent systems based on N,N-dimethylacetamide, tetrahydrofuran, and ionic solids/liquids were optimized for flat sheet asymmetric membrane fabrication via aqueous phase inversion. Resulting membranes exhibited desirable morphologies previously only attainable via phase inversion in toxic coagulants such as acetonitrile, outlining a feasible approach for developing PBI membrane technology at a commercially relevant scale.

1:40 pm (Room 1, Wednesday)

Oral 78 - Polybenzimidazole-Derived Carbon Molecular Sieve Hollow Fiber Membranes with Tailored Oxygen Selective Transport

Jong Geun Seong* (Los Alamos National Laboratory), John Matteson (Los Alamos National Laboratory), Jeremy Lewis (Los Alamos National Laboratory), John Baca (Los Alamos National Laboratory), Alexander Josephson (Los Alamos National Laboratory), Troy Holland (Los Alamos National Laboratory), Joel Kress (Los Alamos National Laboratory), Kathryn Berchtold (Los Alamos National Laboratory), Rajinder Pal Singh Singh (Los Alamos National Laboratory).

Carbon molecular sieve (CMS) membranes have gained much attention due to their viable tunability in gas transport characteristics with polymer precursors selection and tailored carbonization conditions. Despite their demonstrated exceptional potential in CO2 separation processes, the further development of CMS membranes having high O2/N2 selectivity for energy efficient high purity O2 production processes has proven challenging. Herein, we explore a new type of CMS membranes derived from polybenzimidazole (PBI) materials leveraging their tightly packed, rigid, all aromatic molecular structure for high molecular sieving characteristics and exceptional thermal stability resulting in low weight loss upon pyrolysis. In this work, defect-free PBI-based CMS hollow fiber membranes are fabricated for the first time to develop structure-property-performance relationships. With rigorous carbonization protocol optimization, we obtain highly O2 permselective PBI CMS membranes having O2/N2 selectivities approaching 20 enabling high purity (≥90%) O2 production in a 2-stage membrane operation. The influences of operating temperatures and pressures on PBI CMS membrane performance is also explored.

2:00 pm (Room 1, Wednesday)

Oral 79 - Tailoring structural and functional features of high-performance ionenes and ionic composites designed for membrane-based gas separations

Kathryn O'Harra* (University of Alabama), Irshad Kammakakam (University of Alabama), Jason Bara (University of Alabama).

We have developed a library of high-performance (HP) ionenes, or polymers which contain ionic groups along the backbone chain rather than as pendants. These materials combine structural elements typically associated with state-of-the-art gas-separation membranes with functionality utilized in ultra-high-performance engineering polymers (i.e. imide, amide linkages). This work probes the effects of spacing and sequencing of functional features and ionic groups incorporated along the main chain, specifically imidazolium cations paired with fluorinated anions (Tf2N, OTf). Our methods take advantage of established chemistries that impart great control over repeat unit structure and regiochemistry. These ionenes are robust, exhibiting good thermal and mechanical properties, but extremely tailorable based on the modular design and synthetic methods which allow for innumerable potential derivatives. Thus, the thermophysical properties of these materials are highly tunable based on methodical combination of diverse monomers and linkages. These polymeric materials exhibit additional self-assembly and nanostructuring when impregnated with "free" imidazolium-based ionic liquids (IL), which contributes another degree of tunability and increases stability as the IL serves as a non-covalent cross-link between the polymer chains. We are investigating the effects of altering charge density as well as adding functional small molecules and ionic moieties into the polymeric matrix which alter intramolecular interactions and promote further structuring within the framework. These HP-ionenes are thoroughly characterized in order to develop and study structure-property relationships. Composites are similarly analyzed for comparison with the corresponding neat polymers, to gain a better understanding of the coordination between polymer chains and dispersed additives. This presentation focuses on the design and performance of these novel ionenes and composites as gas separation membranes.

2:20 pm (Room 1, Wednesday)

Oral 80 - Azo-UiO-66 MOF for low-energy CO2 capture adsorbents and mixed matrix membranes for CO2/N2 separation

Bradley Ladewig* (Karlsruhe Institute of Technology), Nicholaus Prasetya (Imperial College London).

The synthesis of Azo-UiO-66 was systematically studied regarding the effect of azobenzene ligand content inside the framework [1]. The azobenzene ligand content in the MOF could be rationally tailored by using a simple mixed-ligand approach. This approach also allows quantitative calculation of the ligand constituent in the resulting MOF through 1H-NMR study. The resulting products (Azo(X)-UiO-66, where X denotes the azobenzene ligand percentage in the MOF framework) were then investigated for CO2/N2 separation and CO2 dynamic photo-switching.
The highest CO2/N2 selectivity was obtained in Azo(100)-UiO-66, however with significant reduction in CO2 uptake capacity because of the surface area reduction. Therefore, the MOF with a moderate amount of azobenzene ligand was preferred. Higher azobenzene ligand content inside the framework led to a higher desorption photo-switching capacity. However, the capacity reached a plateau at higher azobenzene loading, likely due to steric hindrance.
The performance of these MOFs in Matrimid and PIM-1 mixed-matrix membranes was evaluated for CO2/N2 separation [1, 2]. Azo(100)-UiO-66 had the best performance, likely due to the excellent CO2/N2 separation performance of the Azo(100)-UiO-66 particles. This study then opens the possibilities to simply tailor light-responsive MOFs which are applicable for post-combustion CO2 capture either as adsorbents or applied in membrane-based separation.

References

1. Prasetya, N. and B.P. Ladewig, Journal of Materials Chemistry A, 2019. 7(25): p. 15164-15172.
2. Prasetya, N., B.C. Donose, and B.P. Ladewig, Journal of Materials Chemistry A, 2018. 6(34): p. 16390-16402.

2:40 pm (Room 1, Wednesday)

Oral 81 - Physical aging of sub-micron PBI membranes at elevated temperatures

Melanie Merrick* (The University of Texas at Austin), Benny Freeman (The University of Texas at Austin).

The promising potential of poly(benzimidazole) (PBI) membranes for high-temperature, hydrogen-selective separations has been documented in the literature for bulk films, composite films, and even hollow fiber membranes. However, the physical aging behavior of sub-micron, spin-cast membranes has never been studied at elevated temperatures (e.g., 200 °C) for PBIs nor any other membrane material. Previous studies of spin-cast membranes at 35 °C have shown that, for membranes at commercially relevant thicknesses, physical aging behavior is highly dependent on the difference between the polymer's Tg and the temperature at which it is aged. For the first time, physical aging studies of spin-cast, 500-nm membranes have been extended to elevated temperatures. PBI membranes cast from commercial grade Celazole® are aged in a high-temperature permeation system while the gas permeabilities are periodically measured over > 1500 hours. When aged at 190 °C, the membranes' gas permeabilities decreased by 60% in the first 300 hours of aging and then plateaued. The observation of a plateau in permeability (i.e., equilibration) has never before been reported for a membrane nor for any polymer more than 200 °C below its Tg. A review of physical aging studies in the polymer physics literature will be presented to give context for this unexpectedly short equilibration time. Qualitatively similar aging behavior was seen at 190 and 175 °C, although the timescale of the observed behavior was longer for the membrane aged at 175 °C. Interestingly, no aging was detected up to 2500 hours for the membrane aged at 150 °C, indicating a very long lag time. These observations cast doubt on the accuracy of both our understanding of physical aging's relationship with temperature and our predictions of membranes' long-term stability in elevated temperature applications.

Oral 17 – MEMBRANE SYNTHESIS AND CASTING I

1:00 pm (Room 2, Wednesday)

Oral 82 - Polyamide Nanofiltration Membrane with Highly Uniform Sub-nanometre Pores for Sub-1Å Precision Separation

Yuanzhe Liang* (Vanderbilt University).

Separating molecules or ions with sub-Angstrom scale precision is important but technically challenging. Herein, we demonstrate that precise solute-solute separation can be achieved using polyamide membranes formed via surfactant-assembly regulated interfacial polymerization (SARIP). The dynamic, self-assembled network of surfactants facilitates faster and more homogeneous diffusion of amine monomers across the water/hexane interface during interfacial polymerization, thereby forming a polyamide active layer with highly uniform sub-nanometre pores compared to those formed via conventional interfacial polymerization. The polyamide membrane formed by SARIP exhibits highly size-dependent sieving of solutes, yielding a step-wise transition from low rejection to near-perfect rejection over a solute size range smaller than half Angstrom. SARIP represents a new approach for the scalable fabrication of ultra-selective membranes with uniform nanopores for precise separation of ions and small solutes.

11:20 am (Room 2, Wednesday)

Oral 83 - Synthesis of Ultrahigh MW Polyvinylamine for Incorporation in Membranes for CO2 Capture

Kai Chen* (The Ohio State University), Yang Han (The Ohio State University), Winston Ho (The Ohio State University).

Membranes for post-combustion CO2 capture are required to have a high CO2 permeance due to the limited driving force. For facilitated transport membranes formed by polyvinylamine (PVAm), a defect-free selective layer of about 100 nm is usually required to render sufficient permeance with high CO2/N2 selectivity. In order to meet such a demand through the knife-coating process, the coating solution needs to have a high viscosity at a relatively low concentration to minimize its penetration into the substrate.

In this study, the challenge was overcome by synthesizing PVAm with an ultrahigh molecular weight (MW) via inverse emulsion polymerization (IEP). Compared to solution polymerization, IEP isolates the reaction in inverse micelles suspended in a continuous organic phase, which allows excellent dissipation of the heat generated by the reaction and reduces gel formation drastically. Hence, the monomer concentration used for the polymerization could be increased to 42.5 wt.%, and the MW, measured by static light scattering technique, was enhanced to as high as 13.5 million. For comparison, the PVAm synthesized via solution polymerization had a MW of only 1.2 million. As a result, the viscosity of a 3 wt.% PVAm solution was improved from 2,000 cp to 30,000 cp, which allowed a lower concentration to formulate the coating solution. Subsequently, a reduction in the selective layer thickness from 170 nm to 120 nm was achieved in a roll-to-roll continuous knife coating process. Besides the PVAm, small molecule amino acid salts were also blended into the membrane to increase the amine content, which enhanced the facilitated transport of CO2. With the ultrahigh MW PVAm to strengthen the polymer matrix, the content of the amino acid salts could be raised from 65 wt.% to 85 wt.%. These two factors led to a ca. 25% increase in the CO2 permeance, i.e., 1077 GPU with >140 selectivity at 57°C.

1:40 pm (Room 2, Wednesday)

Oral 84 - Surface Patterning of Microporous Membranes using Thermally Induced Phase Separation under Confinement

Shouhong Fan* (University of Colorado at Boulder), Yifu Ding (University of Colorado Boulder).

Surface characteristics of microporous membranes significantly impact their performances in liquid-based separations from overall permeance to fouling propensity. Besides chemical modification of the membrane surfaces, surface patterning has been recently shown effective at improving membrane performance. Besides enlarging the active surface area for separation, the presence of the patterns can promote turbulence and increase local shear force, which results in improved mass transfer in boundary layers and fouling resistance. To date, a few patterning strategies have been developed including phase-inversion micromolding, phase-inversion with patterned knife, direct nanoimprint lithography. However, it remains a challenge to precisely replicate microporous membranes with controlled feature size, in a scalable way.
In this presentation, we demonstrate a new membrane casting technique for fabrication of microporous membrane with patterned surface features using lithographically templated thermally induced phase separation (lt-TIPS). Using a PVDF and ATBC as a model system, we successfully replicate a Sharklet □ pattern from pre-patterned PDMS molds onto the PVDF membranes. We will present the impact of dope solution composition and casting temperature on the pattern replication fidelity, membrane morphology (pore size and porosity) and the properties (mechanical, thermal, and permeation).

2:00 pm (Room 2, Wednesday)

Oral 85 - Membrane with cross-linked zwitterionic nanopores achieves sub-nanometer separations

Samuel J Lounder* (Tufts University), Ayse Asatekin (Tufts University).

Self-assembling copolymer membranes have the potential to emerge as the next generation of nanofiltration membranes; however, the sub-nanometer selectivity required for this broad application space has not yet been demonstrated. Random zwitterionic amphiphilic copolymers (r-ZACs) are strong candidates to meet this challenge. These copolymers can self-assemble to form zwitterionic nanopores through which water and small solutes can permeate. Membranes derived from r-ZACs exhibit exceptional fouling resistance and chlorine tolerance, but also low ion retention and a size-based cutoff of ≈ 1.5 nm; they are mere angstroms away from the long sought sub-nanometer selectivity.
In the present work, we employ a novel cross-linking technique to reduce the size of r-ZAC nanopores. Neutral solutes greater than 1 nm are completely rejected by the modified membranes (sucrose rejection > 99.5%; Stokes diameter = 0.94 nm). Furthermore, these membranes demonstrate outstanding divalent-monovalent selectivity when challenged with artificial sea water (sulfate rejection = 99.2%, sulfate-chloride separation factor = 75). In addition, these membranes are extremely resistant to fouling and chlorine damage, which are serious performance limitations for commercial nanofiltration membranes. These results illustrate a promising route to achieving sub-nanometer separations with self-assembling copolymer membranes.

2:20 pm (Room 2, Wednesday)

Oral 86 - Highly Permeable and Selective Crosslinked Polymer Membranes for Energy Efficient Gas Separation

Si Li* (University of Notre Dame), Gregory Kline (University of Notre Dame), Ruilan Guo (University of Notre Dame).

Many glassy polymers used as gas separation membranes suffer from physical aging and plasticization induced by segmental motion during long-term application or sorption of condensable gases. Crosslinking has been demonstrated as one of the most effective approaches, but the much-improved stability upon conventional random crosslinking is inevitably accompanied with limited tunability of intricate network structure and unpredictable performance as well as significant loss of permeability due to densification during crosslinking. Here, to provide a fundamentally new dimension to regulate crosslinked microstructure and consequent membrane properties, we report a new approach of end-linking process to create crosslinked model network with precisely controlled crosslink density and highly tailorable crosslink inhomogeneity from designed telechelic oligomers with well-controlled molecular weight. Specifically, bulky groups are introduced at crosslink junctions to counteract the undesired densification. As results, even the most densely crosslinked membranes (e.g., average inter-crosslink chain length of 3000 g/mol) showed comparable high permeability to their linear counterparts along with expectedly improved selectivity and excellent stability. Via adjusting crosslink density and crosslink inhomogeneity as well as curing conditions, a series of crosslinked membranes with systematically varied yet well-defined model network were prepared and outperformed the 2008 upper bounds for H2 separations, CO2/CH4 and O2/N2 separations. More importantly, it was demonstrated for the first time that crosslink inhomogeneity plays a significant role in regulating gas transport as evidenced by the markedly enhanced separation performance. This talk will discuss the preparation and characterization of these innovative crosslinked model network polymer membranes with a focus on elucidating the fundamental relationship between microscopic structures and macroscopic transport properties.

2:40 pm (Room 2, Wednesday)

Oral 152 - Lyotropic liquid crystal templating for making ultrafiltration membranes

Sahar Qavi (New Mexico State University), Aaron Lindsay (New Mexico State University), Millicent Firestone (Los Alamos National Laboratory), Reza Foudazi* (New Mexico State University).

Self-assembly of amphiphilic molecules in the presence of two solvents (oil/water) leads to the formation of lyotropic liquid crystals (LLCs) with the length scale in the range of 2-50 nm, also called mesophases. LLCs can be used as template to synthesize mesoporous polymers. The key point in this soft templating method is to preserve the structure upon polymerization. Using amphiphilic block copolymers with slow dynamics and crosslinking the monomer (oil) phase to arrest the structure, we have preserved the mesostructure upon thermal polymerization in soft templates. In this work, we show that ultrafiltration (UF) membranes can be produced from the polymerized LLCs. Fabricated membranes have pore sizes in the range of 2-5 nm and exhibit both excellent fouling resistance and high permeance of water, vastly outperforming conventional UF membranes made through non-solvent induced phase separation (NIPS). In addition, the proposed templating method provides flexibility in terms of both the final membrane chemistry and pore size. Therefore, it can be tailored to produce membranes for a wide range of applications including wastewater treatment and food processing.

Oral 18 – DOWNSTREAM BIOPROCESSING

1:00 pm (Room 3, Wednesday)

Oral 87 - Membrane adsorbers for medical isotope purification

Christine E Duval* (Case Western Reserve University), Maura Sepesy (Case Western Reserve University), Priyanka Suresh (Case Western reserve University).

Medical isotopes are radioactive species that can be conjugated to a biomarker and injected into a patient to facilitate localized cancer treatment or imaging. This class of pharmaceuticals has shown promise in treating otherwise pervasive cancers while avoiding full body doses of radiation. Unfortunately, the progression these life-saving drugs to clinical trial has been limited by their availability—a problem that can be traced back to insufficient purification processes. Traditional purification processes for medical isotopes rely on resin-packed columns which are known to suffer from diffusion limitations. This contribution describes our efforts to 1) shorten purification times; 2) minimize elution volumes and 3) improve peak resolution by developing a new class of membrane adsorbers for radiochromatography. These materials are designed to purify medical isotopes, Cu-67 and Ac-225 which have been identified by the US DOE Isotope Program as in-demand and under-supplied. Membranes are prepared by grafting functional polymer brushes from the surface of microfiltration membranes by two methods: 1) UV-induced free radical polymerization and 2) atom transfer radical polymerization (ATRP). The binding groups used in this study are primary amines and phosphonates. Changes in the membrane surface chemistry are monitored with attenuated total reflectance Fourier transform infrared spectroscopy. The number of binding sites are quantified by titration and binding capacities were calculated by fitting equilibrium sorption data with the Langmuir isotherm. Membranes have static binding capacities of 15 mg/g membrane and pure water permeabilities of ~200 LMH/bar. Dynamic binding capacities are measured using membranes loaded in a 0.8 cm2 filter holder at flowrates of1-10 mL/min. Mass transport phenomena are interpreted using a continuum model. The results of this research are laying the foundation for the use of membrane adsorbers in radioanalytical chemistry..

1:20 pm (Room 3, Wednesday)

Oral 88 - New Multimodal Anion-Exchange Membranes for Polishing of Biologics

Joshua Osuofa* (Clemson University), Daniel Henn (Purilogics, LLC), Anna Forsyth (Purilogics, LLC), Jinxiang Zhou (Purilogics, LLC), Scott Husson (Clemson University).

This presentation will highlight findings from a study using new multimodal membranes for high throughput polishing at elevated solution conductivities. Multimodal membranes combine Coulombic interactions, hydrophobic interactions and hydrogen bonding in binding bioprocess impurities, thereby enabling salt-tolerant performance under conditions where conventional ion-exchange media fail. In this study, we evaluated the performance of Purilogics' multimodal anion-exchange membranes (PurexaTM-MQ) under a variety of industry relevant conditions. We will present dynamic binding capacity (DBC) measurements of bovine serum albumin (BSA) and salmon sperm DNA (ssDNA) for a range of salt types, salt concentrations, flowrates, and pH values. Using 1 mg/ml BSA, measured DBC values were greater than 90 mg/ml for the highest conductivities but showed a maximum of 130mg/ml between 7 and 10 mS/cm. The membranes had exceptionally high ssDNA DBC of 50 mg/ml under high conductivity conditions, easily outperforming all other commercial products. Over the 100 cycles, the average binding capacity is 89.78 mg/ml with a standard deviation of 2.7 and there is only a 3% loss in DBC comparing the first and last run. Polishing studies with acidic and basic monoclonal antibodies at >2kg/L loads show improved clearance of aggregated species in comparison to other commercial products. Ultimately, this study shows that Purilogics' PurexaTM-MQ membranes can rapidly purify bioprocess pools, including aggregates, in a single polishing step and without dilution of feed. The extremely high binding capacity for ssDNA may also hold potential applications for plasmid DNA purification in the gene therapy industry.

1:40 pm (Room 3, Wednesday)

Oral 89 - Fouling and Retention Behavior of BioOptimal MF-SL Microfilter during Cell Culture Harvesting

Ranil Wickramasinghe (University of Arkansas), Xianghong Qian* (University of Arkansas), Da Zhang (University of Arkansas).

The rapid advancement in upstream cell culture operations has led to a significant increase in product titers. However, this high-level of productivity is accompanied by a rather high cell density which places a significant burden on the traditional downstream clarification and purification operations, especially during the initial clarification and capture steps. Depth filtration, centrifugation and tangential-flow filtration (TFF) have been used to remove cells, cell debris and other impurities during cell culture harvesting. BioOptimalTM MF-SL from Asahi KASEI operates as a hybrid TFF-depth filter which demonstrates superior performance with high throughput and fast processing time. However, there is very limited understanding on its fouling behavior and product retention characteristic. Our previous studies have demonstrated that a combined pore blockage and cake filtration model is able to describe the fouling behavior of this hybrid filter in the normal flow mode whereas a resistance in series model has been able to predict the performance of the filter operated in the TFF mode. Here the retention of host cell proteins (HCPs), DNA and therapeutic product is investigated during the TFF operation mode. Complementary imaging tools were used to visualize the locations where various components are trapped in the membrane matrix.

2:00 pm (Room 3, Wednesday)

Oral 90 - Development of nanopocket membranes for tangential flow analyte capture (TFAC) of extracellular vesicles

Thomas Gaborski* (Rochester Institute of Technology), Mehdi Dehghani (Rochester Institute of Technology), Shayan Gholizadeh (Rochester Institute of Technology).

The use of extracellular vesicles (EV) for diagnostic and therapeutic applications has seen a major interest increase in recent years because of their capacity to exchange components such as nucleic acids, lipids and proteins between cells. Isolation of a pure population of EVs is the first step in studying their physiological functions since contamination of EV preparations with non-EV proteins can lead to incorrect conclusions about their biological activities. We have recently demonstrated a new method termed tangential flow for analyte capture (TFAC) using ultrathin nanomembranes to purify extracellular vesicles from complex biological fluids such as blood plasma and cell culture media. In order to increase the specificity of EV capture including separation of different EV sub-populations based on size and surface markers, we are developing a series of nanopocket membranes. We use nanosphere lithography (NSL) to pattern and etch nanoscale (50-500 nm) pockets, each with a pore, on the surface of a membrane. Nanopockets can be fabricated with varying radius and depth as well as surface properties akin to size and affinity chromatography.

Oral 91 - Catch and Display Diagnostics: Capturing Non-protein Biologics on Ultrathin Silicon Membranes Using Size and Affinity-based Separations

Michael Klaczko* (University of Rochester), Kilean Lucas (University of Rochester), William Houlihan (University of Rochester), Julia Kuebel (University of Rochester), Jonathan Flax (University of Rochester), Richard Waugh (University of Rochester), James McGrath (University of Rochester).

Point-of-care (POC) diagnostics is of growing interest because of its potential for at-home and field detection as well as surveillance of disease. One of the major challenges of POC technologies is the miniaturization of sample preparation steps needed to purify potential biomarkers from complex biofluids. In central laboratories these steps are often done with some combination of centrifugation, affinity capture, elution, and analysis. We have recently demonstrated the use of ultrathin silicon nanomembranes tangential flow setups to enable the physical capture of diagnostically valuable particles or cells from small volumes of raw biofluids. In this talk we will extend these techniques to include affinity-based capture of bioparticles and cells with silicon nanomembranes. Affinity-based capture is achieved by a simple dip-coating of nanomembranes with functional amphiphilic KODE™ molecules. Detection of target capture is achieved by monitoring transmembrane pressure and direct imaging is made possible by the optical transparency of nanomembranes. We will specifically illustrate the application of KODE™-modified silicon nanomembranes for the capture and detection of therapeutic cells and extracellular vesicles from complex biological mixtures.

Oral 19 – MEMBRANE CHARACTERIZATION

1:00 pm (Room 4, Wednesday)

Oral 92 - Tailoring the Structure and Performance of Model Polyamide-based Membranes for Desalination

William Mulhearn (NIST), Peter Beaucage (NIST), Ryan Nieuwendaal (NIST), Christopher Soles (NIST), Christopher Stafford* (NIST).

The state-of-the-art membranes for desalination are comprised of thin film composites where the permselective layer is a thin but highly crosslinked, interfacially polymerized polyamide. Although effective, the rapid polymerization rate and reaction conditions produce films with rough surface structures and chemical heterogeneity, which precludes some advanced characterization techniques. In 2011, we proposed a paradigm shift in how these types of membranes are fabricated, where the selective layer is created layer-by-layer through a reactive deposition process. By doing so, we can create membranes that are smooth, tailorable, and exceptionally thin (10's nm). In this presentation, I will highlight the development of our automated deposition system for making tailored model polyamide membranes and present recent measurements of performance of these membranes through dead-end filtration. I will also present some recent advanced measurements of network structure and dynamics on these types of membranes that challenge our understanding of the operative transport mechanism in these materials.

1:20 pm (Room 4, Wednesday)

Oral 93 - Applying Transition-State Theory to Model Solute Transport in Membranes with Sub-nanometer Pores

Razi Epsztein* (Technion), Ryan M. Duchanois (Yale University), Cody L. Ritt (Yale University), Menachem Elimelech (Yale University).

Synthetic polymeric membranes with pore radii at the sub-nm scale—such as reverse osmosis (RO), nanofiltration (NF), and ion-exchange membranes (IEX)—are at the core of water purification and desalination processes. While these membrane processes have achieved substantial industrial success, the fundamentals underlying molecular transport in these membranes are not well established. The conventional solution-diffusion model, frequently used to describe transport in dense polymeric membranes, has limited capability to elucidate molecular-level phenomena that govern water and solute permeation in these membranes. Instead, the transition-state theory (TST), which is rarely used to describe transport in synthetic membranes, can highlight important features of molecular diffusion under confinement. More specifically, experimental energy barriers and pre-exponential factors from Arrhenius-type equation can provide molecular-level interpretation of solute transport through membranes in terms of enthalpy and entropy changes that occur during the transport. In this work, we first establish the connection between TST and solute transport in membranes with sub-nanometer pores. We then explore energy barriers and pre-exponential factors of ion transport in NF and IEX membranes and show correlations between these parameters and inherent properties of the ions, highlighting molecular phenomena that occur under extreme confinement, such as ion dehydration and electrostatic repulsion. We conclude with implications to future membranes with improved solute-solute selectivity that are inspired by ion-selective biological channels.

1:40 pm (Room 4, Wednesday)

Oral 94 - Unraveling the Clay-Membrane Interactions during An Osmotically Driven Process via Optical Coherence Tomography

Xin Liu* (Southern University of Science and Technology), Weiyi Li (Southern University of Science and Technology).

This study highlighted the interplay between the clay particles and membrane when employing a membrane with a relatively loose 'active layer' to implement osmotically driven processes. Optical coherence tomography (OCT) was explored to in-situ observe the deposition of the clay particles as the initial permeate flux was varied around the critical point, below which significant accumulation of the clay particles was inhibited. The OCT-based characterization offered a series of tools to quantify the evolution of the distribution of the deposits, thereby providing deeper insights into the underlying mechanisms. It was revealed that stronger clay-membrane interactions could result from the presence of the draw solute in the boundary layer near the feed-membrane interface, especially when divalent salts were exploited to enhance the effective osmotic pressure difference across the active layer. Moreover, the OCT-characterization results indicated that the variation in the clay-membrane affinity could have significant impact on the hydrodynamic stability of the boundary layer. Design heuristics would be developed to optimize the operating conditions for novel osmotically driven processes.

2:00 pm (Room 4, Wednesday)

Oral 95 - Local density and free volume inhomogeneities govern transport properties in reverse osmosis membranes

Michael Geitner* (The Pennsylvania State University), Tyler Culp (The Pennsylvania State University), Abhishek Roy (Dow), Mou Paul (Dow), Steve Jons (DuPont), Jeffrey Wilbur (DuPont Water Solutions), Manish Kumar (The University of Texas at Austin), Enrique Gomez (The Pennsylvania State University).

A mechanistic description of microstructure's effect on membrane properties could lead to the development of next-generation materials for water separation. Quantification of the internal microstructure of fully-aromatic polyamide thin-films, which serve as the active layer in state-of-the-art desalination membranes, is a crucial component to developing such descriptions. Here, we quantify the internal morphology of the polyamide active layers of commercial reverse osmosis membranes via scanning transmission electron tomography. This technique allows 3D reconstructions of the active layer to be obtained and parameters, such as void fraction and surface area, to be measured. Tomogram intensity analysis revealed the nanometer-scale density and free volume distributions, which are used in conjunction with the 3D polyamide models and tracer studies to determine correlations between active layer inhomogeneities and flow. The combination of density and free volume distributions determined from electron tomography with the use of physiochemical markers has provided a robust approach towards the development of structure-property relationships in reverse osmosis membranes.

2:20 pm (Room 4, Wednesday)

Oral 96 - Probing the Structure and Dynamics of Membranes under Realistic Operating Conditions, using Operando Spectroscopy

Casey O'Brien* (University of Notre Dame).

Operando spectroscopy is a general research methodology that combines spectroscopic characterization of a catalytic material with simultaneous analysis of reaction products under realistic operating conditions. Because of its ability to directly correlate catalyst structure—determined by spectroscopic characterization—to catalyst performance, operando spectroscopy is a powerful technique for establishing reliable structure-performance relationships that can guide the rational design of high-performance catalysts. While operando spectroscopy is well-established in catalysis, and has had a significant impact on the field over the past two decades, the operando spectroscopy methodology has not been widely adopted by other fields. We are the first research group to my knowledge to apply the operando spectroscopy methodology to gas separation membranes. The operando spectroscopy methodology applied to membranes involves the spectroscopic characterization of the membrane with simultaneous measurements of trans-membrane permeation rates under realistic operating conditions. By performing both measurements simultaneously, one can correlate membrane structure to performance directly, and elucidate structure-function relationships in great detail. In this talk, I will describe the operando spectroscopy experimental setup, and I will demonstrate with specific examples from our laboratory how operando spectroscopy can reveal unique scientific insights that can accelerate the discovery of high-performance membranes.

Oral 20 – MEMBRANES FOR ELECTROCHEMICAL APPLICATIONS I

1:00 pm (Room 5, Wednesday)

Oral 97 - Modelling and validation of concentration dependence of ion exchange membrane permselectivity: significance of convection and Manning's counter-ion condensation theory

Ryan S Kingsbury* (University of North Carolina at Chapel Hill), Orlando Coronell (University of North Carolina at Chapel Hill).

The electrical efficiency of electrodialysis, reverse electrodialysis, and related electrochemical processes is directly related to the permselectivity of the ion exchange membranes (IEMs) – defined as the extent to which the membrane permits the passage of counter-ions (ions of opposite charge to the membrane, e.g., cations for cation exchange membrane) while blocking passage of co-ions. Permselectivity is not a material constant, but rather depends on the concentration and composition of the electrolyte solutions in contact with the IEM. Thus, even though permselectivity is routinely measured at standardized conditions, the practical utility of such data is limited because we lack an accurate, quantitative way of using it to predict permselectivity under relevant process conditions. Moreover, the concentration dependence of IEM permselectivity has primarily been studied by evaluating the performance of (reverse) electrodialysis stacks rather than individual membranes, which has made it difficult to relate changes in permselectivity to specific membrane characteristics.

In this study, we measured the permselectivity of four commercial IEMs in six different concentration gradients employing 4 M and 0.5 M NaCl as the high salt concentration. We then constructed a predictive model of membrane permselectivity based on the extended Nernst-Planck equation and investigated how accounting for convection and electrostatic effects (via Manning's counter-ion condensation theory) affected model accuracy. We demonstrate that accurate, quantitative predictions of IEM permselectivity as a function of external salt concentrations are possible and require knowledge of only four easily measured membrane properties: water uptake, water permeability, charge, and thickness.

1:20 pm (Room 5, Wednesday)

Oral 98 - The Influence of Ion Association on Ion Solubility in Ion Exchange Membranes

Rahul Sujanani* (The University of Texas at Austin), Oscar Nordness (The University of Texas at Austin), Joshua Saunders (The University of Texas at Austin), Joan Brennecke (The University of Texas at Austin), Lynn Katz (The University of Texas at Austin), Benny Freeman (The University of Texas at Austin).

Ion exchange membranes (IEMs) are widely used in water purification and energy generation systems because the fixed charge moieties on the polymer backbone allow for selective permeation of ionic species. However, a detailed understanding of the impact of polymer structure on ion solubility and diffusivity in IEMs for various electrolytes remains largely incomplete. In prior studies, we developed a thermodynamic modeling framework (i.e., Manning/Donnan model) to predict ion solubility in IEMs. Results for NaCl and MgCl2 solubility in IEMs showed that our framework describes this data remarkably well. However, we recently discovered that our model is unable to predict Na2SO4 solubility in cation exchange membranes (CEMs). Na2SO4 was expected to be less soluble than NaCl, because mobile ions of higher valence are subject to a higher Donnan potential that excludes them more effectively from IEMs. Na2SO4 uptake was higher than expected and was, instead, in quantitative agreement with monovalent salt (i.e., NaCl) uptake. We attribute this to ion association as Na2SO4 forms significant amounts of monovalent [NaSO4]- ion pairs in aqueous solution, that can contribute to enhanced sorption in a CEM. Additionally, we measured the solubility of MgSO4 and results are available for several commercial IEMs. Remarkably, the solubility behavior of MgSO4 in IEMs is similar to that of an electroneutral molecule (i.e., CH3OH). The extent of ion association for Na2SO4 and MgSO4 in aqueous solutions was measured and found to be as high as 50% at higher salt concentrations. These findings and their impact on ion separation using IEMs will be discussed.

1:40 pm (Room 5, Wednesday)

Oral 99 - Functionalized Ceramic Ion Exchange Membranes for Electrodialysis

Gregory M Newbloom* (Membrion, Inc.), Rachel Malone (Membrion, Inc.), Ryan Kingsbury (Membrion, Inc.), Aditya Salunkhe (Membrion, Inc.).

Polymer ion exchange membranes (IEM) suffer from high cost, high ion resistance, dimensional swelling and poor durability which limit their use in applications such as electrodialysis for brackish water desalination. A new class of functionalized ceramic ion exchange membranes have the potential to address all of these challenges simultaneously. In this work, we investigate the use of a surface-modified inorganic membrane created using sol-gel processing of siloxanes without calcination or sintering. Silica is an advantageous material for membranes because of its excellent durability, extremely low cost and lack of swelling in water. Membrane performance can be tuned by size exclusion (i.e., sub-5nm nanometer pores) and surface functionalization (i.e., silane chemistry) to achieve promising anion and cation-selective performance. Ion exchange capacity, permselectivity, water permeance and area specific resistance can be tuned to have comparable performance to commercial polymer IEMs. When scaled and run in an electrodialysis stack, ceramic IEMs outperform their polymer counterparts in current efficiency and areal resistance for brackish water conditions. These promising preliminary results highlight the commercial potential for a new, inexpensive IEM with the potential to elevate the status of electrodialysis as a leading solution to the fresh water crisis.

2:00 pm (Room 5, Wednesday)

Oral 100 - Tradeoff in membranes for artificial photosynthesis: Ion transport and product crossover

Sarah M Dischinger* (Lawrence Berkeley National Laboratory), Dan Miller (Lawrence Berkeley National Laboratory), Blaine Carter (Lawrence Berkeley National Laboratory), Shubham Gupta (Lawrence Berkeley National Laboratory).

Artificial photosynthesis transforms abundant natural resources, namely CO2 and water, into renewable liquid fuels (e.g., ethanol) using sunlight. Polymeric membranes are commonly employed in artificial photosynthesis devices to prevent crossover of CO2 reduction products, which leads to device inefficiencies, while also promoting transport of charge carriers between the electrodes. However, explicit examination of the membrane physicochemical properties required to make these devices successful has been limited because most studies utilize commercial membranes. While many studies have considered either ionic conductivity or organic solute permeation within the framework of membrane free volume fraction, few have considered these transport phenomena simultaneously. Herein, we explore the structure-transport relationships of a membrane platform that are relevant to artificial photosynthesis devices. We systematically varied the water uptake and corresponding microstructure of imidazolium-functionalized poly(phenylene oxide) membranes by controlling the polymer's degree of functionalization. In this manner, we were able to regulate ethanol (a commonly reported CO2 reduction product) permeation and bicarbonate (a common charge carrier) conductivity, both of which increased with increasing free volume fraction. Combined, these results present a tradeoff between membrane properties desirable for artificial photosynthesis (i.e., high charge carrier transport and low CO2 reduction product transport). Amid the wide range of operational characteristics (e.g., current density) reported for artificial photosynthesis devices, optimal membrane properties for this application remain undefined. The systematic control over properties of the material platform discussed herein enables the study of artificial photosynthesis devices of various operational conditions, and could inform rational design of membranes for artificial photosynthesis and other electrochemical applications.

2:20 pm (Room 5, Wednesday)

Oral 101 - Highly Selective Ion Separations with Pressure-driven Flow through Charged Nanoporous Membranes

Chao Tang* (University of Notre Dame), Andriy Yaroshchuk (Polytechnic University of Catalonia), Merlin Bruening (University of Notre Dame).

This presentation explores highly selective separations of Li+ from other monovalent ions during flow of mixed salt solutions through charged nanopores. Remarkably, at ionic strengths from 0.2 to 1 mM, flow through negatively charged, track-etched polycarbonate membranes (30 nm pores) yields Li+/K+ selectivities from 20 to 70 along with Li+ passages (defined as the ratio of permeate to feed concentration) as high as 1. Moreover, both Li+/K+ selectivity and Li+ passage initially increase with flow rate to help overcome the permeability/selectivity trade-off.

Simulations using the extended Nernst-Planck equation both in the membrane and in the unstirred layer next to the membrane help to explain the mechanism of the separations and the high Li+ passage. During flow of electrolyte solutions through membranes containing charged nanopores, a streaming potential spontaneously develops to maintain electroneutrality. This potential gives rise to electromigration, which opposes convection for cations within negatively charged pores. Importantly, with two cationic species electromigration is greater for the more mobile cation. Thus, perhaps counter-intuitively, the membrane shows higher rejections for more mobile ions. The separation becomes very effective if electromigration exceeds convection for the more mobile cation. At high flow rates where diffusion is negligible, rejections of more mobile cations become very high. Li+ passages that approach unity result from concentration polarization as well as the high concentration of Li+ in the nanopores. Control over flow rate and concentration polarization my enable separations with both high selectivity and high Li+ recovery. Future work will aim to extend these separations to higher ionic strengths.

Abstracts – Oral Presentations, Thursday

Presenting authors are indicated by an asterisk*. See the most up-to-date version of the program book with full abstracts at **www.membranes.org/nams-2020/**

Oral 21 – MOLECULAR AND PROCESS MODELING

11:00 am (Room 1, Thursday)

Oral 102 - Effect of packing nonuniformity at the fiber bundle-case interface on flow distribution and separation performance for hollow fiber membrane modules

Lili Sun (University of Toledo), Atabong Etiendem (University of Toledo), Glenn Lipscomb* (University of Toledo, Toledo).

Membrane gas separation processes most commonly utilize hollow fiber membrane modules. Experimental performance of these modules can differ dramatically from theoretical predictions based on independent experimental measurements of membrane transport properties. One of the most significant sources of these deviations is flow maldistribution that prevents uniform counter-current contacting between retentate and permeate. Previous work has demonstrated the impact of fiber size uniformity on flow distribution and performance for lumen-fed modules. While poor packing of the fiber bundle adjacent to the enclosing case has been hypothesized as another significant factor for shell-fed modules, the impact has not been investigated thoroughly.

Simulations of flow through fiber bundles with non-uniform wall packing are reported. Results are reported for fiber bundles possessing a regular square or triangular packing as well as randomly packed bundles. The simulation approach relies on determining the length scale over which the case influences flow distribution and the extent to which flow near the case increases or decreases relative to that expected for a large bundle where the wall has no influence.

The length over which the case influences flow increases dramatically with increasing fiber packing fraction. As the packing fraction approaches 0.6, the impact can extend over the entire bundle diameter for small modules (<5,000 fibers). The impact of poor packing is greater for triangular packings than square packings because irregular triangular packing leads to larger flow regions near the case. The impact is even greater for randomly packed bundles.

The flow distribution results are used to evaluate the performance of a shell-fed hollow fiber module for nitrogen production from air. The results clearly demonstrate the detrimental effect of poor packing along the case especially for small, randomly packed modules.

11:20 am (Room 1, Thursday)

Oral 103 - Estimating fluid pressure gradients within crosslinked aromatic polyamide using molecular dynamics

Riley Vickers* (University of North Carolina at Chapel Hill,) Timothy Weigand (University of North Carolina at Chapel Hill), Casey Miller (University of North Carolina at Chapel Hill), Orlando Coronell (University of North Carolina at Chapel Hill).

The solution-diffusion model is routinely used to analyze reverse osmosis membrane performance at the system scale, or megascale. However, megascale approaches do not provide mechanistic information that can be used to predict the effects of changes in microscale and molecular-scale structure on performance. Previous work has shown that variations in the microscale pore structure of crosslinked aromatic polyamide (XLPA) membranes has a significant effect on water flux (Lin et al., 2018, J. Membr, Sci., 564, 935-944). The goal of this work is to develop a mechanistic understanding of the internal structure of XLPA membranes at the molecular scale and use this understanding to inform a microscale continuum model of fluid flow and salt transport. Closure of the microscale continuum model requires a molecular-scale model to analyze the distribution of fluid pressure within the molecular free volume of an XLPA membrane during operation. Accordingly, we develop a method that utilizes molecular dynamics (MD) simulations to generate representative all-atom XLPA domains and measure the pressure gradient behavior of transported molecules during simulated operation. To achieve this, we utilized experimentally determined XLPA densities and water and solute fluxes during operation as model inputs to verify the accuracy of simulated XLPA structures. The pressure gradient during operation simulation is evaluated by using MD to track average forces exerted by an ensemble of water molecules acting on a plane. The dependence of the pressure gradient on accessible molecular free volume due to entropic fluctuations is also analyzed. This approach provides a basis to simulate water flow and salt transport through dense polymeric membranes at the molecular scale. It will also provide the mechanistic understanding needed for microscale continuum modeling of representatively sized domains and prediction of realistic membrane system performance, dependent upon microscale membrane structure.

11:40 am (Room 1, Thursday)

Oral 104 - High-throughput computational prediction of the cost of carbon capture using mixed matrix membranes

Janice Steckel* (U.S. Department of Energy), Christopher Wilmer (University of Pittsburgh), Samir Budhathoki (DOE/NETL), Olukayode Ajayi (Brewer Science).

Mixed matrix membranes (MMMs) are advanced materials formed by combining polymers with inorganic particles. MMMs based on metal–organic frameworks (MOFs) has been shown to improve selectivity and permeability over pure polymers. We have carried out high-throughput atomistic simulations on 112 888 real and hypothetical metal–organic framework structures in order to calculate their CO2 permeabilities and CO2/N2 selectivities. The CO2/N2 sorption selectivity of 2 017 real MOFs was evaluated using the H2O sorption data of Li et al. (S. Li, Y. G. Chung and R. Q. Snurr, Langmuir, 2016, 32, 10368–10376). Using experimentally measured polymer properties and the Maxwell and Bruggeman models, we predict the properties of all of the hypothetical mixed matrix membranes that could be made by combining the metal–organic frameworks with each of nine polymers, resulting in over one million possible MMMs. The predicted gas permeation of MMMs is compared to gas permeation data from our group as well as published data to evaluate the accuracy of the methodology. A selection of promising materials have been further analyzed. We carried out twelve individually optimized techno-economic evaluations of a three-stage membrane-based capture process. For each evaluation, capture process variables such as flow rate, capture fraction, pressure and temperature conditions were optimized and the resultant cost data were interpolated in order to assign cost based on membrane selectivity and permeability. This work makes a connection from atomistic simulation all the way to techno-economic evaluation for a membrane-based carbon capture process. We find that a large number of possible mixed matrix membranes are predicted to yield a cost of carbon capture less than $50 per tonne CO2 removed, and a significant number of MOFs so identified have favorable CO2/N2 sorption selectivity.

12:00 pm (Room 1, Thursday)

Oral 105 - Opportunities for high productivity and selectivity desalination via osmotic distillation with improved membrane design

Anthony Straub (University of Colorado Boulder), Sangsuk Lee* (University of Colorado Boulder).

Current osmotic membranes rely on dense polymeric materials that are inherently limited in their achievable selectivity and poorly reject low molecular weight contaminants with neutral charge. Osmotic distillation (OD) membranes separate using a phase change and thus offer distillation-quality product water free of all non-volatile solutes. Despite their ultrahigh selectivity, OD membranes have not been widely implemented due to poor water fluxes and detrimental heat transfer effects. In this study, we examine how OD membrane optimization can improve water treatment performance and compare the productivity of simulated OD membrane modules to conventional FO systems. Several OD membrane parameters are optimized, including thickness, thermal conductivity, porosity, and pore diameter. Among the membrane properties explored, we identify that reducing the OD membrane thickness is the most crucial factor in enabling high performance, and a membrane thickness of approximately 0.1

µm is needed to achieve water fluxes exceeding those of current commercial FO membranes. Decreasing the membrane thickness increases the water permeability coefficient of the membrane and also increases the heat conduction through the membrane, which reduces the detrimental impact of temperature polarization. The improved heat conduction offered by thin membranes is critical in large-scale systems, where accumulated heat from the enthalpy of vaporization will otherwise dramatically decrease the water flux. Comprehensive comparison of OD with conventional FO membranes showed that optimized OD membranes can outcompete FO membranes in maximum achievable water flux (25.3 vs. 18.6 kg m-2h-1 for OD and FO membranes, respectively) and module-scale water recovery (0.28 vs. 0.18). Overall, the results of this work demonstrate the promise of OD membranes to overcome the selectivity limitations of conventional polymeric membranes and offer guidelines for future OD membrane design.

12:20 pm (Room 1, Thursday)

Oral 106 - Ion association as key element in RO and NF modeling

Viatcheslav Freger* (Technion – IIT).

Despite wide use of NF and RO membranes, understanding ion transport in these processes still presents a challenge, as current model are unable to reconcile all observations. Mean-field models such as Donnan, Poisson-Boltzmann and Manning's ion condensation have long been the standard, yet they fail to reproduce some observed trends, e.g., permeation of divalent-ion salts and multi-ion mixtures without questionable ad hoc adjustments.
This failure has been related here to the use of mean-field approaches, which were originally devised for high-T* (high-dielectric) media and in which potential variations are small compared to thermal energy kT, thereby charged groups nearly completely dissociate. Unfortunately, this condition is strongly violated in NF and RO, which are low-T*, i.e., the local potentials largely exceed kT. As a sounder treatment, we introduce a solution-diffusion model that adds ion association in a manner of classical Bjerrum theory. Association acts as a charge-regulating and ion-immobilizing mechanism that modifies and, overall, greatly weakens the effect of fixed charges on ion permeation, resulting a largely reduced "effective" membrane charge. In the new treatment, ion uptake and mobility become inter-related thus ion content (e.g., as determined by RBS) no more correlates linearly with ion permeability. Our analysis also highlights the fact that simple continuum electrostatics is unable to accurately predict association and solvation, which requires molecular simulations for determining relevant ion-specific parameters.
The model can explain our recent data on NF of NaCl+CaCl2 mixtures and ion permeation in RO that poorly correlate with very high ion uptake reported for RO membranes, inconsistent with their high ion rejection. The new model thus presents a step towards more physically consistent predictive modeling of ion-selective membrane separations in water-, environment- and energy-related applications.

Oral 22 – MEMBRANE SYNTHESIS AND CASTING II

11:00 am (Room 2, Thursday)

Oral 107 - Atomically Thin Graphene Membranes for Desalination and Molecular Separations

Piran Kidambi* (Vanderbilt University).

Atomically thin graphene with a high-density of sub-nanometer pores could enable the ultimate water treatment/purification/desalination membrane with ultra-high flux and excellent solute rejection. However, the formation of precise and high-density of sub-nanometer pores (0.28-0.66 nm) in the atomically thin graphene lattice over large areas remains extremely challenging.
Here, we report on a facile and scalable process to fabricate large-area graphene membranes with a high density of nanopores (4 - 5.5 × 10^12 cm^-2) for desalination and molecular separations. We demonstrate fully functional centimeter-scale monolayer graphene membranes with water permeance orders of magnitude higher than commercially available desalination membranes, and excellent rejection to salt ions and small organic molecules.
Our work provides a facile and scalable approach to overcome fundamental limitations in the development of nanoporous atomically thin membranes (NATMs) for large area applications.

11:20 am (Room 2, Thursday)

Oral 108 - A Comparison of PolarClean, Gamma-Valerolactone and their Mixture as Bio-derived solvents for Polysulfone Membrane Fabrication

Xiaobo Dong* (University of Kentucky), Tequila Harris (Georgia Tech), Isabel Escobar (University of Kentucky).

Petroleum-derived solvents used for membrane fabrication through the nonsolvent induced phase separation (NIPS) process are known for their toxicity. One big challenge is that by 2020, some common solvents, N-Methyl-2-Pyrrolidone (NMP) and Dimethylacetamide (DMAc) will be banned by Registration, Evaluation, Authorisation and Restriction of Chemicals (REACH) in Europe. Therefore, bio-derived and low-toxicity solvents are a focus of increasing investigations. Examples of bio-derived, low-toxicity solvents include methyl-5- (dimethylamino)-2-methyl-5-oxopentanoate (PolarClean) and gamma-valerolactone (GVL).

PolarClean has been previously used to fabricate polysulfone (PSf) membranes with mixed results, such as having similar flux and rejection as traditional membranes but having membrane pores collapse during membrane cleaning. To address this issue, GVL was investigated as a sole solvent and a co-solvent with PolarClean to fabricate PSf membranes. Membranes prepared using GVL as a sole solvent were observed to be gelatinous, hence not ideal for filtration.

GVL and PolarClean were used as co-solvents, viable membranes were cast with surface charge and hydrophilicity not being significantly different from membranes made using PolarClean alone. Specifically, the solvent mixture produced PSf membranes with zeta potential values of about -20mV and contact angles at around 65°. Furthermore, the average pore size of membranes decreased from 48nm to lower than 1nm as adding GVL in dope solutions. The membranes prepared using solvent mixtures showed pure water flux at 300-500 LMH and over 99% BSA rejection. Therefore, the use of PolarClean/GVL as co-solvents showed promise for the fabrication of PSf membranes. After optimizing the dope solution recipe, the slot die technique was used to successfully fabricate the membranes at a large scale.

11:40 am (Room 2, Thursday)

Oral 109 - Electrospun Pretreatment Membranes

Husain Mithaiwala* (Arizona State University), Matthew Green (Arizona State University).

With the ever-growing population that nearly tripled in the 20th century, managing water resources has become one of the most pressing concerns of scientists both in academia and industry. The reverse osmosis (RO) water treatment process is a well-researched technology among the pressure driven processes to produce potable water. RO is an energy intensive process and often RO membranes are susceptible to fouling and scaling that drives up operational cost and hinder the efficiency. Thus, to increase the performance of RO membranes feed water is pretreated to remove pollutants before desalination. This work aims to fabricate pretreatment membranes to prevent the effects of fouling and scaling by introducing hydrophilic character to membrane surface.

This work explores electrospinning, a cost-effective and scalable technique, to blend two polymers into a nonwoven membrane comprised of fibers ~100 nm –100 μm in diameter. A rotary drum collector holding the mat simultaneously collects the hydrophobic poly(vinyl chloride) and hydrophilic poly(vinyl alcohol) (PVA) fibers being electrospun from two separate solutions. The work aims to tune the hydrophilicity of the resulting membrane by controlling the relative deposition rate of PVA onto the mat. After electrospinning, the mats are cross-linked with a poly(ethylene glycol) diacid to impart mechanical strength and tune the porosity. Fiber morphologies were characterized by scanning electron microscopy and the elemental composition across the membrane was mapped by energy dispersive X-ray spectroscopy. Additionally, membrane characterization tests (e.g., differential scanning calorimetry, thermogravimetric analysis, and Fourier-transform infrared spectroscopy) confirmed the presence of both polymers on the mat. A rigorous analysis method was developed combining the above results to estimate mass deposition rate, relative concentration of both polymers on the mat, and correlate PVA concentration with hydrophilic character.

12:00 pm (Room 2, Thursday)

Oral 110 - Aqueous phase separation of responsive copolymers for sustainable and mechanically stable membranes

Wouter Nielen* (University of Twente), Joshua D. Willott (University of Twente), Wiebe M. de Vos (University of Twente).

Membranes are often used in environmentally friendly applications and as a sustainable alternative to conventional processes. Unfortunately, the vast majority of polymeric membranes are produced via an unsustainable and environmentally unfriendly process that requires large amounts of harsh reprotoxic chemicals like N methyl 2 pyrrolidinone (NMP) and dimethylformamide (DMF). In this work, we investigate an aqueous phase separation (APS) system that uses weak polyelectrolytes, whose charge is dependent on the pH (weak polyelectrolytes) to produce membranes. Specifically, the copolymer polystyrene-alt-maleic acid (PSaMA) is used. PSaMA contains responsive monomers required for the aqueous phase separation and also unresponsive hydrophobic monomers that provide mechanical stability to the resultant membranes. This work demonstrates that by controlling the precipitation of PSaMA, it is possible to prepare a wide range of membranes, from open microfiltration membranes capable of treating oily waste water, to dense nanofiltration-type membrane with excellent micropollutant retentions and high mechanical stability. The only solvents used in this APS system are water and the green solvent acetic acid, thus making our APS process significantly more sustainable and environmentally friendly compared to conventional membrane fabrication methods.

12:20 pm (Room 2, Thursday)

Oral 111 - Stepwise synthesis of oligoamide film on porous supports for preparing new types of membranes

Paramita Manna (Ben-Gurion University of the Negev), Roy Bernstein (Ben Gurion University), Ron Kasher* (Ben Gurion University of the Negev).

Porous polymeric membranes are widely used in potable water purification, wastewater treatment, the food and the pharmaceutical industries, and haemodialysis. However, producing specialized membranes with diverse transport properties is challenging. A method for fabricating membranes with controllable transport properties is described here by stepwise synthesis of aromatic oligoamide on a porous polymeric support. The use of aromatic oligoamide affords good water permeance due to its hydrophilic character. Alternate couplings of trimesoyl chloride and meta-phenylenediamine yielded an oligoamide dendrimer that was covalently bonded to the support. The water permeance and molecular weight cutoff (MWCO) of the synthesized membranes were controlled (with values of 4.6 to 543 L·m−2·h−1·bar−1 and 22.6 to 332 kDa, respectively) by adjusting the number of oligoamide synthesis cycles in the range of 2.5 – 20.5. The oligoamide membrane with 5.5 synthetic cycles showed a high rejection of the negatively charged rose bengal dye (95% rejection) with high flux (126.4 ± 4.2 L·m−2h−1 at 5.2 bar), as compared with other membranes reported in the literature. The fabricated membranes are potentially highly useful for the separation of macromolecules with specific ranges of molecular weight, for industrial separations that require membranes with tunable MWCO ranges, or for the separation of charged macromolecules.

Oral 23 – MEMBRANE DISTILLATION AND PERVAPORATION

11:00 am (Room 3, Thursday)

Oral 112 - Elucidating the Trade-off between Membrane Wetting Resistance and Water Vapor Flux in Membrane Distillation

Chenxi Li (University of British Columbia), Xuesong Li (Tongji University), Xuewei Du (Colorado State University), Ying Zhang (University of British Columbia), Tiezheng Tong (Colorado State University), Arun Kota (North Carolina State University), Jongho Lee* (University of British Columbia).

Membrane distillation (MD) has been receiving attentions as a promising approach for desalinating high-salinity industrial wastewater. Generally, high surface hydrophobicity is preferable for membranes used in MD. However, when a commonly available hydrophobic membrane is modified to increase surface hydrophobicity, a sizable reduction of water vapor flux has been widely reported. Understanding the relation between the membrane wettability and water flux is crucial to determining the balance of long-term operation stability and water production yield in MD.

In this study, we elucidate the trade-off relation between the membrane wetting resistance (i.e., surface hydrophobicity) and the water flux in MD. We prepared microporous membranes that possess different surface wettabilities through coating quartz fiber (QF) substrates with silane molecules of different hydro- or fluoro-carbon chain lengths. With almost identical thermal conductivities and fiber structures of all membranes, we observed a decreasing trend of water vapor flux as the wetting resistance increases. An electrochemical impedance spectroscopy revealed that the membrane with a lower hydrophobicity exhibits a larger area of liquid-vapor interface inside the membrane pores. We then employed nanoporous alumina membranes that consist of straight nanopores, modified with identical sets of silane molecules used for QF membranes. No trade-off relation was observed for the alumina membranes. We conclude that an increased wetting resistance of hydrophobic membranes of irregular pore structures would result in a reduced liquid-vapor interface area available for evaporation and condensation of water, and therefore a reduced water vapor flux. Our study can provide a guideline to design MD membranes to ensure a maximized water flux for given feed water properties that require a certain wetting resistance.

11:20 am (Room 3, Thursday)

Oral 113 - Cost optimization of gap membrane distillation

Timothy V Bartholomew* (National Energy Technology Laboratory), Alexander Dudchenko (Stanford University), Nicholas Siefert (National Energy Technology Laboratory), Meagan Mauter (Stanford University).

Estimating the cost of membrane distillation (MD) processes at scale is confounded by the vast array of possible process configurations, module configurations, and operating conditions and the extreme sensitivity of MD costs to these parameters. This work develops a comprehensive MD cost optimization framework for minimizing the levelized cost of water (LCOW) as a function of feed salinity and temperature. Our gap MD model fully captures solution properties, heat transfer, and mass transfer relationships in the modules to determine the cost optimal design as a function of multiple decision variables, including process configuration (feed location and connectivity between the hot and cold-side), the gap type (air, permeate, conductive), equipment sizes (membrane, heater, chiller, and heat exchanger units), and operating conditions (flow rates, hot and cold side temperatures, and heating and cooling duties). We estimate that the minimum LCOW of conductive gap MD ranges between $10 and $16 per m3 for feed salinities of 25 g/L to 200 g/L and water recoveries of 30% to 75%. We update several heuristic assumptions for cost-optimal system operation, including: increased thermal conductivity of the gap decreases system costs; cost-optimal MD operates at the highest allowable temperature; and the LCOW is a strong function of brine salinity. Finally, we find that improving membrane properties and lowering heating and cooling costs will provide only modest reductions in costs (<25%).

11:40 am (Room 3, Thursday)

Oral 114 - Modeling Pilot-Scale Air Gap Membrane Distillation as a Special Case of Permeate Gap Membrane Distillation

Mukta Hardikar* (The University of Arizona), Phakdon (University of Arizona), Itzel Marquez (Central Michigan University), Eduardo Saez (University of Arizona), Andrea Achilli (The University of Arizona).

Air gap is considered one of the most energy efficient membrane distillation (MD) configurations as it recovers heat from the feed stream while using the air gap to maintain a transmembrane temperature difference/driving force. To design an optimal module, air gap membrane distillation (AGMD) performance at pilot-scale needs to be modelled accurately and verified using experimental data. Uncertainty in volume of air gap occupied by distillate presents a hindrance in accurately modeling AGMD. The water flux through the membrane affects the rate at which distillate drains from the gap and the volume fraction of distillate water in the air gap. This influences the rate of heat transport as the distillate provides an additional path for heat conduction across the air gap . Hence, determining extent of flooding in an air gap is essential to model performance of pilot-scale AGMD. To isolate this phenomenon in the air gap, a comparative study between permeate gap MD (PGMD) and AGMD was done using a pilot-scale module. Both configurations produced similar water-fluxes and temperature profiles which indicates partial flooding of the air gap even in AGMD configuration. However, this similarity in flux values steadily decreases with increase in water flux in both configurations. The pilot experimental results are used in this project to model AGMD as a special case of PGDM where the extent of flooding of the gap is analyzed as a function water flux. The model is further corroborated by comparing model predictions with experimental results using modules of different membrane material, feed solution properties and operating conditions. This method will help model the flooding effect in the air gap and select parameters that optimize AGMD performance.

12:00 pm (Room 3, Thursday)

Oral 115 - Economic Evaluation of Membrane Distillation as a Competitive Brackish Water Desalination Technology

Haamid S Usman* (Concordia University), Md. Saifur Rahaman (Concordia University), Khaled Touati (Concordia University).

Membrane Distillation (MD) is an emerging technology that can provide water to remote areas and arid regions. Intensive works have been performed to optimize the operating parameters of MD. However, few studies have investigated its economic feasibility. This study provides a systematic analysis evaluating the economics of producing water using the MD system considering the direct and indirect effects of key parameters (plant size, use of waste heat, membrane permeability). For MD, the results show that utilizing solar energy, waste heat recovery, and reducing the plant size has a significant impact on the cost of produced water (WP). The analysis showed that, for a standalone MD plant, the price was seven times higher than that of reverse osmosis. Similarly, the effect of energy consumption was prominent in the comparison between standalone MD and MD plant assisted with the waste heat recovery. The results show that the effect of using waste heat decreases the WP from 6.8 to 4.63$/m3, compared to the MD standalone system (WP decreases by around 32%). Finally, a sensitivity analysis showed that improving the operating and membrane parameters, the energy consumption, and identifying the process scalability effects significantly improve the economic aspects of MD for brackish water desalination..

12:20 pm (Room 3, Thursday)

Oral 116 - Conducting thermal energy to the membrane/water interface for the enhanced desalination of hypersaline brines using membrane distillation

Jingbo Wang* (University of California, Los Angeles), Yiming Liu (University of California, Los Angeles), Unnati Rao (University of California, Los Angeles), Navid Ebrahimi (University of California, Los Angeles), Eric Hoek (University of California, Los Angeles), Tzahi Cath (Colorado School of Mines), Nils Tilton (Colorado School of Mines), Craig Turchi (NREL), Yongho Sungtaek Ju (University of California, Los Angeles), David Jassby (University of California, Los Angeles).

Membrane distillation (MD) is a membrane-based thermal desalination process capable of desalinating hypersaline brines. Standard MD systems rely on the feed stream as the thermal energy carrier to drive the desalination process. However, the reliance on the feed to carry thermal energy has limitations including thermal polarization and feed stream temperature drop (and subsequent driving force drop). Providing thermal energy to the membrane/water interface has the potential of eliminating thermal polarization, reduce limitations on single-pass recoveries, and decrease the number of heat exchangers in the system. Recent developments in solar collection technologies enabled solar desalination due to its availability and low cost, so long as this heat can be delivered to where it's needed. We developed an MD process using advanced layered composite membrane materials that are thermally and electrically conducting. The integrated polymeric and metallic layers address needs for high-salinity desalination, such as the ability to conduct heat from an outside source to the membrane/water interface and drive electrochemical anti-scaling reactions. The composite membranes were tested in a vacuum MD configuration using feed water containing 100g/L NaCl. The new MD system showed stable performance with water flux up to 9 LMH, and salt rejection >99.9% over hours of desalinating hypersaline feed, with the only heat provided through the thermally conducting membrane layers. The system exhibited >75% energy efficiency in a single pass. System configuration, vacuum levels and crossflow velocities all contribute to the overall system performance. Through capturing solar heat outside of the membrane module and delivering it directly to the membrane/water interface, this process has the potential of solving several problems associated with thermally-driven desalination processes, such as minimizing heat losses, increasing thermal efficiency, and limiting the number of heat exchangers.

Oral 24 – MODULE MODELING AND DESIGN

11:00 am (Room 4, Thursday)

Oral 117 - Rough or wiggly? Membrane topology and morphology for fouling control

Bowen Ling* (Stanford University), David Ladner (Clemson University), Ilenia Battiato (Stanford University).

During filtration in reverse osmosis membranes (ROM), the system performance is dramatically affected by membrane fouling which causes a significant decrease in permeate flux as well as an increase in the energy input required to operate the system. In this work, we develop a model, able to dynamically capture foulant evolution, that couples the transient Navier–Stokes and the advection–diffusion equations, with an adsorption–desorption equation for the foulant accumulation. The model is validated against unsteady measurements of permeate flux as well as steady-state spatial fouling patterns. For a straight channel, we derive a universal scaling relationship between the Sherwood and Bejan numbers, i.e. the dimensionless permeate flux through the membrane and the pressure drop along the channel, respectively, and generalize this result to membranes subject to morphological and/or topological modifications, i.e. whose shape (wiggliness) or surface roughness is altered from the rectangular and flat reference case. We demonstrate that a universal scaling can be identified through the definition of a modified Reynolds number, that accounts for the additional length scales introduced by the membrane modifications, and a membrane performance index, an aggregate efficiency measure with respect to both clean permeate flux and energy input required to operate the system. The proposed approach is able to quantitatively investigate, optimize and guide the design of both morphologically and topologically altered membranes under the same framework, while roviding insights into the physical mechanisms controlling the overall system performance.

11:20 am (Room 4, Thursday)

Oral 118 - Characterization of laminar, transitional, and turbulent flow regime in membrane modules

Alexander Dudchenko* (Stanford University), Meagan Mauter (Stanford University).

Mass transfer, heat transfer, and pressure drop in membrane modules are commonly approximated using Sherwood, Nusselt, and friction factor correlations that are derived explicitly for laminar, transitional, or turbulent flow regimes. However, the transition point from laminar to transitional flow regime and transitional to turbulent flow regime are not quantified for membranes modules and instead are assumed to occur at 2,300 Re and 4,000 Re, respectively. These assumed transition points are only valid for flow in smooth pipes without entrance or exit effects and are not representative of conventional membrane module designs. Herein, we present an experimental study of flow transition in conventional membrane module designs with the open channel and spacer filled channel configuration. We use a combination of tracer experiments and pressure drop measurements to determine the transition point from laminar to transitional flow regime, and transitional to turbulent flow regime. Finally, we compare the experimentally acquired flow transition points to those observed in CFD simulations. The results of this work demonstrate that transitional and turbulent regimes occur at much lower Re value than commonly assumed and have resulted in erroneous estimates of mass transfer rates, heat transfer rates, and membrane process performance parameters..

11:40 am (Room 4, Thursday)

Oral 119 - 3D Numerical simulations of temperature polarization in vacuum membrane distillation systems with active membrane heating

Mark Dudley* (Colorado School of Mines), Nils Tilton (Colorado School of Mines), David Jassby (UCLA), Eric Hoek (UCLA), Craig Turchi (NREL), Tzahi Cath (Colorado School of Mines), Michael Heeley (Colorado School of Mines).

Vacuum membrane distillation (VMD) is a process in which warm feed flows over a vacuum chamber, separated by a hydrophobic membrane. Feed water evaporates from the membrane surface, flows through the membrane due to the difference in partial vapor pressure, and is later condensed. VMD is attractive for its ability to treat high saline brines and leverage the use of solar heat. One challenge for VMD is temperature polarization, which is the cooling of the feed flow near the membrane surface. We perform numerical simulations of a new approach to VMD that counteracts temperature polarization by actively heating the feed directly at the membrane surface, by applying a heat flux through a thermally conductive membrane. We develop an in-house 3D numerical code that simulates fully-coupled heat and mass transport within the feed channel and actively heated membrane. For that purpose, we also develop volume-averaged transport equations governing heat and mass transport within the membrane and at the interface between the membrane and feed flow. We use the code to perform a parametric study of how operating conditions, flow cell geometry, and membrane properties affect temperature polarization and vapor production in the system.

12:00 pm (Room 4, Thursday)

Oral 120 - Direct numerical simulations of unsteady mixing in direct contact membrane distillation systems with different spacer blockages and vertical offsets

Jincheng Lou (Colorado School of Mines), Jacob Johnston (Colorado School of Mines), Denis Martinand (University of Aix Marseille), Nils Tilton* (Colorado School of Mines).

Direct contact membrane distillation (DCMD) is a thermal process in which warm feed and cool distillate flow on opposite sides of a hydrophobic membrane. Water evaporates from the feed, travels through the membrane, and condenses in the distillate. To date, CFD studies of DCMD focus primarily on the challenge of temperature polarization. For high concentration brines, however, concentration polarization is another major challenge that leads to mineral scaling. In addition, unsteady mixing triggered by membrane spacers is also not well understood. To study coupled temperature polarization, concentration polarization, and unsteady mixing in the DCMD treatment of high concentration brines, we develop an in-house CFD method for simulating 2-D heat and mass transport in plate-and-frame DCMD systems with cylindrical spacers. The coupled momentum, energy, and mass transport equations are solved using a finite-volume method with an efficient non-iterative unsteady scheme. Spacers are simulated using recent advances in immersed boundary methods. We perform a parametric study of polarization under a variety of operating conditions, and investigate the unsteady variation of temperature, concentration, and vapor flux along the membrane surface. We explore the transition from steady flow to unsteady vortex shedding, and show that the transition strongly depends on the blockage ratio and vertical offset of the spacers. We also find that though vortex shedding increases transmembrane flux, it also creates regions of preferential solute accumulation on the membrane. This suggests that the dual objectives of maximizing permeate production flux and minimizing concentration polarization can be at odds.

12:20 pm (Room 4, Thursday)

Oral 121 - Hydrodynamic load exerted on a moving bundle of hollow fibers using constraint dissipative hydrodynamics: hydrorattle simulation

Albert S. Kim* (University of Hawaii at Manoa), Kwang Jin Lee (Kolon Industries, Inc.), Moo Seok Lee (Kolon Industries, Inc.), Hyeon-Ju Kim (Korea Research Institute of Ships and Ocean Engineering), Jung-Hyun Moon (Korea Research Institute of Ships and Ocean Engineering).

For municipal and industrial wastewater treatment, membrane bioreactor (MBR) is widely used as a combined process of a membrane process such as microfiltration or ultrafiltration and biological wastewater treatment. Hollow fiber membranes are preferred in many membrane applications due to the high packing ratio, including submerged MBR processes using feed suction processes. As interstitial spaces between packed fibers are of the same order of the fiber thickness, membrane fouling due to the accumulation of chemical and biological solid materials on the fiber surfaces has limited the MBR performance of wastewater treatment. A standard cleaning method is to diffuse small bubbles from the bottom of the vertically packed membranes and sweep out weakly bound foulants on the fiber surfaces, but steady generation and distribution of small, equal-sized bubbles are usually costly. Instead, a new approach is to reciprocate (periodically) submerged membrane cassettes using a motor and crank in a stationary wastewater tank. The relative motion of the cassettes to the fluid generates inertial forces on attached foulants and therefore reduces, minimizes, and prevents contact probabilities of suspended solids on the membrane surfaces. This new method is called Low Energy No Aeration (LENA), recently developed by Kolon Industry in South Korea. Counter-intuitively, LENA consumes 25% less energy than conventional MBR systems that use bubble diffusers or blowers. A potential concern is the mechanical stability of the hollow fibers attached to cassettes in periodic reciprocation dynamics. Although the fiber motion resembles practical wave motion (e.g., of a musical instrument like guitar strings), the finite fiber length mandates sophisticated theoretical approaches of flexible body dynamics, which is computationally challenging. In this light, we mimic the flexible hollow fiber (under periodic motion) as a linear collection of linked spheres, i.e., a particle chain, by incorporating constraint dynamics developed in traditional molecular dynamics. Forces exerted on small local segments of a hollow fiber are calculated as those acting on a sphere whose diameter is equal to that of the fibers' outer diameter. Gravitational, hydrodynamic, and constraint forces are separately calculated and observed at various instances during a reciprocation period. Several suggestions for hollow fibers manufacturing and operating cassette reciprocating are provided based on simulation results. This research is supported by Kolon Industry and Korea Research Institute of Ships and Ocean Engineering (grant no. PES9060)

Oral 25 – MEMBRANES FOR ELECTROCHEMICAL APPLICATIONS II

11:00 am (Room 5, Thursday)

Oral 122 - Understanding water-splitting thermodynamics and kinetics in micropatterned bipolar membranes

Subarna Kole* (Louisiana State University), Christopher Arges (Louisiana State University).

The palette of applications for bipolar membranes (BPMs) has expanded in recent years to numerous electrochemical applications including fuel cells, electrolyzers, and photoelectrochemical cells. BPMs have historically been deployed in electrodialysis setup for mineral acid and base production. The need to have disparate pH environments in electrochemical cells, and prevent species crossover, have motivated researchers to examine BPMs as an electrolyte separator. BPMs have the unique capability to split water into protons and hydroxide ions in addition to conducting those ions in opposite directions in an electrochemical setup. A BPM consists of a cation exchange membrane (CEM) appended to an anion exchange membrane (AEM) that have intimate contact and feature a water dissociation catalyst. Although BPMs have been in existence for a half-century, there are still many open-ended questions about these materials. For instance, 'How interfacial area impacts the kinetics and thermodynamics for water-splitting?' To answer this, we have prepared micropatterned BPMs with precisely controlled interfacial areas using soft-lithography techniques. Our results show that increased interfacial area is more effective for reducing the on-set potential (almost 250 mV) for water-splitting rather than boosting the rate for water-splitting (10-15% improvement in current density at 2V). Interfacial area affects the activation barrier for water-splitting and the rate of water-splitting is under mixed kinetics-diffusion control due to water delivery to the bipolar junction interface and migration of ions away from the interface. This latter attribute explains, in part, the limited improvement in current density. Finally, the soft-lithography approach devised by our group has been successful in making BPMs with different chemistries ranging from perfluorinated polymers to poly(arylene) hydrocarbon polymers vital for fuel cell demonstrations. This work was supported from NSF Award # 1703307.

11:20 am (Room 5, Thursday)

Oral 123 - Co-permeation behavior of methanol and acetate in polyether-based cation exchange membranes

Jung Min Kim* (Auburn University), Bryan Beckingham (Auburn University).

The permeation of multiple solutes in ion exchange membranes is crucial for innovative energy conversion applications including photoelectrochemical CO_2 reduction cells, which reduces CO_2 to valuable chemicals such as methanol and acetate. One of the primary concerns with such devices is the lack of ion exchange membranes with the requisite conductivity but limited crossover of CO2 reduction products. Towards developing such membranes, understanding the permeation, and co-permeation behavior of these products in ion exchange membranes is critical. In previous work, co-permeation behavior of methanol and acetate across the cation exchange membrane Nafion® 117 was investigated and differences in transport behavior relative to their single solute permeation behavior were observed. Here, to further understand this emergent transport behavior, we investigate a series of polyether-based membranes of varied charge density. Model membranes are synthesized by UV photopolymerization with varying poly(ethylene glycol) diacrylate to 2-acrylamido-2-methylpropanesulfonic acid content and characterized by measuring water uptake and in-plane ionic conductivity. Diffusive permeabilities of these membranes to methanol, acetate, and the mixture of two were measured by a diffusion cell coupled with in situ attenuated total reflectance Fourier transform infrared spectroscopy to monitor the time-resolved concentration of each permeant in the downstream receiver cell. We find the inclusion of AMPS in these membranes has a distinct increase in permeability to acetate in the presence of co-permeating methanol. This behavior is attributed to the screening of electrostatic interactions (Donnan exclusion), in which methanol interferes with the electrostatic repulsion between membrane-bound sulfonate anions and transporting acetate anions.

11:40 am (Room 5, Thursday)

Oral 124 - Carbon nanotube thin film deposition on hollow fiber membranes: addressing the scalability challenges of conductive membranes

Melissa J Larocque* (McMaster University), David Latulippe (McMaster University), Charles De Lannoy (McMaster University).

Electrically conductive membranes (ECMs) are of significant interest for their ability to mitigate fouling and biofilm growth. Until recently, most ECM research had focused on flat sheet membranes suitable for bench-scale studies. Other formats such as hollow fiber (HF) are often preferred in industrial applications for their high packing density. While ECMs in HF format are emerging in research, existing approaches tend to deposit excess conductive material on HFs with little consideration for the trade-off between membrane conductivity and permeance [1]. We aim to optimize the conductive material deposition process on HF membranes by studying the effect of crossflow induced shear forces.

We developed a crossflow deposition approach for adhering a carbon nanotube (CNT) thin film onto commercial HF membranes. Functionalized single wall CNTs were suspended with sodium dodecyl sulfate and polyvinyl alcohol then filtered in crossflow along the active surface of polyether sulfone (PES) HF membranes (Repligen, 0.2 μm). Coated membranes were then cured to adhere the CNT thin film layer to the surface. The impact of design parameters affecting shear forces (feed pressure, crossflow velocity) and CNT thin film density (CNT concentration, volume filtered) were quantified by composite membrane permeance and conductivity in a design-of-experiments statistical approach. A Robeson-like plot was generated to compare composite membrane permeance and conductivity across the various experimental conditions studied. From this comparison, we have noted composite membrane performance limits with conductivities as high as 2500 S/m and permeance as high as 6000 LMH/bar. These metrics are several orders of magnitude higher than that of a previous HF ECM study [1]. The CNT coated HF membranes were tested with a bacterial feed solution to assess fouling mitigation under an applied potential.

Reference

[1] G. Wei, J. Dong, J. Bai, Y. Zhao, Y. Li. Environ. Sci. Technol., vol. 53, 2019.

12:00 pm (Room 5, Thursday)

Oral 125 - Selective ion conducting membranes for non-aqueous redox flow battery applications

Patrick McCormack (University of Virginia), Hongxi Luo (University of Virginia), Gary Koenig (University of Virginia), Geoffrey M Geise* (University of Virginia).

Providing sustainable supplies of clean energy is a critical global challenge for the future. As intermittent renewable energy sources are increasingly deployed, grid-scale energy storage solutions are increasingly needed. One approach to addressing this challenge is to use flow battery technology to store and deliver grid-scale amounts of energy. Non-aqueous redox flow batteries can be operated at higher voltages and energy densities compared to aqueous systems and, as such, may offer high volumetric energy density compared to other flow batteries. A significant challenge facing non-aqueous flow batteries, however, is the lack of selective membrane separators engineered for the unique challenges of non-aqueous electrochemical systems. Suitable membranes for non-aqueous redox flow battery applications must be stable in aggressive solvent environments, offer high conductivity, and provide selectivity to prevent cross-over of redox active molecules.

Here, we report the synthesis and characterization of a series of negatively charged ion conductive polymeric membranes for non-aqueous flow battery applications. Fixed negative charges were added to a poly(phenylene oxide) backbone via a custom sulfonate group-based side chain. Lithium ion conductivity and ferrocene permeability properties were characterized to evaluate the selectivity of the membrane for ion transport relative to redox active molecule cross-over (as ferrocene is a representative redox active molecule). The polymers exhibited combinations of conductivity and selectivity that are favorable compared to other Nafion-based materials, and the materials appear to be dimensionally stable in a non-aqueous electrolyte over a period of several months. This presentation discusses the transport properties and stability of these materials that show promise for non-aqueous flow battery applications.

12:20 pm (Room 5, Thursday)

Oral 126 - Graphene/Nafion Based Membrane Structure for Crossover Mitigation in Energy Storage and Conversion Systems

Saheed A Bukola* (National Renewable Energy Laboratory), Zhaodong Li (National Renewable Energy Laboratory), Christopher Antunes (National Renewable Energy Laboratory), Jason Zack (National Renewable Energy Laboratory), Glenn Teeter (National Renewable Energy Laboratory), Jeffrey Blackburn (National Renewable Energy Laboratory), Bryan Pivovar (National Renewable Energy Laboratory).

Polymer electrolyte membranes such as Nafion® are the most preferred cation membranes/ separators for energy storage (e.g. redox flow batteries) and conversion (e.g. fuel cells and electrolyzers) systems. They demonstrate incredibly high ionic conductivity, chemical stability, and mechanical strength. However, the ion-conducting pores of Nafion® membrane are not selective to allow only charge balancing ion through it while simultaneously inhibiting any unwanted electroactive species. This results in significant crossover related issues that have detrimental effect on membrane and catalyst degradation, poor coulombic efficiency, and battery self-discharge. Herein, we report research findings on vanadium ion diffusion and migration through membrane with and without single-layer graphene in both electrically driven H-cell and static H-cell. Single-layer graphene completely inhibits vanadium ion diffusion/ migration at current density below 200 mA cm-2. Proton transmission rate through single-layer graphene was found to occur 300 times faster than vanadium ion. This results in a remarkable selectivity which may have significant impact on flow batteries application and other electrochemical devices in which proton conductivity is required, and transport of other electroactive species is detrimental.

Oral 26 – ORGANIC SOLVENT SEPARATIONS

1:00 pm (Room 1, Thursday)

Oral 127 - Molecular origin of flux non-linearity in Organic Solvent Nanofiltration

Kelly P. Bye (University of Oklahoma), Michele Galizia* (University of Oklahoma).

Organic solvent nanofiltration (OSN) is attracting interest due to compactness, modularity and superior energy efficiency. However, while recent OSN research efforts mostly focused on the synthesis and fabrication of solvent resistant composite polymer membranes, fundamental understanding of chemical and physical aspects that govern solvent and solute transport in OSN membranes remains entirely unexplored. The transport mechanism itself in OSN membranes is poorly understood: some researchers have hypothesized a solution-diffusion mechanism,

others a pore-flow mechanism. Finally, others have considered a combination of the previous two mechanisms. A peculiar OSN feature is the flux non-linearity vs. □p, that is, a negative departure of flux from linearity is observed starting from □p = 10 atm, which has been the subject of a long standing debate in the literature. Despite membrane compaction has been invoked to explain this phenomenon, this hypothesis has no quantitative support.
In this study, we critically discuss the hypothesis of membrane compaction. To demonstrate that the molecular origin of flux non-linearity is purely thermodynamic, we propose, for the first time, a thermodynamic-diffusion framework which describes solvent transport in OSN membranes in terms of the concentration gradient produced by the applied pressure across the membrane. The model predictions agree very well with experimental data. Solvent diffusion coefficient in the membrane increases with increasing □p, which further confirms that flux decline is not related to membrane compaction. The developed framework allows to quantify both frame of reference and non-ideal thermodynamic effects on solvent diffusion coefficients in OSN membranes.
This study demonstrates that the solution-diffusion model provides a satisfactory description of small molecule transport in OSN membranes, without the need to resort to pore-flow or more complicated transport models.

1:20 pm (Room 1, Thursday)

Oral 128 - Thin-film nanocomposite membranes with enhanced permeability and selectivity for organic solvent forward osmosis

Bofan Li* (National University of Singapore), Susilo Japip (National University of Singapore), Tai-Shung Chung (National University of Singapore).

Organic solvent forward osmosis (OSFO) is relatively immature among all organic solvent separation processes. In this study, thin-film nanocomposite (TFN) membranes have been fabricated with not only an enhanced permeability, but also a dramatically improved draw-solute/solvent selectivity for OSFO by functionalizing the polyamide network with two macrocyclic molecules, sulfothiacalix[4]arene (STCAss) and sulfocalix[4]arene (SCA). Both STCAss and SCA are found to be homogenously dispersed inside the polyamide layer thanks to the formation of ionic bonding between them and the polyamide network. Their synergistic effects on membrane formation and separation performance have been examined in terms of morphology, chemistry, and microstructure as well as solvent transport properties and OSFO performance. In OSFO processes, both STCAss and SCA enable the TFN membranes to selectively reject the draw solutes, but only the STCAss-functionalized membrane has an ethanol flux doubling the pristine one due to the enhanced molecular sieving as confirmed by measuring the diffusivities of water, methanol and ethanol. Comparing with other TFN membranes for FO processes which generally cannot simultaneously improve the solvent flux and reduce the reverse solute flux, our molecularly-engineered TFN membranes show much superior performance and provide new insights to fabricate TFN membranes for sustainable separation processes.

1:40 pm (Room 1, Thursday)

Oral 129 - Studies in separation of organic solvent mixtures during reverse osmosis and nanofiltration by a perfluorodioxole copolymer membrane

John Chau (New Jersey Institute of Technology), Kamalesh Sirkar* (New Jersey Institute of Technology).

Using a 1.67 μm thick film of a particular variety of a glassy amorphous copolymer, perfluoro-2,2-dimethyl-1,3-dioxole copolymerized with tetrafluoroethylene, (PDD-TFE), supported on an e-PTFE support, organic solvent reverse osmosis (OSRO) separations of a number of types of binary organic solvent mixtures were studied. A few studies employed a 0.6 μm thick film of the same copolymer variety designated CMS-7. Different types of binary organic solvent mixtures were studied with feed pressure going up to 5000 kPa in some cases. The following systems were studied (similar classes mentioned together): toluene--N-Methyl-2-pyrrolidone (NMP), toluene--dimethylformamide (DMF); ethanol--NMP, ethanol--dimethylsulfoxide (DMSO); methanol—DMSO, methanol--NMP; toluene--methanol; ethanol--n-butanol; NMP--n-heptane; heptane--dodecane; heptane--ethanol. In many such systems, a pure solvent was obtained as permeate. Explanations for the observed separation behavior have been provided. In most cases, it required experimental determination of sorption behavior of various solvent mixtures in

the copolymer membrane and the extent of sorption selectivity. It also included considerations of various existing models: Spiegler-Kedem model involving solute reflection coefficient used in desalination; the irreversible thermodynamics model of Johnson et al. showing the relation between salt rejection and its partition coefficient at high flux; Don Paul's model for solution-diffusion for organic solvent mixtures; Wijmans and Baker's considerations for ultramicroporous membranes.
To illustrate OSRO-facilitated solvent-exchange in organic solvent nanofiltration (OSN), studies were carried out also to retain the dye Oil Blue N during OSN using the following binary solvent mixtures: toluene-NMP; methanol-NMP. Additional applications have been considered as well..

2:00 pm (Room 1, Thursday)

Oral 130 - Complex aromatic hydrocarbon mixture OSRO separation using defect-engineered Torlon® hollow fiber membranes

Hye Youn Jang (Georgia Institute of Technology), Ryan Lively (Georgia Institute of Technology).

The fractionation of complex liquid hydrocarbon mixtures is an important emerging area of membrane science. Polymeric asymmetric hollow fiber membranes can enable this concept, especially if size and number of defects in the membrane skin layer can be precisely engineered. Here, we fabricated various "defect-engineered" hollow fiber membranes by modifying the composition of the volatile solvents of the polymeric solution as well as spinning parameters to improve the organic solvent reverse osmosis (OSRO) separation performance of complex aromatic hydrocarbon mixtures. The quality of the membranes was probed using both gas and liquid permeation tests, which provided the data for a useful fundamental analysis that estimates the pore size and surface porosity of the asymmetric hollow fiber membranes. That information, combined with organic solvent permeation data, provides an engineering platform to estimate the relative permeation rates of organic molecules within a complex mixture. We utilized these experiments to test simplified hypotheses regarding the quantitate estimation of the pore size and pore size distribution information on the hollow fiber membranes of any arbitrary membranes as well as separation performance which have never been conducted so far. Theses gas-liquid analysis enables extension of our predictive capabilities of various organic solvent separation of defect-engineered membranes with simple gas permeation test. Such that this study can be a platform for OSRO separation/fractionation of complex aromatic hydrocarbon mixture.

2:20 pm (Room 1, Thursday)

Oral 131 - Organic solvent nanofiltration (OSN) membranes for separation of close-boiling mixtures and high-boilers

Petrus Cuperus* (SolSep BV), Ingrid Wienk (SolSep BV).

A main driving force for developing OSN technology for separation of components in mixtures with close boiling points is the potential energy saving. For current "easy" solvent recovery processes that incorporate distillation, energy reductions of 50-60 % have been calculated when using straight-forward NF. Mixtures that have close boiling points may benefit even more from the use of membranes as the distillation processes are intensive. This has also a techno-economic advantage.
In this paper possibilities for organic solvent nanofiltration (OSN) in the lower molecular mass regions (<300 Da) are evaluated. This was combined by work particularly at higher temperatures (close to 100°C and beyond) in real process streams. Typically possibilities in close-boiling mixtures and high-boilers were looked at. In these cases, the molecular weight difference between the molecules is relatively small. Such membranes that can achieve an efficient separation between components with small differences (size, shape, polarizability, complexibility) are not standard and definitely not available commercially at this moment.
A very important issue in OSN is to translate work on lab-scale – model- systems into the real system. During membrane development it is attractive to use simple solvent-solute systems that give straight-forward results. However, in industrial environments these become less relevant because of the typical interactions in those other systems. Vast amounts of research have already been directed towards getting better characterization procedures

as well as transport models but progress is relatively small. We will elaborate on the suggestion that an experimental is the only way out to develop industrial relevant OSN membranes and systems.

2:40 pm (Room 1, Thursday)

Oral 150 - Polymer Membrane-based Liquid Hydrocarbon Fractionation

Ronita Mathias* (Georgia Institute of Technology), Kirstie Thompson (Georgia Institute of Technology), Daeok Kim (Imperial College London), Jihoon Kim (Imperial College London), Neel Rangnekar (ExxonMobil), JR Johnson (ExxonMobil), Scott Hoy (ExxonMobil Research and Engineering), Irene Bechis (Imperial College London), Andrew Tarzia (Imperial College London), Kim Jelfs (Imperial College London), Benjamin McCool (ExxonMobil Research and Engineering), Andrew Livingston (Imperial College London), M.G. Finn (Georgia Institute of Technology), Ryan Lively (Georgia Institute of Technology).

The separation of multicomponent, complex liquid mixtures, which are mixtures without a clear singular solvent, is an important, emerging area of membrane science. In this work, novel spirocyclic polymers were incorporated into thin-film composite membranes with the aim of separating liquid hydrocarbons that are typically found in the light fractions of crude oil. These cover a range of classes such as alkanes, cycloalkanes, alkyl aromatics, and polyaromatics. These hydrocarbon species were utilized not only to create a molecular weight cut-off (MWCO) curve but also to assess the performance of this class of membranes in the fractionation (as opposed to purification) of complex mixtures. A MWCO of ~ 250 Da was obtained for dilute mixtures, accompanied by high selectivities amongst different classes of hydrocarbons as well as molecules within the same class but with sub-nm differences in size. The polymer membranes also demonstrated the fractionation of light crude oil through a combination of class- and size-based "sorting" of molecules, where the permeate is enriched with molecules lighter than 170 Da equivalent to a boiling point less than 200°C. Such scalable, selective membranes offer potential for the hybridization of energy-efficient technology with conventional processes like distillation.

Oral 27 – CONTAMINANT REMOVAL FROM WATER

1:00 pm (Room 2, Thursday)

Oral 132 - Adsorption of organic micropollutants to polymer surfaces probed by second harmonic scattering laser spectroscopy

Will Cole (University of California Berkeley), Hoyun Wei (University of California Berkeley), Son Nguyen (University of California Merced), Charles Harris (University of California Berkeley), Dan Miller* (LBNL), Richard Saykally (University of California Berkeley).

Modern thin-film composite RO membranes suffer from poor rejection of organic micropollutants, including pharmaceuticals, endocrine disruptors, and disinfection byproducts. The physico-chemical interactions between organic solutes and the membrane surfaces are poorly understood, but may play a significant role in determining membrane rejection.

The purpose of this study was to quantitatively measure the adsorption of several organic micropollutants and surrogate molecules to a model polymer surface. Angle-resolved second harmonic scattering laser spectroscopy was used to probe the adsorption of resonant aqueous cationic and anionic dye molecules to polystyrene surfaces. Calculation of the Gibbs free energy for adsorption of malachite green to negatively charged polystyrene and naphthol yellow S to positively charged polystyrene revealed highly favorable interactions. The Gibbs free energy of adsorption for the adsorption of malachite green to negatively charged polystyrene was in agreement with other independent measurements.

For non-resonant organic solutes, including the micropollutants caffeine and pentoxifylline, a competitive displacement methodology was employed to obtain values for the Gibbs free energy of adsorption. The mechanism of adsorption for these organic micropollutants appears to be dominated by van der Waals interactions, with surface

charge playing a relatively minor role, even for ionic adsorbers. Preliminary results on the adsorption of organic solutes to polyamide surfaces will also be presented.
This experimental technique provides a straightforward way of quantifying the adsorption of critical organic micropollutants to polymer membrane surfaces. These measurements could inform the design of membranes with improved rejection of organic micropollutants, reducing the need for costly secondary treatment of purified wastewater.

1:20 pm (Room 2, Thursday)

Oral 133 - Laser-induced graphene polymer composite membranes as electrically active filters for contaminant removal

Christopher J Arnusch* (Ben Gurion University of the Negev).

The control/elimination of microorganisms, viruses and micropollutants is relevant in many water treatment systems from many water sources. Laser-induced graphene (LIG) is a three-dimensional porous electrically conductive graphene material generated by irradiation of polymer substrates with a 10.6 μm carbon dioxide laser. Previously, we showed that LIG was antifouling, and with an applied voltage, the electrically conductive properties of the surface led to active antimicrobial effects.[1] Also, LIG was made on porous supports such as polyethersulfone ultrafiltration (UF) membranes, and the as-prepared LIG-membranes were used as porous electrodes during filtration of contaminated waters, leading to decontamination.[2] However, the carbon layer was relatively brittle and the UF separation properties were lost. Here we developed LIG-polymer composites, which improved both the robustness of the surface and separation properties of the membrane.[3] Also, carbon and metal nanoparticle LIG composites will be presented, which led to improvements in the antifouling and antimicrobial characteristics of the surfaces.[4,5] The multi-functionality of these surfaces was also demonstrated in the electrochemical degradation of micropollutants and virus destruction. This method to "laser-print" electrically conductive antifouling graphene coatings on membranes might enable many advanced applications in water treatment and separations including micropollutant degradation and other contaminant removal from various water sources.

References

1. Singh, S. P. et al. ACS Applied Materials and Interfaces 2017, 9 (21), 18238-18247.
2. Singh, S. P. et al. ACS Nano 2018, 12 (1), 289–297.
3. Thakur, A. et al. ACS Applied Materials and Interfaces 2019, 11 (11), 10914-10921
4. Gupta, A. et al. Chemical Communications, 2019, 55, 6890.
5. Thakur, A. et al. Journal of Membrane Science 2019, 591, 117322.

Oral 134 - Printing zwitterionic copolymer thin film composite (TFC) membranes: Enabling Tunability of Thickness and Remarkable Permeances for Nanofiltration Applications

Xin Qian* (University of Connecticut), Jeffrey McCutcheon (University of Connecticut), Ayse Asatekin (Tufts University), Samuel Lounder (Tufts University), Tulasi Ravindran (University of Connecticut).

Zwitterionic copolymer self-assembled membranes have been shown to have impressive selectivity for nanofiltration applications. In this work, we evaluate the use of electrospray to print thin layers of amphiphilic zwitterionic selective layers onto UF membrane substrates. Dyes with different sizes are used to probe the selectivity of these membranes while their thickness is reduced by orders of magnitude compared to casting methods. As thickness is decreased, water permeance was found to proportionally increase while dye selectivity was maintained. While it was found that a threshold minimum thickness was required to maintain selectivity for some molecules, the intrinsic permeability of the polymer films changed with decreasing thickness . Notably, our best performing membrane exhibited a water permeance value of 205 LMH/bar and a Chlorophyllin rejection at 99.67%. This permeance is indistinguishable from the supporting UF membrane permeance, suggesting that even higher permeance without selectivity loss is possible with more permeable support layers.

2:00 pm (Room 2, Thursday)

Oral 135 - Enhancing antimicrobial properties of Chitosan/ Graphene Oxide membrane for waste water treatment application

Raheleh Daneshpour* (University of Arkansas), Lauren Greenlee (University of Arkansas).

Biofouling is one of the major challenges in wastewater treatment by membranes. So, enhancing antimicrobial properties of the membrane is critically needed to avoid biofilm growth which leads to flux reduction and more energy consumption. Graphene oxide (GO) and Chitosan (CS) are promising candidate materials for antimicrobial membranes due to their abundancy, low cost, scalability, stability in water and their oxygen-containing functional group.

The ultimate goal of this project is to achieve an optimum composition of CS/GO and PVA (Polyvinyl alcohol) to create a membrane with a broad chemical and biochemical contaminant rejection. The anti-microbial properties of the CS/GO-PVA membrane allow this type of membrane to be used broadly for disinfection in wastewater treatment processes and demonstrate contaminant rejection performance that is comparable to RO/NF rejection and we will demonstrate the resulting membrane effectively decreased the biofilm formation in cross-flow systems.

2:20 pm (Room 2, Thursday)

Oral 136 - Early-stage membrane fouling detection for improving membrane cleaning efficiency

Thomas Schäfer* (Polymat, University of the Basque Country), Iliane Rafaniello (POLYMAT, University of the Basque Country).

Membrane fouling can seriously affect membrane filtration processes. As fouling cannot be completely avoided, focus lies on optimizing membrane cleaning intervals in order to lengthen membrane lifetime and reduce process downtimes and cleaning-costs. Interviews with some of the major Spanish desalination companies revealed that until today operators lack a method sufficiently sensitive for detecting membrane fouling at an early-stage which would avoid significant drops in water flux and permit optimizing process operating conditions such as membrane cleaning protocols. Several methods have been proposed for membrane fouling detection but until today none has proven to be sensitive enough to handle membrane fouling in an anticipative way.

This work presents an approach of overcoming the limitations of conventional methods by combining quartz crystal microbalance with dissipation monitoring (QCM-D) and multi-parameter surface plasmon resonance (MP-SPR). The resulting technique yielded an accurate, fast, and non-invasive fouling monitoring of unprecedented sensitivity. Real-time monitoring of adsorption of different foulants on modified polyamide (PA) films was studied and correlated with an independent filtration system in order to prove the concept of the combined device. We could successfully demonstrate that QCM-D/MP-SPR enables characterizing the build-up of even the very first fouling layers at the nanoscale and therefore possessed a sensitivity high enough such as to anticipate the need of membrane cleaning before the filtrated water flux decreases. QCM-D/MP-SPR furthermore allowed to test different membrane cleaning strategies in a time-efficient and material-extensive manner, maximizing in this way cleaning efficiency and prolonging membrane lifetime.

The potential of the QCM-D/MP-SPR-based monitoring device for increasing the process efficiency and mitigate membrane fouling will be critically discussed.

Oral 28 – INDUSTRIAL APPLICATIONS

1:00 pm (Room 3, Thursday)

Oral 137 - Produced water desalination using high temperature membranes

Cheng Chen* (Chevron), Prakhar Prakash (Chevron), Xiaofei Huang* (Hydranautics), Rich Franks (Hydranautics).

Medical isotopes are radioactive species that can be conjugated to a biomarker and injected into a patient to facilitate localized cancer treatment or imaging. This class of pharmaceuticals has shown promise in treating otherwise pervasive cancers while avoiding full body doses of radiation. Unfortunately, the progression these life-saving drugs to clinical trial has been limited by their availability—a problem that can be traced back to insufficient purification processes. Traditional purification processes for medical isotopes rely on resin-packed columns which are known to suffer from diffusion limitations. This contribution describes our efforts to 1) shorten purification times; 2) minimize elution volumes and 3) improve peak resolution by developing a new class of membrane adsorbers for radiochromatography. These materials are designed to purify medical isotopes, Cu-67 and Ac-225 which have been identified by the US DOE Isotope Program as in-demand and under-supplied. Membranes are prepared by grafting functional polymer brushes from the surface of microfiltration membranes by two methods: 1) UV-induced free radical polymerization and 2) atom transfer radical polymerization (ATRP). The binding groups used in this study are primary amines and phosphonates. Changes in the membrane surface chemistry are monitored with attenuated total reflectance Fourier transform infrared spectroscopy. The number of binding sites are quantified by titration and binding capacities were calculated by fitting equilibrium sorption data with the Langmuir isotherm. Membranes have static binding capacities of 15 mg/g membrane and pure water permeabilities of ~200 LMH/bar. Dynamic binding capacities are measured using membranes loaded in a 0.8 cm2 filter holder at flowrates of1-10 mL/min. Mass transport phenomena are interpreted using a continuum model. The results of this research are laying the foundation for the use of membrane adsorbers in radioanalytical chemistry..

1:20 pm (Room 3, Thursday)

Oral 138 - Membrane based upgrading of Biogas: Using Realtime Controls and Selective Distribution of Membrane Types to Effectively Respond to Varying Gas Flow and Gas Quality

Steven K Pedersen* (Evonik Canada Inc.), Erik Hoving (Evonik Cooperation).

Evonik Industries has been a membrane supplier in the RNG (renewable natural gas) market since 2011 under the tradename SEPURAN® GREEN. During this time Evonik has partnered with numerous local equipment manufacturers and engineering firms to design and install over 250 membrane based Biogas upgrading units that achieve up to 99.5% methane yield. One of the challenges our customers face, particularly in the Landfill gas space, is how to design and operate systems that can easily accommodate changes in gas flow rates due to barometric conditions, seasonality, and well field management issues. Evonik engineers have discovered a process configuration that uses multiple (new) membrane types in different membrane steps within the membrane plant to solve this problem. With this patent pending concept, it is possible to meet pipeline gas specifications during 20% swings in raw Biogas flow without sacrificing efficiency. This can be done in real-time without operator intervention by using only one extra control valve. The result is a lower overall capital cost, less compression energy and higher sales gas flows with more uptime.

1:40 pm (Room 3, Thursday)

Oral 139 - Industrial Applications of Membranes in the Membrane Science, Engineering and Technology (MAST) Center at the University of Arkansas

Ranil Wickramasinghe* (University of Arkansas).

The Membrane Science, Engineering Engineering and Technology (MAST) Center is a National Science Foundation Industry and University Cooperative Research Center. It consists of University of Arkansas, University of Colorado Boulder, New Jersey Institute of Technology and Pennsylvania State University, an international site and numerous industrial sponsors. The center focuses on industrially relevant research. The Center has four research themes: water purification, biopharmaceutical processing, chemical separations and membrane fundamentals. This presentation provides an overview of MAST Center projects that relate to water treatment and biopharmaceutical processing. Though much of the research was conducted at the University of Arkansas, the presentation highlights the fact that projects often involve multiple sponsors and faculty at more than one institution. Research in the area of wastewater treatment focuses on the development of membrane processes for direct potable reuse of wastewater, treatment of hydraulic fracturing produced water, and poultry production wastewaters. Membrane based processes are attractive for a number of reasons. They can be used to validate virus clearance

as they provide an absolute barrier to pathogens. Further they are light weight and have a small footprint. A major challenge is membrane fouling. Here we have developed electrocoagulation as a pretreatment step.
In the area of biopharmaceutical processing we have investigated bioreactor harvesting and validation of virus clearance. In all cases industrially relevant feed streams have been used. The use of commercially available hollow fibers, that combine depth filtration and tangential flow filtration for bioreactor harvesting, has been investigated. Flux decline has been modeled. The performance of a number of commercially available virus filters over a set of bracketed conditions has been investigated. The results provide insights into the performance of these filters.

2:00 pm (Room 3, Thursday)

Oral 140 - High Throughput Development of Carbon Molecular Sieve Membranes and Adsorbents for Olefin/Paraffin Separations

Jay (Junqiang) Liu* (The Dow Chemical Company).

Pressure swing adsorption (PSA) and membrane systems have the potential to significantly reduce the energy consumption for olefin/paraffin separations. The heart of future successful PSA or membrane systems will be advanced materials. The high throughput development of novel Carbon Molecular Sieve (CMS) materials with tunable pore sizes for olefin/paraffin separations derived from cation exchange resins and polyvinylidene chloride copolymers (PVDC) will be discussed.
PSA using CMS adsorbent can potentially be commercialized sooner than membrane technology to provide an interim solution. However, PSA has much more complex set of transport processes and is an inherently dynamic process. The engineering of the PSA process using benchtop PSA experiments coupled with ASPEN modelling will be discussed. PSA processes with additional steps from the basic Skarstrom cycle, i.e. co-current blowdown and effluent rinse, were investigated. Propylene purity of 89% at 89% propylene recovery was demonstrated using a 25% propylene-75% propane feed.
CMS membrane offers steady state operation. A thin discriminating layer of the expensive material can save the CMS membrane cost. However, industrially relevant defect-free CMS membranes has been a challenge due to the melting and softening of most of the precursors during pyrolysis. Novel self-standing CMS membranes were made from low cost PVDC films and hollow fibers. PVDC crosslinks before melts, which "locks" the morphology from the polymer precursor to the final carbon membrane. The new PVDC CMS membranes have about 50% more micropore volume than those derived from Matrimid® polyimide. Because of that, the gas permeabilities were 5-10 times higher in the new PVDC CMS membrane than those in the Matrimid® CMS membranes. Therefore, despite the dense thick wall ~ 30-50 μm, the novel PVDC CMS membranes have exceptional separation performance for multiple industrial separations.

Oral 29 – AWARDS SESSION

1:00 pm (Room 4, Thursday)

Oral 141 - 3D Printed Adsorber for Capturing Chemotherapy Drugs before They are Released in the Body

Hee Jeung Oh* (Pennsylvania State University), Mariam Aboian (Yale University), Michael Yi (University of California, Berkeley), Jacqueline Maslyn (University of California, Berkeley), Whitney Loo (University of California, Berkeley), Xi Jiang (Lawrence Berkeley National Laboratory), Dilworth Parkinson (Lawrence Berkeley National Laboratory), Mark Wilson (University of California, San Francisco), Teri Moore (UCSF), Colin Yee (UCSF), Grecory Robbins (Carbon, Inc.), Florian Barth (Carbon, Inc.), Joseph DeSimone (Carbon, Inc.), Steven Hetts (University of California, San Francisco), Nitash Balsara (University of California, Berkeley).

Due to longer life expectancies, the prevalence of age-related diseases is increasing rapidly, and the need for developing biomedical devices that can solve big health problems is similarly greater. Inspired by absorption columns, which are routinely used in industry to remove pollutants from chemical streams, my research focuses on the design of biomedical membranes for capturing unwanted toxins in the body.

Cancer is becoming the leading cause of death in most developed nations. Despite efforts to develop increasingly targeted and personalized cancer therapeutics, dosing of drugs in cancer chemotherapy is limited by systemic toxic side effects. During intra-arterial chemotherapy infusion to a target organ, excess drug that is not trapped in the target organ passes through to the veins draining the organ, and is then circulated to the rest of the body, causing toxicities in distant locations. Typically, more than 50-80% of the injected drug is not trapped in the target organ and bypasses the tumor to general circulation.

In the context of reducing the toxicity of chemotherapy, we have designed, built, and deployed porous adsorbers for capturing chemotherapy drugs from the blood stream after these drugs have had their effect on a tumor, but before they are released into the body where they can cause hazardous side effects. The porosity was obtained by 3D printing of lattice structures within a cylinder. The surface of porous cylinders was coated with an ion-containing nanostructured block polymer which is responsible for capturing doxorubicin, a widely used chemotherapy drug with significant toxic side effects. Using a swine model, we show that our initial design enables the capture of 69 % of the administered drug without any immediate adverse effects. This development represents a significant step forward in minimizing toxic side effects of chemotherapy..

1:20 pm (Room 4, Thursday)

Oral 142 - Mechanical properties of thin-film composite membranes and the roles they play on transport in osmotic processes

Jaime Idarraga-Mora* (Clemson University), Anthony Childress (Clemson University), Parker Friedel (Clemson University), Michael Lemelin (Clemson University), Alton O'Neal (Clemson University), Morgan Pfeiler (Clemson University), Steven Weinman (The University of Alabama), Apparao Rao (Clemson University), David Ladner (Clemson University), Scott Husson (Clemson University).

Mechanical properties (e.g. strength and stiffness) play a role in the transport (i.e. water permeance and salt reverse flux) of polymer-based thin-film composite (TFC) membranes used for Osmotic Processes (OP). Our goal has been to create guidelines for the design of TFC membranes for OP applications, given the trade-off between membrane mechanical integrity and productivity. We studied the mechanical properties of the three individual polymeric layers (active, porous support, and backing) that comprise TFC membranes, and developed models to correlate these mechanical properties with TFC membrane performance in OP.

Firstly, we will discuss a study on carbon nanotubes that were used as additives to create porous nanocomposite supports with improved stiffness. Results showed that increasing the stiffness of TFC membrane nanocomposite supports using low carbon loadings was effective in enhancing water production in desalination operations. Secondly, we will present findings from an analysis of woven polyester mesh backing layers, in which we studied the role of mesh size on burst strength and mass-transfer resistance of TFC membranes used for OP. The findings showed that mass-transfer resistances in OP are an additive effect and can be reduced simply by using more open (yet functional) backing layers such as polyester woven mesh. Thirdly, we will discuss the effect of alcohol contact on the transport properties of commercial TFC membranes and the correlation of these properties to the reduction in stiffness of the active layer after alcohol contact. The correlation suggests that it is the mixing of water with the alcohol that facilitates penetration of the alcohols into the active layer, thus increasing the active layer free volume for water permeation. Finally, I will provide research opportunities that I envision would improve TFC membrane productivity in OP applications.

1:40 pm (Room 4, Thursday)

Oral 143 - Engineering Selective Desalination Membranes by Controlling Functional Group Configuration

Hongxi Luo* (University of Virginia), Kevin Chang (University of Virginia), Kevin Bahati (University of Virginia), Geoffrey Geise (University of Virginia).

Membrane-based desalination processes have been developed to mitigate stress on the global freshwater supply. Continuous population growth and socioeconomic development will increase demand for freshwater in the future. As such, the design of highly selective membranes with improved desalination performance is a pressing engineering challenge. Precise functionalization of membrane materials is becoming increasingly viable due to recent developments in polymer chemistry, yet little is currently known about how the functional groups can be used to engineer membrane water/salt transport properties. In developing this understanding, an important, but sometimes overlooked, challenge stems from changing membrane water content, which affects the membrane water and salt transport properties and can obscure the influence of changes to the polymer functional groups. Therefore, to effectively study the influence of functional groups on water and salt transport properties, it is important to control membrane water content over the range of functionalization.

We recently prepared and reported the properties of a series of equivalent water content model membranes using different ratios of glycidyl methacrylate (GMA), 2-hydroxyethyl methacrylate (HEMA), and glycerol methacrylate (GMAOH) to study the effects of hydroxyl group configuration on membrane water and salt transport properties. Spreading out the distribution of hydroxyl groups in the polymer by shifting the membrane composition from a vicinal diol (GMAOH rich) configuration to a mono alcohol (HEMA rich) configuration led to improved water/salt sorption, diffusion, and permeability selectivity properties, which are crucial for desalination membranes, at equivalent membrane water content. The results of our study suggest that engineering the configuration/distribution of functional groups in polymers used to prepare desalination membranes could be a reasonable approach to design highly selective desalination membranes.

2:00 pm (Room 4, Thursday)

Oral 144 - More than wetting resistance: How membrane surface wettability regulates water vapor permeability and mineral scaling in membrane distillation

Tiezheng Tong* (Colorado State University), Xuewei Du (Colorado State University), Wei Wang (North Carolina State University), Arun Kota (North Carolina State University), Yiming Yin (Colorado State University), Kofi Christie (Vanderbilt University), Shihong Lin (Vanderbilt University).

Membrane distillation (MD) is a sustainable desalination technology suitable for the treatment of hypersaline wastewater. But MD is constrained by membrane wetting and mineral scaling, which compromise its function of effective desalination. Very recently, omniphobic membranes have attracted great interests as functional materials to mitigate membrane wetting in MD. However, the design framework for MD membranes is still incomplete, due to the typically low water vapor flux of omniphobic membranes and the lack of fundamental relationship among membrane surface wettability, water vapor permeability, and scaling resistance. In the first part of my presentation, I will present the investigation of membrane wetting resistance-water permeability relationship in MD, which is enabled by the fabrication of monolithic omniphobic membranes. We discovered a trade-off between membrane wetting resistance and water vapor permeability in MD desalination, in which an increase of wetting resistance leads to a decrease of water vapor flux. We also elucidate the underlying mechanism of such trade-off, which indicates the important role of water-air interfacial area in regulating water vapor transport within microporous membranes. In the second part of the presentation, I will discuss the effect of membrane surface wettability on mineral scaling in MD. We showed that superhydrophobic membranes delayed induction time, reduced membrane wetting, and enhanced scaling reversibility associated with gypsum scaling. Such mitigating effects were not observed for silica scaling, which displayed distinct behavior from gypsum scaling. In sum, our results collectively demonstrate that membrane surface wettability regulates not only wetting resistance, but also water vapor permeability and scaling resistance of MD membranes. Our work provides valuable insights on the smart design of "fit-for-purpose" membrane materials for diverse water purification applications.

2:20 pm (Room 4, Thursday)

Oral 145 - Comprehensive study of dissolved methane harvesting using omniphobic membrane contactor

Abhishek Dutta* (University of British Columbia).

Recovery of dissolved methane from anaerobic effluents has long been a challenge, particularly at low temperatures, due to an increased solubility of the gas in the effluent. We propose a novel approach, solvent-based membrane contactor (SMC), for dissolved methane recovery from wastewater. In this approach, an omniphobic membrane which traps air in its pores when placed in solvent, is placed between a methane-rich feed solution (e.g., anaerobic effluent) and an organic draw solvent that has an order of magnitude higher methane solubility than water. Driven by the solubility difference, methane is then extracted from the feed into the draw, while water transport is deterred due to the low water solubility of the draw solvent. Through this study, we successfully demonstrate over 90% recovery at three different temperatures and flow conditions, using methane-saturated deionized water. We also develop a mass transfer model to investigate the impact of temperature, flow condition, and presence of dissolved carbon dioxide on methane recovery. Our experimental data and predicted values from the model are in excellent agreement, showing that the energy gain from the recovered methane is higher at a lower temperature, the recovery kinetics is mainly influenced by feed flow condition, and the recovery rate is nearly unaffected by presence of carbon dioxide. An energy analysis comparison of the SMC also shows the possibility of the process to contribute towards net energy generation. We further extend our work using methane saturated AnMBR effluent, to study effects of different foulants and fouling on methane transfer in the SMC process. Our experiments showed continued high methane recoveries (>90%) with minimal fouling effects on the recovery process. We conclude that our SMC process, when combined with a low-temperature anaerobic treatment process, has great potential for harvesting dissolved methane as a renewable energy source from anaerobic effluents.

2:40 pm (Room 4, Thursday)

Oral 146 - Quantification of the CO2-induced plasticization of glassy polymer and CMS film membranes from the behavior of CH4 diffusivity, and how observations of multicomponent sorption and permeation rectified recurring

Giuseppe Genduso* (King Abdullah University of Science and Technology), Ingo Pinnau (KAUST).

The replacement of conventional absorber-stripper units with membrane systems can drastically scale down the environmental and economic impact of natural gas pretreatment. The classical literature recommended CO2/CH4 membrane permeability selectivities higher than 40 for maximal economic return. However, under real mixture conditions, CO2-induced plasticization restricts the number of materials that fulfill this specification. Hence, the quantification and understanding of thermodynamic (i.e., competitive sorption) and kinetic (i.e., plasticization) effects are necessary to design tailored materials for membranes for gas separation.

We present the behavior of different classes of polymer films based on experimental data of permeation and sorption of CO2-CH4 equimolar mixtures at 35 °C and up to total pressures of 20 atm. In all cases, CO2 vs. CH4 mixed-gas solubility coefficients regressed linearly independently from the concentration of the sample atmosphere, and this translated in mixed-gas selectivities that increase with pressure.

Mixed-gas diffusion coefficients were determined from experimental solubility and permeability coefficient data and revealed that CO2 diffuses in the same manner both in pure- and in mixed-gas conditions. Methane, instead, displayed higher diffusion coefficients when in mixture with CO2 than during unary permeation. This CO2-induced plasticization effect was quantified precisely by calculating the ratio between mixed- and pure-methane diffusion coefficients. This analysis showed how, for example, intra-chain rigidity (i.e., restricted chains' building blocks mobility) does not prevent CO2-induced plasticization. In the final part of this work, we compared the mixed-gas CO2/CH4 permeation, solubility and diffusion properties of CTA with 6FDA-mPDA — by introducing a mixed-gas diffusion upper bound — and we investigated the CO2/CH4 mixture transport behavior of a carbon molecular sieve film.

Oral 30 – MEMBRANES FOR ELECTROCHEMICAL APPLICATIONS III

1:00 pm (Room 5, Thursday)

Oral 147 - Tailoring Non-Aqueous Electrolyte Concentration enables Optimum Ion Transport in Pentablock Terpolymer Cation Exchange Membranes

Michelle Lehmann* (University of Tennessee), Jameson Tyler (University of Tennessee), Ethan Self (Oak Ridge National Laboratory), Jagjit Nanda (Oak Ridge National Laboratory), Tomonori Saito (Oak Ridge National Laboratory), Thomas Zawodzinski (University of Tennessee).

Medium to large scale energy storage devices, such as redox flow batteries, will be imperative for the implementation of smart grids and to accommodate the increasing use of renewable energy sources. Sodium (Na) based electrochemical energy storage is of interest for large-scale applications due to sodium's low cost and high abundance compared to lithium and vanadium. The membrane plays a critical role in a flow battery, separating the catholyte and anolyte by preventing transport of the redox active species through the membrane. For this purpose, mechanically robust, chemically resistant membranes are being developed. A pentablock terpolymer with a sulfonated center block was successfully converted to trifluoromethanesulfonimide (TFSI) form and evaluated for use in a Na-based non-aqueous flow battery. The pentablock copolymer provides robust mechanical properties, and the membranes maintain their mechanical integrity, flexibility, and dimensional stability when immersed in tetraethylene glycol dimethyl ether (TEGDME). The storage modulus of the dry membrane is ~2 GPa, while the plasticized membrane is 300 MPa, at 25 °C. Membranes were equilibrated in TEGDME solutions, with and without the addition sodium triflate (NaTf). The ionic conductivity significantly increased with the addition of NaTf, with an optimum occurring at a salt concentration of 0.5M NaTf in TEGDME. The ionic conductivity increased from 2.4 × 10-6 S/cm in TEGDME only to 4.1 × 10-5 S/cm (30 °C) for 0.5 M NaTf in TEGDME. The robust pentablock copolymer can be fabricated into thin (20-30 μm) membranes, which allows for high ion conductance. In addition to these results, this presentation will discuss the membranes' selectivity for flow battery applications.

1:20 pm (Room 5, Thursday)

Oral 148 - Powering up Biomedical Devices by Harnessing Energy from Reverse Electrodialysis using Sodium Concentrations from the Flow of Human Blood

Efecan Pakkaner* (University of Arkansas), Jessica Orton (University of Arkansas), Jamie Hestekin (University of Arkansas), Christa Hestekin (University of Arkansas).

With the growth of the healthcare industry, some groundbreaking advancements in implantable biomedical devices, including cardiac pacemakers, artificial organs, drug pumps, and sensors, have been achieved. However, finding a safe, reliable, and stable power source for these devices continues to be a significant challenge for researchers and medical doctors. The present work proposes a miniaturized reverse electrodialysis (RED) biofuel cell design that could potentially be used as a power supply for implantable medical devices by taking advantage of the salinity differences between different bloodstreams that continuously flow inside the human body. Moreover, the effects of different system parameters, including spacer thickness and salinity gradient difference on the electrochemical performance of the biofuel cell, were also investigated. Preliminary results showed that decreasing the spacer thickness from 1 mm to 500 μm increased the maximum power density delivered by 47% from 0.55 to 0.9 mW/m2. By using an external salinity source to enhance the concentration difference, the power density was increased by almost 120%. By the aid of 3-D printing and nano-lithography techniques, the bio-fuel cell will be further miniaturized to optimize the power output and to develop a compact and implantable design.

1:40 pm (Room 5, Thursday)

Oral 149 - Ionic conductivity of ion exchange membranes: Measurement techniques and salt concentration dependence

José Carlos (University of Michigan, Ann Arbor), Jovan Kamcev* (University of Michigan, Ann Arbor).

Ion exchange membranes (IEMs) are required for the efficient operation of various electrochemical processes, including electrodialysis, electrolysis, and fuel cells, among others. Ion transport across IEMs in such processes is driven by an externally applied electric field and is typically quantified by the membrane ionic conductivity (or membrane resistance). Accurate measurements of the membrane ionic conductivity are necessary to improve the fundamental understanding of electric field-driven ion transport in charged polymer membranes. Various techniques for measuring the ionic conductivity of IEMs equilibrated with salt solutions have been proposed, but none has emerged as a widely accepted, standard method. Moreover, there is frequent disagreement between the different techniques when measuring the ionic conductivity of the same membrane under similar conditions, which frustrates the advancement of fundamental understanding of transport phenomena in membranes. In particular, there appears to be disagreement between researchers on the salt concentration dependence of membrane ionic conductivity in dilute solutions. Some studies have reported membrane ionic conductivity values that sharply decrease with decreasing salt concentration, sometimes by several orders of magnitude. This observation is often explained by invoking the microheterogeneous model, which presumes that IEMs consist of two distinct phases – a charged gel phase and a solution phase – both of which contribute to the overall conductivity of a membrane. The solution phase, which has properties analogous to the bulk solution, dominates the overall membrane resistance at low salt concentrations, leading to sharply decreasing membrane ionic conductivity values. In this study, we demonstrate, via several independent techniques, that ionic conductivity measurements could be significantly influenced by resistances at the membrane/solution/electrode interface, particularly at low salt concentrations, which could also explain the sharply decreasing "apparent" ionic conductivity values in dilute salt solutions. After accounting for these interfacial resistances, ionic conductivity values of dense (i.e., nonporous) IEMs plateau with decreasing salt concentration, which agrees well with the Nernst-Einstein framework for a single conducting phase.

Abstracts – Poster Presentations, Monday

Presenting authors are indicated by an asterisk*. See the most up-to-date version of the program book with full abstracts at **www.membranes.org/nams-2020/**

Poster Session 1: Materials

Poster 1 - Influence of aliphatic and aromatic fluorine groups on the gas permeability and morphology of fluorinated polyimides

Albert Wu* (MIT), James Drayton (MIT), Katherine Mizrahi Rodriguez (MIT), Zach Smith (MIT).

Partially fluorinated polymers show exceptional separation performance for a large variety of gas pairs, in part due to inefficient packing caused by the comparatively large size of fluorine versus hydrogen atoms. While the incorporation of aliphatic fluorine groups in the polymer backbone is relatively commonplace (e.g., CF3 groups as a bulky side group), there are very few studies on the impact of aromatic fluorine groups on gas transport. Here, we present an analysis of two pairs of structurally-analogous, partially fluorinated polyimides – one pair altered via aliphatic fluorination, and the other via aromatic fluorination. Challenges related to the synthesis of highly fluorinated polyimides are discussed and the effects of differing fluorine group connectivity on solid-state morphology are related to observed changes in permeability and permselectivity, diffusion and diffusion selectivity, and their energetics. Developing an understanding of structure-property relationships related to fluorination can allow for improved rational design of high-performance polymers in the future.

Poster 2 - Interlayer Spacing and Separation Performance of Graphene Oxide Membranes in Organic Solvent

Sunxiang Zheng* (University of California, Berkeley), Baoxia Mi (University of California, Berkeley).

Ultrathin membranes formed by stacking two-dimensional (2D) graphene oxide (GO) hold great promises for enhanced permeability and selectivity in organic solvent nanofiltration (OSN). The nanochannels formed between neighboring nanosheets serve as the solvent transport paths (functionally equivalent to nanopores in porous membrane materials), which needs to maintain a stable interlayer spacing in order to achieve desired separation. In this study, we applied liquid-phase ellipsometery to precisely quantify the interlayer spacing of GO when the membrane is completely soaking in the organic solvents. We found that solvents with high dipole moments (e.g., dimethylformamide, n-methyl-2-pyrrolidone) tends to cause significant swelling of GO with the interlayer spacing increasing up to 2.74 nm, whereas the GO layers maintain a stable interlayer spacing (i.e., ~ 0.8 nm) in non-polar solvents such as hexane and toluene. To understand such unique swelling behavior, the solubility parameter of GO was experimentally determined to quantitatively evaluate the affinity between GO and the testing solvents. Our results suggest that the solubility parameter difference between GO and the solvents is a good indicator of the interlayer spacing. Additionally, it was found that solvents with different polarity shows distinct transport resistance, which was further investigated by molecular simulation. Separation performance of the GO membranes in different organic solvents were tested to understand the mass transport in confined GO nanochannel. The layer-stacked GO membrane could achieve > 95% rejection of small organic dye molecules (e.g., rhodamine B and methylene blue) in acetone and superior solvent flux (more than four times faster than polymeric membranes) without additional stabilization steps owing to its low interaction with acetone. These findings shine lights on the design and application of GO membranes in OSN.

Poster 3 - Fluoride- and Hydroxide-Containing Facilitated Transport Membranes for CO2 Removal From Solid Oxide Fuel Cells

Kai Chen* (The Ohio State University), Witopo Salim (Membrane Technology & Research Inc.), Yang Han (The Ohio State University), Mike Gasda (Bloom Energy Corporation), Winston Ho (The Ohio State University).

In order to utilize the H2 remaining in the anode exhaust of a solid oxide fuel cell (SOFC), a CO2-selective membrane was developed to remove CO2 from the anode exhaust and recycle the concentrated H2 back to the SOFC. As the cooled exhaust still has an average temperature of 120ºC, it is challenging for polymeric membranes to maintain both high CO2/H2 selectivity and CO2 permeance. Previously, facilitated transport membranes were developed using tetramethylammonium hydroxide (TMAOH) as the mobile carrier to selectively remove CO2. However, TMAOH has been reported to degrade via a nucleophilic substitution mechanism, and a possible higher temperature of the anode exhaust in practice may exacerbate the degradation.
Herein, more stable fluoride- and hydroxide-containing species were investigated for their potentials of replacing TMAOH. The thermal stabilities of the quaternaryammonium-containing compounds were characterized by TGA. Moreover, quantitative results from NMR showed that tetramethylammonium fluoride (TMAF) was 14 times more stable than TMAOH at 130ºC. Next, membranes with the optimal composition containing TMAF were synthesized, and their transport performances at different temperatures and feed CO2 partial pressures were measured, respectively. The membrane showed a CO2 permeance of 108 GPU and a CO2/H2 selectivity of 106 at 120ºC, and it was stable for 120 h. The TMAF-containing membrane also displayed an improved stability of about 4 times better than the TMAOH-containing membrane at 130ºC. To date, scale-up of 21"-wide membranes with a total length of 720 ft has been demonstrated, and the membranes exhibited consistent transport results as the lab-scale samples. Additionally, a techno-economic analysis showed that, when the H2 recovery of the membrane process was set at 99%, the CO2 removal cost was calculated to be $62.9/tonne. Lastly, a CO2 removal of 76.9% could be obtained when the air sweep was replaced by vacuum on the permeate side.

Poster 4 - Advancing NF Membranes: Role of Surface Charge and Support

Francisco Leniz* (University of Kentucky), Dibakar Bhattacharyya (University of Kentucky), Isabel Escobar (University of Kentucky).

Nanofiltration (NF) membranes have been reported to selectively separate charged small molecules and ions providing a variety of applications, as for example rejection of negatively charged ions/molecules or recovery of valuable metals (multivalent cations), simultaneously reducing the osmotic pressure of the system by letting pass non-desired ions. Therefore, tuning and understanding the role of surface charge will allow better NF design as well as suitable applications.
Moreover, the ultrafiltration (UF) membrane as support for NF membranes has been reported to impact both the formation and performance of the NF layer. In this study, we use 3 different UF supports (two polyethersulfone (PES) and one polysulfone (PSf – commercial PS35), and we synthesize the NF layer by rapid interfacial polymerization using two (commercially available) amine groups: traditional negatively charged NF membrane with piperazine (PIP); and another with modified chemistry by addition of polyallylamine hydrochloride (PAH) – more positively charged.
It was observed an impact on the performance of the synthesized NF membranes from both differences in the substrate as well as by the addition of PAH, making them part of the variables (direct or indirect) of the final NF performance. An increase in the zeta potential (around 2 points in the isoelectronic point) of the membranes is obtained from the addition of PAH for all substrates used. In terms of performance, on one hand, water permeabilities were affected: up to 7% decrease by addition of PAH, and a decrease of 2 and 3 folds by varying support. On the other hand, rejection of salt increased: up to 20 – 30% points of PAH addition, and 20 – 40% from different support. Deep characterization of substrates, as well as the synthesized membranes, is performed for correlation with the mentioned variables. Finally, a study in lanthanides is included, allowing a better understanding of surface charge relation with Ln+3 separation by NF membranes.

Poster 5 - The development of honeycomb-graphene oxide coated porous polymeric membranes

Feihu Wang* (Ben-Gurion University of the Negev).

Graphene oxide (GO) has been studied as an anti-bacterial material in many applications. In membrane science, GO was incorporated into membranes using various approaches including binding, blending, and laser-induced methods. A recent study demonstrated that honeycomb structured graphene films coated on glass, exhibit high

antibacterial activity. Here, we report the fabrication of microporous patterned GO films coated on polymeric membranes that may use as an anti-biofilm pre-membrane.
PTFE membranes were coated with patterned GO using the "breath figure" method. The GO was transferred to an organic solvent with the aid of a surfactant and placed on the membrane surface until the solvent was evaporated. The influence of various parameters (such as humidity, solvents, and concentrations) on the patterned morphologies was investigated. The GO coated membrane was then analyzed for stability and permeability.
Microporous structures were obtained at 54% ≤ humidity ≤ 98%. Chloroform was the most efficient solvent to produce a honeycomb structure (1-4μm, Figure 1) when compared with other solvents. Homogenous patterned GO films were obtained at 3 mg/ml GO and 6 mg/ml surfactant. The structure of the films can be divided into three regions: 1. a compacted GO layer on the membrane surface 2. vertically oriented GO with cavity structures in the middle acting as pillars 3. a surface microporous structured layer. The structures and permeability of the microporous GO films were stable after washing three hours with 50% ethanol.
The results show that a microporous GO can form on a polymeric membrane. This can open new directions in modifying membranes with thin graphene films for the development of remediation platforms of the aquatic environment. The performance and antifouling properties of these membranes are currently being investigated.

Poster 7 - Activated Carbon as a Photothermal Absorber for Solar Driven Air-Gap Membrane Distillation

Maryam AlQaydi* (Khalifa University), Arwa Alshareif (Khalifa University), Ibrahim Mustafa (Khalifa University), Faisal AlMarzooqi (Khalifa University), Hassan Arafat (Khalifa University).

1. Introduction
There have been recent studies demonstrating the use of activated carbon (AC) additives in the synthesis of membranes for membrane distillation (MD) applications, related to inducing higher membrane surface area, hydrophilicity and preferable pore structure. However, no previous work was performed on the utilization of AC additives in solar-driven MD. Herein, we investigated for the first time, the effects of AC additives on the performance of polymeric polyvinylidene fluoride (PVDF) membranes, when utilized as photoabsorbing nanoparticles for solar-driven MD.
2. Methods
We designed an AGMD setup, as shown in (Figure 1). Our in-house fabricated membranes using the phase inversion method were utilized within the device, in which temperature, membrane flux, solar irradiation, and humidity were measured. We also performed physical characterizations to provide theoretical insights and understandings related to the membrane performance results.
3. Results
The results show that the PVDF-9%AC resembled the highest flux values, which is attributed to its excellent photo-thermal properties, highest wettability and highest porosity (Figure 2).
4. Discussion
In our novel approach, we demonstrated a solar MD process without external heating (external heating is a typical approach in literature), which was solely driven by direct solar heating, enabling the applicability of such technology as a standalone, scalable AC modified PVDF membranes fabricated via simple and facile methods. Furthermore, efficiencies and energy demands are investigated to assess the commercial criteria.
We show that the utilization of small amounts (5 to 9 %) of AC nanoparticles can significantly boost the flux, directly driven by solar energy. PVDF-AC9% yielded the lowest energy demand and the highest solar efficiency (30.1 %).

Poster 8 - Application of 3D Printed Minimal Surface Area Spacers on PVDF Membranes for Treating Fracking Wastewater via Membrane Distillation

Shahriar Habib* (The University of Alabama), Steven Weinman (The University of Alabama).

Fracking wastewater is considered one of the principle waste streams from the oil and gas industry. Along with a high concentration of salts, it contains a wide range of harmful contaminants such as heavy metals, emulsified oil, and other organics, which pose serious environmental and remediation challenges. The main technique used to treat this water is deep well injection, which is suspected to cause earthquakes. Recently, membrane distillation (MD) has emerged as a promising candidate for treating fracking wastewater. MD is a thermal-driven separation process where water from the feed side (hot temperature side) transports to the permeate side (cold temperature side) through a hydrophobic membrane as vapor molecules, leaving behind non-volatile components. Though MD is a promising way to treat fracking wastewater, flux reduction and pore wetting due to fouling of the hydrophobic membrane is an issue because flux reduction reduces the productivity and pore wetting reduces the separation efficiency of the membrane. As such, an efficient and cost-effective strategy is required to mitigate membrane fouling and will have significant economic and environmental impacts. One approach for fouling mitigation is to use 3D printed spacers in the feed and permeate flow channels. Studies suggest that spacers can improve the water flux performance and membrane fouling resistance. However, the performance of the system depends on the design of the spacers as it governs the fluid hydrodynamics through the feed channel. In this work, three different 3D printed spacers have been used to test for performance in MD. Two are minimal surface area spacers and one is a mimic of a conventional mesh spacer. Commercial PVDF membranes have been tested in a custom made MD system with and without the spacers to compare the pressure drop across the feed channel, water flux performance, and fouling resistance.

Poster 9 - Molecularly Engineered 6FDA-based Polyimide Membranes for Sour Natural Gas Separation

Zhongyun Liu* (Georgia Institute of Technology), Yang Liu (Georgia Institute of Technology), Wulin Qiu (Georgia Institute of Technology), William Koros (Georgia Institute of Technology).

Glassy polyimide membranes are attractive for industrial applications in sour natural gas purification. This polymer family benefits from excellent separation performance in some cases under aggressive natural gas streams and well-known spinnability into thin-skinned hollow fibers. Most prior studies focused on CO2 removal alone and overlooked H2S/CH4 separation applications; however, co-existence of CO2 and H2S in many realistic sour natural gases makes simultaneous removal of CO2 and H2S particularly important. Moreover, the lack of fundamental understanding of relationships between chemical structures of polyimides and their gas transport properties, especially for H2S, constrains the design and engineering of advanced membranes for such challenging applications. In this study, 6FDA-based polyimide membranes with engineered structures were synthesized to tune the separation performance and plasticization properties of membrane for sour natural gas purification. Pure gas (CO2, H2S, and CH4), binary (CO2/CH4) and ternary (H2S/CO2/CH4) mixed gas permeation tests were performed to understand the effects of polyimide membrane chemical structure on their gas transport properties. Controlling polymer chain packing and plasticization tendency of such polyimide membranes under aggressive sour natural gas were found to offer better combined H2S and CO2 removal efficiency compared to conventional polymer. Fundamental insights into the observed "structure-to-property" relationships will be discussed to guide next-generation membranes for sour natural gas separation.

Poster 10 - Membrane with cross-linked zwitterionic nanopores achieves sub-nanometer separations

Samuel Lounder* (Tufts University), Ayse Asatekin (Tufts University).

Self-assembling copolymer membranes have the potential to emerge as the next generation of nanofiltration membranes; however, the sub-nanometer selectivity required for this broad application space has not yet been demonstrated. Random zwitterionic amphiphilic copolymers (r-ZACs) are strong candidates to meet this challenge. These copolymers can self-assemble to form zwitterionic nanopores through which water and small solutes can permeate. Membranes derived from r-ZACs exhibit exceptional fouling resistance and chlorine tolerance, but also low ion retention and a size-based cutoff of ≈ 1.5 nm; they are mere angstroms away from the long sought sub-nanometer selectivity.

In the present work, we employ a novel cross-linking technique to reduce the size of r-ZAC nanopores. Neutral solutes greater than 1 nm are completely rejected by the modified membranes (sucrose rejection > 99.5%; Stokes diameter = 0.94 nm). Furthermore, these membranes demonstrate outstanding divalent-monovalent selectivity when challenged with artificial sea water (sulfate rejection = 99.2%, sulfate-chloride separation factor = 75). In addition, these membranes are extremely resistant to fouling and chlorine damage, which are serious performance limitations for commercial nanofiltration membranes. These results illustrate a promising route to achieving sub-nanometer separations with self-assembling copolymer membranes.

Poster 11 - Tuning Pore Size and Robustness of Membranes Formed by Scalable Self-Assembly of Random Copolymer Micelles

Luca Mazzaferro* (Tufts University), Ilin Sadeghi (Massachusetts Institute of Technology), Ayse Asatekin (Tufts University).

Broader use of membrane separations is limited by their capabilities, typically confined to size-based separations. Self-assembly of functional polymeric materials is a powerful method for designing membranes capable of new separations, including the separation of organic compounds of similar size from each other. This can potentially be achieved by membranes that mimic biological pores such as porins, with pores that are only slightly larger than the target solute and functionalized with groups that selectively interact with one compound over another. Our group has developed a method for preparing such membranes by the spontaneous self-assembly of random amphiphilic copolymer in methanol into micelles that are then coated onto a support to form a selective layer of tightly packed micelles. These membranes have ~1-4 nm pores lined with carboxylic acid groups that enable charge-based separation of organic solutes. The carboxylic acid groups can further be converted to other functional groups. This can provide selectivity towards solutes by affinity. The goal of this study was to tune the pore size and increase the robustness of these thin film composite membranes. The pore size was tuned by changing the composition of the copolymer and also by changing the casting solvent. The robustness of the system was altered by different crosslinking procedures, which further altered pore size. We observed a 7% to 99.5% range in the rejection of a neutral solute with a hydrodynamic diameter of 1.7 nm, though these changes were poorly correlated with the size of micelles in solution. This study was essential to better understand and prove the versatility of this membrane manufacturing approach and its' potential to address various separation challenges in pharmaceutical and bioseparations.

Poster 12 - Electron tomography reveals the 3D microstructure of virus filtration membranes

Kaitlyn Brickey* (Pennsylvania State University), Andrew Zydney (Penn State University), Enrique Gomez (Penn State University).

Virus filtration membranes are instrumental in ensuring the safety of potable water supplies and high value biopharmaceuticals like monoclonal antibodies. Virus filtration is particularly challenging in biopharmaceutical applications because the filter must provide high transmission of the drug product, which can be as large as 10-12 nm for monoclonal antibodies, while still providing >99.9% retention of small parvovirus (around 20 nm). Several studies have shown that the performance of some virus filters can be compromised under certain conditions including low pressure operation, high throughput filtrations, and after process disruptions. However, the relationship between the virus filtration performance and the underlying membrane pore structure is still largely unknown. To evaluate the pore size, porosity, and 3D morphology of the pores within commercial virus filtration membranes, high-resolution cryogenic focused ion beam scanning electron microscopy (cryo-FIB) and scanning transmission electron microscopy (STEM) tomography were utilized. Cryo-FIB allowed for large volume imaging while STEM tomography provided reconstructions with higher resolution for both symmetric (Pall – Pegasus™ SV4) and highly asymmetric (MilliporeSigma – Viresolve® Pro) virus filters. Initial analysis of volume reconstructions for the Viresolve® Pro reveal highly interconnected pores with a porosity of 29-33% at the filter exit. The average pore size decreases in the direction of flow until reaching a membrane depth that was previously identified as the primary region for virus capture. Other membrane parameters critical to performance such as closed void fraction and tortuosity were also quantified. These results demonstrate the capabilities of electron tomography for revealing the pore structure of various virus filtration membranes and provide key relationships between the membrane pore morphology and virus filter performance needed for both water and bioprocessing applications.

Poster 13 - Tuning the binding strength of membrane adsorbers in radiochromatography

Priyanka Suresh* (Case Western reserve University), Christine Duval (Case Western Reserve University).

Radiochromatography plays a key role in the fields of medical isotope purification and nuclear forensic in which one element must be quickly and selectively purified from its neighbors on the periodic table. Radiochemists still rely on resin-based chromatography columns which are known to suffer from diffusion limitations—leading to long purification times. An alternative to resins is membrane adsorbers (MAs) which have been successful in reducing mass transport limitations in affinity-based purification of proteins. In radiochemistry, phosphate and phosphonate ligands are used to sequester lanthanides and actinides due to their high polarizability. At low pH, the phosphoryl oxygen (hard base) will chelate with hard acids (polarizable ions like lanthanides and actinides). The hardness (ability to donate electrons) of the oxygen plays a key role in determining affinity of the ligand for the cation.

To control selectivity, we aim to tune the hardness of the phosphoryl oxygen by controlling the local chemical environment around the ligand by changing the composition of the polymer brush. This work describes a simple system that has been designed to study the effect of H-bond (accepting and donating) co-monomers on the basicity of an amphoteric (H-bond donating and receiving) phosphate ligand for La3+. Membranes are functionalized using UV-initiated free radical polymerization to graft random heterogeneous copolymers of ethylene glycol methacrylate phosphate (ligand). The co-monomer is either H-bond donating or H-bond receiving and does not bind with La3+. The effect of brush composition on membrane capacity, affinity and binding kinetics are derived from batch sorption experiments. Introducing an H-bond donating co-monomer, reduces the binding capacity of the MA from 17.6 to 4.2 mg of La3+/g of MA while increasing the water uptake capacity by 16%. Understanding the effect of molecular-level design of MAs will aid in the future design of radiochromatography materials.

Poster 14 - Resilient Hollow Fiber Nanofiltration Membranes Fabricated from Copolymers

Michael Dugas* (University of Notre Dame).

The need has emerged for membrane platforms that can be tailored to accommodate the highly varied applications, feed compositions, and treatment demands of various separation processes. Nanofiltration (NF) membranes based on copolymer materials are a promising separations platform because they can be engineered at the molecular scale to address an array of process needs. Here, for example, a resilient NF membrane is developed through the molecular design of a poly(trifluoroethyl methacrylate-co-oligo(ethylene glycol) methyl ether methacrylate-co-glycidyl methacrylate) copolymer that can be dip-coated onto hollow fiber supports. By exploiting the microphase separation of the copolymer repeat units and by elucidating the processing-structure-property relationships for the dip-coating process, membranes with pores 2 nm in diameter that exhibit a hydraulic permeability of 15.6 L m-2 h-1 bar-1 were generated. The membranes were functionalized post-coating with hexamethylene diamine to incorporate positively-charged moieties along the pore walls, yielding membranes that rejected 98% of the MgCl2 from a 1 mM feed solution. Moreover, the diamine crosslinked the copolymer such that the membranes could operate stably in ethanol, an organic solvent that otherwise damaged the parent membranes. These results demonstrate the ability to develop resilient, valence-selective NF membranes through the molecular engineering of the copolymer. The stability of these membranes has led to investigations of how they can withstand exposure to various organic solvents. Preliminary results suggest that as the length of the crosslinking agent increases, the extent of crosslinking also increases. This increase in crosslinking, in turn, leads to an increased resistance to solvent degradation. By combining the versatility of molecularly designed copolymers with functionalities to increase the integrity of the membrane can further the use of membrane technologies into emerging applications.

Poster 17 - Effect of Metal Ions Dissociated in Cross-linked Poly(Ethylene Oxide) (XLPEO) on Physical Properties and Gas Transport Characteristicss

Taliehsadat Alebrahim* (The State University of New York at Buffalo), Alisa Chakraborty (The State University of New York at Buffalo), Haiqing Lin (The State University of New York at Buffalo).

Metal-organic frameworks (MOFs) have been dispersed in polymers to improve gas separation performance, and favorable interactions between the MOFs (usually metal centers) and polymers are preferred to improve interfacial compatibility and eliminate any non-selective voids. However, there lacks an understanding of the effect of polymer-metal ion interactions on gas separation properties. In this study, we investigate the effect of metal ions dissociated in amorphous cross-linked poly(ethylene oxide) (XLPEO) on physical properties and gas transport characteristics. Polymer electrolytes were prepared by photo-polymerization of solutions containing poly(ethylene glycol) diacrylate (PEGDA), poly(ethylene glycol) methyl ether acrylate (PEGMEA), and different salts such as LiClO4, Ni(BF4)2, and Cu(BF4)2. The salts can be fully dissociated, and the Li+, Ni2+, and Cu2+ ions form conjugations with the ethylene oxide in XLPEO, as validated by FTIR and XRD. Increasing the salt content increases the density (following the additive model) and glass transition temperature (following the Gordon-Taylor equation) except for Cu(BF4)2. For instance, adding 17 wt.% LiClO4 in PEGDA-co-PEGMEA increases the Tg from -65 ºC to -36 ºC and decreases CO2 permeability from 720 Barrer to 20 Barrer. Additionally, the CO2/N2 selectivity decreases from 48 to 31 because of the competing effect for the polar ether to interact with metal ions, instead of CO2. The effect of the salt loading on the gas permeability can be described using the Vogel-Fulcher-Tammann (VTF) equation. The effect of the salt loading on gas solubility and diffusivity is investigated. The elucidation of the effect of the polymer and metal ion interactions on gas transport properties may provide guidance in designing mixed matrix materials containing MOFs.

Poster 19 - High Density Membrane Protein-Polymer Nanosheets-Based Biomimetic Membranes

Yu-Ming Tu* (The University of Texas at Austin), Woochul Song (University of Texas at Austin), Benny Freeman (The University of Texas at Austin), Manish Kumar (University of Texas at Austin).

Biomimetic and bioinspired membranes containing membrane protein (MP) channels and MP-mimic functionalized materials have emerged as significant platforms to develop filtration membranes with specific solute selectivity. These membranes typically consist of highly permeable biological or bioinspired pore structures or channels with well-defined pore geometries embedded in a relatively impermeable synthetic membrane matrix. Nevertheless, membranes integrating MP-based membranes into current manufactured technology for separations has remained challenges. These challenges are possibly due to the amount of time needed to create these membranes, the use of vesicular morphologies and the low density of MPs in resulting membranes, which have resulted in much lower than anticipated improvements. In this presentation, we will describe a comprehensive approach to fabricate high-performance biomimetic filtration membranes that are based on densely packed MP channel in block copolymers (BCPs) to create robust and flat nanosheets. These highly packed crystalline structures and well-ordered nanosheets were constructed by a two-hour organic solvent evaporation method and further deposited on a porous substrate as high-throughput selective layers to form composite membranes. Three different types of β-barrel MP channels with pore sizes of 0.8 nm, 1.3 nm, and 1.5 nm were utilized in this work. The resulting three different MP-BCP nanosheet membranes created demonstrated high water permeability of ~300-1,000 (l m-2 h-1 bar-1) and still maintained a sharp and tunable selectivity with three different pore sizes of MP channels. The approaches described could be adapted with the insertion of MP channels of different pore sizes or chemistries into polymer matrices to conduct specifically designed separations.

Poster 20 - Polyvinyl Alcohol (PVA) Hydrogel Underlayer to Support CVD Graphene on Ultrafiltration Membranes as a Support for Water Desalination

Mansour Saberi* (Clemson University), Scott Husson (Clemson University), Stephen Creager (Clemson University).

Fresh water is a primary concern for all the people around the world. So far, seawater desalination has been performed via membrane technologies. One of the most important and recently studied candidates for use in Reverse Osmosis (RO) or nanofiltration (NF) membranes is graphene which has excellent chemical and mechanical stability. Graphene is the thinnest possible membrane with its one atom thickness acting as the membrane. Small thickness is desirable because it gives high flux. But high flux must be accompanied by high salt rejection. The application of plasma methods can make the graphene membrane and its substrate controllably nanoporous to give the desired high flux and high salt rejection. In a further benefit, graphene manifests greater resistance to chlorine than current polyamide membranes. In the fabrication of an efficient graphene NF/RO membrane, it is critical to have a suitable flexible substrate (hydrogel) that can offer support to the one-atom-thick graphene membrane while still allowing rapid water transmission. In this project, the process of transferring CVD graphene on to a hydrogel-modified NF membrane will be discussed. CVD graphene is fragile and can easily be torn if it is directly transferred to a membrane support structure. To allow transference without damage, a hydrogel underlayer—polyvinyl alcohol (PVA)—has been synthesized and cross-linked and applied onto a membrane support. The degree of cross-linking and the thickness of the PVA layer can affect the permeability of a PVA / support membrane.
The graphene is transferred onto the PVA support by a simple but unique approach to decreasing the chance that defects will form. Experimental work combined with membrane characterization methods (SEM, AFM, and, FTIR) and filtration studies using a home-made RO system to examine the graphene as RO membrane will be presented. These results will be compared with the other synthesized membrane and different types of supports.

Poster 21 - Ion Transport Improvement by Surface Modified Beads

Jamie Hestekin (University of Arkansas), Leticia Santos de Souza* (University of Arkansas), John Moore (University of Arkansas), Christa Hestekin (University of Arkansas).

Human organs, such as kidneys, have been developed and improved for thousands of years to achieve their current level of efficiency. Although it might be audacious trying to aim the same level of selectivity as the human body, several medical applications have been widely studied to develop devices and procedures that will reasonably mimic nature's job. Selective ion removal at low concentration is one of the important subjects to be improved upon for many biomedical applications. Electrodialysis and electrodeionization are examples of suitable technologies for those applications, but still have limitations. Therefore, the newer method of wafer-enhanced electrodeionization combines the sensitivity of these methods, while reducing drawbacks such as leaking and channeling by giving a fixed structure for the resins. However, there is still a demand for more selective units in biomedical and other applications. Therefore, ion transport improvement by surface modification of resin beads was the objective of this study. Unmodified and modified beads were tested in solution as well as incorporated into the wafer format for different ions in blood. We found characteristics of charged groups that works well in electrodeionization and improves selectivity towards that of a kidney.

Poster 22 - Surface Nano-structuring with Tethered Poly(acrylic acid) Chains for Tuning Ultrafiltration Membrane Performance

Yian Chen* (University of California, Los Angeles), Yoram Cohen (University of California, Los Angeles).

Polysulfone (PSf) based ultrafiltration (UF) membranes have gained popularity in filtration/treatment of aqueous streams in various industries. However, in selecting UF membranes, one is typically constrained by the solute rejection-permeability tradeoff which restricts the permeability that can be attained for a given required solute molecular weight cutoff (MWCO). Accordingly, this study presents an approach to fine-tuning the performance of PSf UF membranes based on a two-step surface nano-structuring (SNS) process. The PSf membrane surface was first activated (to form free radicals) via exposure to atmospheric pressure plasma (APP) composed of He/O2, followed by graft polymerization at different reaction conditions to form on the PSf base membrane a layer of tethered poly(acrylic acid) (PAA) chains. Membrane surface topography, chemistry and surface charge were evaluated along with determination of the tethered polymer layer thickness and its swelling characteristics in an aqueous environment. Fine tuning of membrane performance was assessed with respect to different GP conditions such as initial monomer concentration, graft polymerization (GP) pH, time, and temperature. Both APP surface activation and GP conditions impacted SNS-PAA-PSf membrane performance as quantified based on membrane permeability and MWCO. Surface structuring with a tethered PAA layer enabled tuning of membrane performance to achieve membranes of the same MWCO over a wide permeability range or a given permeability over a range of MWCO. For example, SNS-PAA-PSF membranes of permeability of 13.2 L/m-2·h-1·bar-1 and 28.4 L·m2·h/bar were synthesized to have a MWCO range of 4-9 kDa and 11.5-17 kDa, respectively. The above results demonstrated that one can achieve membrane performance that overcomes the commonly reported permeability-solute rejection tradeoff.

Poster 23 - Membrane Capacitive Deionization using Quaternized Polymer Blends

Robert McNair* (University of Manchester), Robert Dryfe (University of Manchester), Gyorgy Szekely (King Abdullah University of Science and Technology).

Membrane capacitive deionization (MCDI) for water desalination has emerged as an innovative technique to help tackle global water scarcity. The employment of ion selective membranes with high ion-exchange capacity has the potential to markedly increase salt removal during MCDI desalination. Membranes based on polybenzimidazole (PBI) and polyethylenimine (PEI) of various mass loadings were prepared via phase inversion. Both polymers possess a high number of nitrogen atoms; which can undergo quaternization to impart anion-exchange character. Membranes were surface modified with methyl groups and their structure and composition were fully characterized. Electrochemical properties (ion-exchange capacity, permselectivity, area resistance) were determined and the membranes were employed as an anion-exchange membrane (AEM) in a bench-scale MCDI configuration using brackish water. The desalination performance was quantified in terms of salt adsorption capacity and charge efficiency. Owing to the combination of nitrogen-rich PBI/PEI, all blend membranes displayed high quaternization degree as confirmed by NMR and elemental analysis. Further, all membranes displayed high ion-exchange capacity (outperforming some commercial membranes) and permselectivity; making them ideal candidates for MCDI anion-exchange membranes. The membranes produced a significantly higher salt adsorption capacity and charge efficiency than conventional CDI without membranes, whilst outperforming similar membranes in literature. For the first time, this study fabricated quaternized membranes based on blends of PBI/PEI and were applied as AEMs in MCDI. The membranes exhibited over a 4x increase in salt adsorption capacity and charge efficiency compared to CDI. This was attributed to greatly enhanced chloride ion removal through the highly quaternized anion-exchange membranes. The use of such membranes will help to establish the industrial application of MCDI for water desalination.

Poster 24 - Ion-Stabilized Organic Solvent Nanofiltration Membranes from PIM/PBI Polymer Blends

Gergo Ignacz* (King Abdullah University of Science and Technology), Gyorgy Szekely (King Abdullah University of Science and Technology).

Robust, ion-stabilized membranes from polybenzimidazole (PBI) and its blends with polymers of intrinsic microporosity (PIM-1) were fabricated for organic solvent nanofiltration in polar aprotic solvents (PAS)[1]. Despite their drawbacks such as toxicity and high boiling point, PASs are widely used in the chemical industries due to their excellent solubility properties. Green alternatives solvents to substitute the undesired conventional PASs are emerging in the literature; however, their industrial viability is yet to be demonstrated. Therefore, it is necessary to develop membranes that are stable in PAS, which could facilitate the implementation of OSN by the industrial sector. Chemical cross-linking aims to increase the stability of polymer membranes under harsh environments such as PAS. In this work, flat sheet membranes comprised of PBI and 4–12 wt% PIM-1 were fabricated via conventional phase inversion followed by the reduction of nitrile PIM-1 to amine PIM-1. Stabilization of the membranes was achieved via ion formation under acidic conditions. Such simple ion-stabilization methodology could replace cumbersome cross-linking. The aging of the membranes and their performance at room and elevated temperatures, in a wide range of pH, in the presence of organic bases, in six conventional and two green PAS, were investigated. The novel stabilization approach eliminates the use of cumbersome chemical cross-linking. The incorporation of PIM-1 into the membranes improved the permeance up to 4 times and simultaneously decreased the MWCO. Filtration studies covering 3 months revealed a moderate membrane aging and a comprehensive thermal treatment study showed a stable but anomalous performance up to 125 °C.

Reference

[1] Ignacz, G.; Fei, F.; Szekely, G. Ion-Stabilized Membranes for Demanding Environments Fabricated from Polybenzimidazole and Its Blends with Polymers of Intrinsic Microporosity. ACS Appl. Nano Mater. 2018, 1, 11, 6349-6356. https://doi.org/10.1021/acsanm.8b01563.

Poster 25 - Synthesis and Gas Transport Properties of CO2-philic Anionic Poly(ionic liquid) Composite Membranes

Irshad Kammakakam* (University of Alabama), Jason Bara (University of Alabama).

Herein, we report the development of, to the best of our knowledge, the first anionic poly(IL) membranes via a photopolymerization for enhanced CO2 separation, in which the poly(IL) component has delocalized sulfonimide anions pendant from the polymer backbone and imidazolium cations as mobile counterions. Two types of photopolymerizable methacryloxy-based IL monomers (MILs) with highly delocalized anions (–SO2–N(-)–SO2–CF3 and –SO2–N(-)–SO2–C7H7) and mobile imidazolium ([C2mim]+) counter cations were successfully synthesized and photopolymerized with two distinct amounts of free IL containing the same structural cation ([C2mim][Tf2N]) and 20 wt% PEGDA crosslinker, to serve as a composite matrix. The structure-property relationships of the newly developed anionic poly(IL)-IL composite membranes were extensively characterized by FT-IR, TGA, DSC, and XRD analysis. All of the newly developed anionic poly(IL)-IL composite membranes exhibited superior CO2/CH4 and CO2/N2 selectivities together with moderate CO2/H2 selectivity and reasonable CO2 permeabilities. This study provides a promising strategy to explore the benefits of anionic poly(IL)-IL composites to separate CO2 from flue gas, natural gas, and syngas streams with strong candidate materials for practical applications.

Poster 26 - Synthesis and Characterization of Chlorinated RO Membrane Model Compounds – Insights into the Role of N-Cl Species in the Chlorination of Polyamide Membranes

Dean Welsh* (DuPont / FilmTec Corporation), Rachel Ehlert (DuPont), Kyoungmoo Koh (DuPont), Mou Paul (Dow), Tom Peterson (Dow), Nipon Pothayee (Dow), Mark Rickard (DuPont), Abhishek Roy (DOW), David Wilson (Dow).

Polyamide reverse osmosis (RO) membranes may be inadvertently exposed to chlorine-based bleaching agents or purposely exposed to clean biofouling from the membrane. This exposure may chlorinate the polyamide layer, and the chlorinated RO membrane can show poor performance. It has been reported that the generation of N-Cl species in polyamide by exposure to bleaching agents causes decreased performance. However, there is no clear evidence to support the proposed mechanism of chlorination of polyamide. In this study, four model compounds were prepared by amidation or chlorination to understand the chlorination process in amide structures. The chemical structures of chlorinated model compounds were successfully identified with LC/MS, IR, NMR, and X-ray crystallographic analysis. The chloro-substitution selectively occurred on the phenyl rings of m-phenylenediamine units, however, N-Cl species were not observed in our model experiment conditions. The other critical finding was that ring-chlorination also causes a significant change of the Amide II band in the IR spectrum which has been used as evidence of N-Cl generation in the literature.

Poster 27 - Spray-Coated Graphene Oxide Hollow Fibers for Nanofiltration

Gheorghe Falca* (King Abdullah University of Science and Technology)), Lakshmeesha Upadhyaya (King Abdullah University of Science and Technology), Valentina Elena Musteata (King Abdullah University of Science and Technology), Suzana Nunes (King Abdullah University of Science and Technology).

Process intensifications in the pharmaceutical, oil and food industry require high energy to recover solvents and separating molecules in the range of nanofiltration. Recently, the development of new membranes manufacture techniques, material together with their mechanical and thermal resistance has facilitated the membrane implementation in such sectors. However, simple, scalable, low-cost fabrication methods for GO deposition on hollow fibres, are still in an early phase. We propose here a simple spray-coating method for deposition of GO sheets on cross-linked hollow fibre supports. We first fabricated polyetherimide hollow fibre membranes and cross-linked them with hexamethylene diamine. These supports have a strong tolerance to various organic solvents. Thereafter, the spray-coating of GO dispersions led to the formation of a stable, selective layer on the hollow fibres. Excellent adhesion between GO and the substrate was achieved. Overall, this simple method enables the application of GO hollow fibre membranes in organic solvent nanofiltration, with high performance demonstrated in water and acetone.

Poster 28 - Laser-Induced Graphene for Charged Membrane Applications Enabled by Sequential Infiltration Synthesis

David Bergsman* (Massachusetts Institute of Technology), Beza Getachew (MIT), Jeffrey Grossman (Massachusetts Institute of Technology).

Membrane-based processes are becoming increasingly popular for water treatment due to their relatively high energy efficiency and low cost. However, these advantages are mitigated by the need for additional pre-treatment steps that are required to maintain effective membrane operation. The treatment and prevention of membrane fouling constitutes a large fraction of typical operational costs. One potential approach to combat fouling is to design conductive membrane coatings that can prevent the attachment and growth of biofoulants, both electrostatically and via electrochemical generation of reactive oxygen species. Despite their potential, these coatings are often expensive, requiring additional chemicals and non-scalable methods to produce, e.g. carbon nanotube mats or other graphitic coatings deposited by vacuum filtration. In this work, we explore the use of laser-induced graphene (LIG) for the creation of conductive ultrafiltration membranes. Porous polyethersulfone (PES) membranes are first treated with alumina using sequential infiltration synthesis (SIS) before being irradiated with an infrared laser. We show that this alumina, which can be scalably produced using spatial SIS, can localize LIG formation to the surface of the membrane, preventing the buried, un-lased areas of PES from melting and losing their porosity during the lasing process. This allows the top-most layer of the PES to be a conductive coating that can be used to charge the membrane surface and used to improve membrane performance (e.g. fouling mitigation). The formation of LIG is verified by scanning electron microscopy and Raman spectroscopy. The conductive layer is also shown to possesses relatively high conductivity, which is important for reducing power consumption in devices. Insight into the mechanism behind the improved stability to melting provided by SIS is provided by thermogravimetric analysis, differential scanning calorimetry, and Fourier-transform infrared spectroscopy.

Poster 31 - Nafion based mosaic nanofiltration membranes with tuneable charge and permeability

Michael Zelner* (Technion - IIT), Philipp Jahn (Universität Duisburg-Essen), Mathias Ulbricht (Universität Duisburg-Essen), Viatcheslav Freger (Technion - IIT).

Desalination of wastewater for irrigation needs tunable rejection of different ions to minimize scaling and maximize benefits for crops and soil. Ideally, while NaCl is rejected, multivalent ions should remain in the water. Available

commercial nanofiltration (NF) membranes are not optimal for this application and novel membranes with tunable charge and high permeability are desired. We report here novel NF membranes with two-component (mosaic) polyelectrolyte active layer of tunable characteristics. The membranes were prepared by coating a polysulfone UF support first with an anionic layer of Nafion and, thereafter, applying a second "caulking" layer of a different polymer. The permeability was tuned by varying the thickness of first Nafion layer, while charge and rejection properties were tuned by varying the type and amount of the second polymer, which could be either neutral (poly(vinyl alcohol), PVOH) or cationic (poly(vinyl amine), PVAm).
A pure Nafion membrane showed poor rejection, apparently, due to an excessive number of defects. Coating with PVOH strongly reduced the defect rate and produced a strongly negatively charged membrane, as confirmed by rejection of different single salts. In contrast, PVAm coating sealed the defects and counter-balanced the Nafion charge, producing a neutral + with a more uniform rejection of different ions. By varying its average thickness of Nafion in the range 0.25 to 2.5 µm, the water permeability could be tuned in a wide range of 0.5 to 9.5 LMH/bar with only a moderate variation of rejection. The proposed method shows potential for developing tunable membranes for an application that requires optimization of rejection rates for different ions, such as wastewater desalination. It may also be useful for developing membranes with a mosaic structure, which is currently under examination.

Poster 32 - Graphene Oxide (GO) Nanocomposite Anion Exchange Membranes for Improved Electrodialysis

Levente Cseri* (University of Manchester), Peter M. Budd (University of Manchester), Gyorgy Szekely (King Abdullah University of Science and Technology).

Electrodialysis (ED) is the second most common desalination technique in the number of operational plants serving many communities worldwide. Recently, commercial anion exchange membranes (AEMs) have struggled to develop at the same rate as cation exchange membranes, due to insufficient selectivity, chemical- and dimensional stability. The incorporation of nanomaterial fillers has been proposed to improve membrane properties [1]. In this work, mechanically robust and highly permselective AEMs were developed based on polybenzimidazolium and graphene oxide (GO) nanocomposite [2]. GO powder was modified for better dispersibility via diazonium chemistry and used to fabricate dense, flat sheet AEMs with various GO loadings. The AEMs exhibited excellent mechanical strength, high ion exchange capacity, good to exceptional permselectivity and relatively low area resistance. A classical trade-off was observed at low GO loadings (0.25–2.5%) between high selectivity and low resistance. Key properties of the AEM such as mechanical strength and permselectivity were improved with the introduction of a small amount of GO filler. As little as 1% GO content resulted in unprecedented permselectivity of 0.99. Nevertheless, filler overloading may result in somewhat lower mechanical strength and high area resistance. The membrane having 1% GO loading performed best in brackish water electrodialysis tests with the highest salt removal, lowest power consumption and best current efficiency. A permselectivity – reciprocal area resistance plot with an empirical upper bound was proposed as a reference for future developments. The results of this work show that fine-tuning of membrane properties may be achieved even with a low amount of GO filler leading to improved ED processes.

References

[1] Alabi et al., npj Clean Water, 2018, 1, 10. (DOI: 10.1038/s41545-018-0009-7)
[2] Cseri et al., J. Mater. Chem. A, 2018, 6, 24728-24739. (DOI: 10.1039/C8TA09160A).

Poster 33 - Large-Area Composite-Membrane based on Ultra-Thin Carbon Nanomembranes

Nikolaus Meyerbroeker* (CNM Technologies GmbH), Polina Angelova (CNM Technologies GmbH), Henning Vieker (CNM Technologies GmbH), Albert Schnieders* (CNM Technologies GmbH).

Recently scientists at Bielefeld University demonstrated that nanometer-thin carbon nanomembranes (CNMs) made by cross-linking of self-assembled monolayers of aromatic molecules possess an extremely high pore density of one sub-nm channel per square nanometer [1]. Thus, this two-dimensional (2D) material combines an exceptionally high permeance (~ 800 LMH) with close to full rejection of salts [2] and small organic molecules like urea.
We have succeeded to implement nanometer-thin CNMs as active layers in large-area composite membranes consisting of a polymer support layer with pores in the micrometer-range giving access to the free-standing active

CNM-layer. These composite membranes are mechanically stable and can be mounted in modules. Currently, we can produce sheets with areas of up to 20 x 20 cm^2 in the laboratory. A concept for a pre-industrial pilot production with a capacity of up to several 10.000 m^2 is in place.
We will demonstrate in laboratory experiments applications of our composite membranes as semipermeable membranes in forward osmosis for the cold concentration of watery solutions like juices, milk, beer, wine, urine etc. Another possible application is in the provision of ultrapure water, where the membrane can filter last amounts of salt, heavy metals or small organic molecules from otherwise pure water.

References

[1] Y. Yang, P. Dementyev, N. Biere, D. Emmrich, P. Stohmann, R. Korzetz, X. Zhang, A. Beyer, S. Koch, D. Anselmetti, A. Gölzhäuser, ACS Nano 12, 4695 (2018).
[2] Y. Yang, R. Hillmann, Y. Qi, R. Korzetz, N. Biere, D. Emmrich, M. Westphal, B. Büker, A. Hütten, A. Beyer, D. Anselmetti, A. Gölzhäuser, Adv. Mater. 32, 1907850 (2020).

Poster 34 - Biomimetic carbon nanotube-containing membranes for water purification

Mauricio Dantus (Technion), Yun-Chiao Yao (Lawrence Livermore National Laboratory), Aleksandr Noy (Lawrence Livermore National Laboratory), Viatcheslav Freger (Technion - IIT).

Membrane technology is constantly evolving to address a wide range of industrial needs. Nevertheless, there is still room for improving and expanding this technology. For example, water purification membrane manufacturing has not seen a major change for decades, undergoing only incremental improvements in permeability and selectivity, while removal of certain non-charged species, such as boron, still awaits a breakthrough.
Biomimetic membranes rely on a different approach of membrane design, imitating biological cell membranes. Instead of employing a single material to perform all tasks, Nature allocated separate functions to support (matrix) and selective channels. By following this design, we aim to develop and characterize such a synthetic membrane built of an amphiphilic lipid or block-copolymer matrix and short (~10 nm) carbon nanotube porins (CNTPs) as selective water transport channels. We use a surface-modified anodized aluminum oxide (AAO) membrane with 100-200 nm pores as a support for such biomimetic membrane.
Previous results demonstrated the possibility of incorporating single CNTPs into planar lipid bilayer and multiple CNTPs into liposomes and polymersomes. Our present goal is to upscale this membrane by developing a macroscopic array of pore-spanning lipid or polymer membranes with embedded CNTPs. Following earlier strategies developed in our group, we optimized a layer-by-layer approach to assemble a pore-spanning lipid-like matrix membrane on top of the support. As part of the upscaling process we tested the approach using a patch clamp on single pore chips in an osmotic cell to verify the possibility of forming the lipid layer as well as embedding the CNTPs in it. The nanocomposite membrane was tested in terms of adhesion to the support, integrity, and defect rate. We will present proof-of-concept results assessing water permeability, ion selectivity and electrical properties of the pore-spanning membranes before and after incorporation of CNTPs.

Poster 35 - Preparation of organic solvent and thermal resistant polytriazole membranes with enhanced mechanical properties

Stefan Chisca* (KAUST), Gheorghe Falca (KAUST), Valentina Elena Musteata (King Abdullah University of Science and Technology), Suzana Nunes (King Abdullah University of Science and Technology).

The membrane technology is considered a sustainable process that can partially replace conventional separations, such as evaporation and distillation, in chemical, pharmaceutical, and petrochemical industries. However, most of these industries are highly organic solvent-based, and one of the major challenges is to prepare membranes which are stable in a wide range of pH's and solvents. In addition to solvent and pH resistance, many of OSN applications require membranes stable at temperatures in the range of 60–90 °C or even higher.
In this study, we combine the synthesis of the crosslinked polytriazole membranes by a sustainable phase inversion process with the requirements of OSN applications. In the first step, the polymer was dissolved in an ionic liquid for the membrane casting and immersion in water, followed by exposure to two non-toxic diepoxy crosslinkers with different lengths. The membranes were tested from 25 °C to 105 °C, in strong polar solvents, such as DMF. The

permeance values are in the range of 3.7–10.6 L m−2 h−1bar−1, and the molecular weight cut-off is around 1000 g mol−1. To correlate the physical aging and compaction during filtration we measured the creep recovery. Additionally, we studied the polymer relaxations at high temperatures by using dynamic mechanical analysis.

Poster 36 - Hydrogen-sieving single-layer graphene membranes obtained by crystallographic and morphological optimization of catalytic copper foil

Mojtaba Rezaei* (Ecole polytechnique fédérale de Lausanne), Kumar Varoon Agrawal (Ecole polytechnique fédérale de Lausanne).

Gas separation membranes based on single-layer-graphene are highly attractive because the size of graphene nanopores can be tuned to separate gases by the size-sieving mechanism. A prerequisite for this is synthesis of high-quality polycrystalline single-layer graphene film by chemical vapor deposition (CVD) is crucial. The quality of graphene in the context of membranes is reflected by the size and the density of the intrinsic vacancy defects, and is affected by the catalytic metal substrate and the CVD environment. Generally, expensive high-purity Cu foil is used to obtain gas-sieving performance from single-layer graphene. For the eventual scale-up of graphene membranes, it is highly attractive to use low-cost Cu foils, however, as we show here, these Cu foils are rough and graphene membranes derived from these foils do not yield gas-sieving performance. Herein, we conduct a systematic high-temperature annealing study on two separate, commercial, low-cost Cu foils leading to their transformation to Cu(111). The annealing process smoothened the Cu surface, decreasing the root mean square (RMS) surface roughness from 206 to 99 nm. The RMS roughness on the individual Cu step, measured using the scanning tunneling microscopy (STM), was only 0.23 nm. The oriented Cu grains yielded single-layer graphene with a significantly lower defect density with ID/IG ratio decreasing from 0.18 ± 0.02 to 0.04 ± 0.01. Finally, single-layer graphene films, synthesized on the annealed low-purity Cu foil, yielded H2-selective membranes with H2 permeance reaching 1000 gas permeation units (GPU) in combination with attractive H2/CH4 and H2/C3H8 selectivities of 13 and 26, respectively.

Poster 37 - Nanoporous, Ionic, Lyotropic Liquid Crystal Polymer Membranes: permanent pore modification, characterization of ion-exchange properties, and fabrication of ultrathin films

Michael McGrath* (University of Colorado at Boulder).

Thin polymer membranes made from cross-linked, self-assembled lyotropic liquid crystals (LLC) exhibit uniform, sub-1-nm ionic pores and show promise as membranes in water filtration and ion-selective transport. This work presents a facile method to change the pore environment of a bicontinuous cubic phase LLC membrane by in-situ polymerization of polyelectrolytes within the confined, highly ionic pores. The presence of the polyelectrolyte significantly changes the net charge and ion transport properties of the membranes. The demonstrated pore modification approach paves the way for facile, rational tuning of the LLC pore environment for a particular separation. Secondly, this work develops new techniques to fabricate supported ultrathin (100 nm) LLC polymer membranes that exhibit a high density of hexagonally-packed aqueous sub-1-nm ionic pores that are vertically-aligned in the membrane transport direction. Improvements in thin film fabrication of LLC polymer membranes that contain vertically-aligned, straight-channel pores will allow LLC polymer membranes to achieve both high permeance and high selectivity in nanofiltration.

Poster 38 - Porous, high surface area polymers with controlled pore size <20 nm, via 1-step controlled radical polymerization

Mahati Chintapalli* (PARC).

Membranes from porous polymers with controlled pore size below 20 nm, narrow pore size distribution, and high porosity would enable a range of applications in size- and interaction-selective separations, catalysis, and thermally insulating materials. These new porous polymers could cover a range of pore sizes not accessible by MOFs or polymers of intrinsic microporosity. Because the pore size is well below the wavelength of visible light, the materials are optically transparent, which could enable new applications, such as photocatalytically-active membranes.

We have demonstrated two novel methods based on controlled radical polymerization to produce high porosity polymers with narrow pore size distribution and average pore size below 20 nm. Monoliths, powders, and films were produced using scalable, ambient pressure processing conditions (i.e. without supercritical fluids). We will discuss the effects of polymerization control, monomer rigidity and processing conditions on the pore structure, optical properties, and permeability. Materials with porosities over 60 % and Brunauer Emmett Teller surface areas over 1000 m2/g were synthesized.

Poster 39 - Oxone®-Mediated TEMPO-Oxidized Cellulose Nanomaterials Form I and Form II

John Moore* (University of Arkansas).

The 2,2,6,6-tetramethylpiperidin-1-oxyl (TEMPO) oxidation of cellulose, when mediated with Oxone® (KHSO5), can be performed simply and under mild conditions. Furthermore, the products of the reaction can be isolated into two major components: Oxone®-mediated TEMPO-oxidized cellulose nanomaterials Form I and Form II (OTO-CNM Form I and Form II) [1] . This study focuses on the characterization of the properties of OTO-CNMs. Nanoparticle-sized cellulose fibers of 5 and 16 nm, respectively, were confirmed through electron microscopy. Infrared spectroscopy showed that the most carboxylation presented in Form II. Conductometric titration showed a two-fold increase in carboxylation from Form I (800 mmol/kg) to Form II (1600 mmol/kg). OTO-CNMs showed cellulose crystallinity in the range of 64%–68% and crystallite sizes of 1.4–3.3 nm, as shown through XRD. OTO-CNMs show controlled variability in hydrophilicity with contact angles ranging from 16 to 32°, within or below the 26–47° reported in the literature for TEMPO-oxidized CNMs. Newly discovered OTO-CNM Form II shows enhanced hydrophilic properties as well as unique crystallinity and chemical functionalization in the field of bio-sourced material and nanocomposites.

Poster 40 - 6FDA-based Ionic Polyimides-Ionic Liquid Composites for Advanced Gas Separation

Grayson Dennis* (University of Alabama).

Herein, we report our recent findings on three constitutional isomeric ionic polyimides derived from 4-4'-hexafluorophenyldianhydride (6FDA) using a facile synthetic method. First, 6FDA and 1-(3-aminopropyl)imidazole (API) were reacted via condensation reaction to produce a bisimidazole diimide monomer called as "6FDA-API". Later, this 6FDA-API was polymerized with equimolar amount of three α-α'-dichloro-xylenes (para, meta, or ortho) via the Menshuntkin reaction, forming three novel 6FDA-based ionic polyimide with xylene-linked imidazolium ionic polymer backbone having chloride as the counter anion for each imidazolium cation. This resultant halide 6FDA-basaed ionic polyimide was precipitated into water and lithium bistriflimide to form [6FDA-API(p,m,o)-xylene] [Tf2N]. Further, one molar equivalent of 1-butyl-3-methyl-imiadazolium bistrifimide ionic liquid was added to plasticize the materials by means of improving their gas separation abilities, particularly, CO2 separation over other light gases such as H2, N2, and CH4. Since the para substituted polymer has the highest CO2 permeability of 6.5 barrer, two other ionic liquids, [1-ethyl-3-methyl-imidazolium] [Tf2N] and [1-benzyl-3-methyl-imidazolium] [Tf2N], were used to determine the effects of different ionic liquid substituents on this ionic 6FDA-based polyimide systems to create a full picture of ionenes and their IL composites. All the gas permeabilities were correlated to the physical and thermal properties of all neat and composite materials.

Poster 136 - Single-layer graphene membranes by crack-free transfer for gas separation with sub-angstrom resolution

Shiqi Huang* (Ecole polytechnique fédérale de Lausanne), Jing Zhao (Nanjing Tech University), Kumar Varoon Agrawal (Ecole polytechnique fédérale de Lausanne).

Porous single-layer graphene has been regarded as the ultimate gas separation membrane, capable of yielding ultrahigh gas permeance and an attractive molecular selectivity, due to their atomic thickness.1 However, the attractive performance of atom-thick graphene membranes in the size-selective separation of gas mixtures has not been realized. The development of these membranes face two major bottlenecks: 1) crack-free fabrication of large-area membranes; 2) develop a controllable lattice etching technique mitigating the trade-off between the pore density and the pore-size distribution (PSD).

Herein, we reported several novel approaches to achieve record-high gas separation performance from single-layer graphene-based membrane. Firstly, a novel nanoporous-carbon-assisted graphene transfer technique was developed, enabling the transfer of relatively large area single-layer graphene onto a macroporous support without inducing cracks or tears.2 For the first time, gas-sieving from the intrinsic defects in the chemical vapor deposition derived graphene was observed. Furthermore, a synergistic strategy of decoupling pore-nucleation and pore-expansion on graphene lattice resulted in high-performance single-layer graphene membranes (H2 permeance of 1340 to 6045 gas permeation units (GPU); H2/CH4 separation factor of 15.6 to 25.1).3 Moreover, millisecond graphene gasification was developed to realize high pore-density and narrow PSD nanoporous single-layer graphene, yielding record gas separation performance CO2/N2 selectivity of 24.4 with corresponding CO2 permeance over 9550 gas permeation units.4 Overall, we demonstrate that graphene-based membranes are indeed capable of reaching the predicted high performance in gas separation.

References

(1) Wang. L. et al. Nature Nanotechnology 2017, 12, 509–522.
(2) Huang. S. et al. Nature Communications 2018, 9, 1–11.
(3) Zhao. J. et al. Science Advances 2019, 5, eaav1851.
(4) Huang, S et al. Submitted.

Poster 137 - Centimeter-Scale Gas-Sieving Nanoporous Single-Layer Graphene Membrane

Wan-Chi Lee (Ecole polytechnique fédérale de Lausanne), Kumar Varoon Agrawal (Ecole polytechnique fédérale de Lausanne).

Nanoporous single-layer graphene (NSLG) membranes are promising for molecular separation, capable of yielding large yet selective molecular permeance based on molecular-sieving. However, the scale-up of the gas-sieving NSLG membrane for gas separation remains challenging. Typically, cracks and tears are introduced in the suspended graphene during the membrane fabrication, and dominate the overall gas transport. Herein, we demonstrate the fabrication of centimeter-scale crack-free NSLG membranes for gas separation by mechanically reinforcing the graphene film with a two-layer composite carbon film that prevents crack generation and propagation in the graphene film. The bottom layer of the composite film is a 100-nm-thick block-copolymer film derived nanoporous carbon film with a pore size of 20–30 nm. The top layer is a 500-nm-thick multi-walled carbon nanotube film with a pore size of 200–300 nm. The reinforced graphene film is successfully suspended on a low-cost macroporous nonwoven metal wire mesh with a pore opening of 20 μm on a centimeter-scale while completely avoiding cracks and tears. As a result, H2/CH4 and H2/CO2 selectivities of 11–23 and 5–8, respectively, higher than the corresponding Knudsen selectivities of 2.8 and 4.7, respectively, could be obtained from the centimeter-scale nanoporous graphene membranes. The membranes are mechanically robust and could be successfully pressurized to 5 bar without inducing failure. Overall, the approach reported here helps accelerate the adoption of NSLG membranes for gas separation.

Poster 138 - Synthetic saponite clays as additives for reducing aging effects in PIM-1 membranes

Federico Begni (Università degli Studi del Piemonte Orientale), Geo Paul (Università degli Studi del Piemonte Orientale), Elsa Lasseuguette* (University of Edinburgh), Enzo Mangano (University of Edinburgh), Chiara Bisio (Università degli Studi del Piemonte Orientale), Maria-Chiara Ferrari (U. Edinburgh), Giorgio Gatti (Università degli Studi del Piemonte Orientale).

Polymers of Intrinsic Microporosity represent one of the most promising polymeric materials for gas separation applications. Their very rigid and contorted backbone induces unusually high free volumes and high internal surface area, with high gas permeabilities and moderate ideal selectivity, especially for O2/N2, CO2/N2 pairs with values lying above the Robeson's upper bound. However, the high FFV of PIM1 tends to be shortlived, soon collapsing to leave fewer transport pathways and reduce gas permeability. One way to tackle this problem is the addition of fillers within the polymeric matrix. Here we report the use of synthetic clays named saponites as fillers, to slow down the physical aging of PIM1 membranes. Mixed Matrix Membranes based on two different saponites samples (one completely inorganic and one functionalized with a surfactant) have been obtained and their permeation performances have been studied in the course of one year in order to explore physical aging effects over time. Without filler, PIM1 exhibits the classical aging behaviour of polymers of intrinsic microporosity, namely a progressive decline in gas permeation. In contrary, with saponites, MMMs present a plateau after one week within the aging showing that the fillers slow down the aging of PIM1 membranes in the long term. After one year of aging, the total reduction for CO2 permeability for native PIM1 was 80% whereas for the MMMs it was 49% and 56% for the inorganic and the functionalized saponite respectively. Interactions between the fillers and the polymeric matrix in addition to aging effects have been also monitored through ss-NMR spectroscopy. The 13C spin–lattice relaxation time (T1) measurements reveal that PIM1 chains intercalation between T-O-T lamellar sheets could be one of the mechanisms responsible of PIM1 slowing down aging. Chains confinement between lamellar sheets could play a significant role in reducing chains densification, while maintaining small free volumes.

Poster 139 - Free volume manipulation of a 6FDA-HAB polyimide using a solid-state protection/deprotection strategy

Sharon Lin* (Massachusetts Institute of Technology), Taigyu Joo (Massachusetts Institute of Technology), Francesco Maria Benedetti (Massachusetts Institute of Technology), Laura Chen (Massachusetts Institute of Technology), Albert Wu (MIT), Katherine Mizrahi Rodriguez (MIT), Qihui Qian (Massachusetts Institute of Technology), Cara Doherty (CSIRO), Zach Smith (MIT).

Polymer membranes have shown great promise for applications in gas separations as they are both energy-efficient and easily processable. In order to be suitable for the gas separation industry, membrane materials should be solution-processable, as well as highly permeable and selective. It has been shown in recent years that polymers with rigid backbones and contorted structures possess both high permeability and selectivity due to their inefficient chain packing that leads to the formation of larger free volume elements, allowing gas molecules to pass through more easily. However, free volume elements are formed via a “bottom-up” method in which careful polymer chemistry design is required, which can lead to limited opportunities to rationally and selectively manipulate free volume without changing polymer chemistry.

In this study, we report a free volume modification (FVM) method through the use of solid-state deprotection chemistries. Tert-butoxycarbonyl (t-BOC), a common chemical protecting group, was appended onto a polyimide consisting of 2,2'-bis-(3,4-dicarboxyphenyl) hexafluoropropane dianhydride (6FDA) and 3,3'-dihydroxy-4,4'-diamino-biphenyl (HAB). This polymer (6FDA-HAB-t-BOC) was formed into a self-standing film, after which, different thermal treatments were performed to selectively remove t-BOC. However, despite performing thermal treatments well below the glass transition temperature of 6FDA-HAB (~300 °C), this particular FVM approach only produced subtle changes relative to the unprotected polyimide in terms of polymer density, fractional free volume, average free volume element size, and gas transport properties. While these findings suggest that thermal deprotection of functional groups in glassy polymer films can be used to selectively manipulate free volume and gas transport performance, more robust systems than linear polyimides are required to preserve free volume elements that are generated via this FVM method.

Poster 140 - Facile and time-efficient carboxylic acid functionalization of PIM-1: effect on molecular packing and gas separation performance

Katherine Mizrahi Rodriguez* (MIT), Albert Wu (MIT), Qihui Qian (Massachusetts Institute of Technology), Gang Han (MIT), Sharon Lin (Massachusetts Institute of Technology), Hyunhee Lee (MIT), Won Seok Chi (Chonnam National University), Francesco Maria Benedetti (Massachusetts Institute of Technology), Cara Doherty (CSIRO), Zach Smith (MIT).

Current industrial gas separations, such as CO2 removal from natural gas, rely primarily on energy-intensive and environmentally unfriendly processes. Polymer membranes offer a promising alternative due to their potentially lower energy costs and ease of operation, but are infrequently deployed because of performance limitations of currently available polymers. In this work, the influence of backbone chemistry and polymer packing structure on gas transport performance was investigated for a high-performance polymer of intrinsic microporosity, PIM-1. A facile and time-efficient PIM-1 hydrolysis method was developed, yielding solution-processable PIM-COOH with greater than 89% conversion in 48 h. Due to densification of PIM-COOH in comparison to PIM-1, permeability decreases while selectivity simultaneously increases by more than 100% for CO2-related and H2-related gas pairs. Transport trends are mechanistically evaluated through a suite of materials characterization tests including surface area measurements, wide-angle X-ray scattering, and positron annihilation lifetime spectroscopy (PALS). Additionally, a deconvolution of sorption and diffusion is presented to assess how changes in backbone chemistry and packing structure influence transport behavior.

Poster 141 - Strategy for acquiring high CO2 permeance performance for dense ceramic-carbonates membranes by total conductivity modification

Oscar Ovalle* (Arizona State University), Jerry Lin (Arizona State University).

In the last decade, ceramic-carbonate dual-phase (CCDP) membranes have been studied for CO2 separation at high temperatures. To contribute a new approach for CCDP comprehension, the present study is focused on the effects of membrane microstructure on CO2 permeance. For that purpose, the microstructure of the samarium doped ceria (SDC)-carbonates membrane was modified using the sacrificial template method using NiO and graphite as sacrificed phases. The supports uniaxially pressed were sintered over 1100°C, the graphite was removed in-situ, and the NiO was removed using H2 and HNO3. Finally, the disks were infiltrated with a eutectic mixture of Li/Na/K2CO3. Three different membranes were synthesized with different microstructure: pure-SDC (p-S), graphite-templated-SDC (gt-S), and NiO-templated-SDC (Nt-S). SDC supports prepared by different methods have different pores volume fraction to tortuosity ratio. Because of this, different carbonate conductivity was observed for SDC-MC membranes prepared by different methods. Also, dense and porous supports conductivity was measured to calculate the solid volume fraction to tortuosity ratio. The performance of the membrane is limited by this parameter determined by the microstructure of the solid phase. The total conductivity of membranes at 750°C was 0.024, 0.0089, and 0.0052 S/cm for p-S, gt-S, and Nt-S, respectively. p-S presents the highest conductivity because it has the highest solid geometrical factor. CO2 permeance experiments were performed for CCDP membranes between 700-750 °C, with a feed gas of CO2 and N2 (50/50 vol%), and helium as the sweep gas. At 750 °C CO2 permeation flux is observed 0.44, 0.32, 0.30 ml/cm^2min for p-S, gt-S and Nt-S, respectively. These data are consistent with the total conductivity results. The SDC-MC membrane total conductivity can be modified by varying the microstructure. The CO2 permeance can be correlated to the total conductivity data.

Poster 149 - Synthesis and in situ functionalization of microfiltration membranes via high internal phase emulsion templating

Ryan Zowada (New Mexico State Univeristy), Muchu Zhou* (New Mexico State University), Reza Foudazi (New Mexico State University), Anna Malakian (Clemson University).

Typical membrane production requires solvents to induce phase separation that results in open channels for fluid flow. However, these solvents are typically harmful to the environment, and can be costly as the porosity of the membrane increases the amount of solvent required increase. In this work, membranes are produced using high internal phase emulsion (HIPE) templating, which do not require organic solvents. HIPEs are emulsions where the internal phase makes up more than 74% of the total emulsion volume. The benefits of using a water-in-oil HIPE templating system is to use water instead of organic solvents, and the high internal volume results in an interconnected porous structure. The aim of this work is to utilize HIPE templating as an effective and eco-friendly means of membrane production, and to utilize functionalization techniques of HIPE templating in order to enhance membrane performance. The hydrophobic membrane was functionalizing so that its surface was hydrophilic in order to increase permeability. The functionalization of the membrane surface was studied using conductometric titration and Fourier transform IR spectroscopy. Our study shows HIPE templated microfiltration membranes have both higher intrinsic permeability and higher rejection efficiency than conventional microfiltration membranes.

Abstracts – Poster Presentations, Tuesday

Presenting authors are indicated by an asterisk*. See the most up-to-date version of the program book with full abstracts at **www.membranes.org/nams-2020/**

Poster Session 2: Processes

Poster 41 - Polyelectrolyte complexation induced Aqueous Phase Separation for the next generation of sustainable membranes

Muhammad Irshad Baig* (University of Twente), Elif Nur Durmaz (University of Twente), Joshua D. Willott (University of Twente), Wiebe M. de Vos (University of Twente).

Polymeric membranes are typically prepared using a technique called Non-solvent induced phase separation (NIPS). Unfortunately, this technique relies on large amounts of aprotic organic solvents such as N-methyl-pyrrolidone (NMP). NMP is reprotoxic and unsustainable and recently, its use has been restricted within the European Union. Here, we report a new and sustainable approach to produce polymeric membranes. The new Aqueous Phase Separation (APS) technique relies on polyelectrolyte complexation for the phase inversion. A homogeneous solution of a strong polyanion, poly(sodium 4-styrenesulfonate) (PSS), and a weak polycation poly(allylamine hydrochloride) (PAH), is prepared at pH ~14. At this pH, PAH is uncharged and hence, cannot complex with PSS. The homogeneous solution is then cast as a thin film and immersed in a pH ~ 1 coagulation bath. At this low pH, PAH becomes charged and complexes with PSS to form a water-insoluble polyelectrolyte complex film, a membrane. The structure and morphology of the resulting membranes is tuned by varying the polyelectrolyte solution concentrations, molecular weights of the polyelectrolytes, and the salinity of the coagulation bath. This allows the production of microfiltration membranes that can remove dispersed oil from waste-water, ultrafiltration membranes that are able to concentrate dilute protein solutions, and nanofiltration membranes that effectively remove small organic molecules from water streams. The new APS technique also provides a new type of separation where it is important to retain organics without retaining salts. With the demonstrated control over pore size and membrane structure that allows the production of membranes with excellent separation properties, the completely water-based APS technique is truly an alternative to the now dominant solvent-based NIPS approach. Furthermore, different types of polyelectrolytes could be used for this approach to further optimize the process.

Poster 42 - Impact of Post-Synthetic Modification Routes on Filler Structure and Performance in Metal–Organic Framework Based Mixed-matrix Membranes

Qihui Qian* (Massachusetts Institute of Technology).

This work systematically investigates an exemplary mixed-matrix membrane (MMM) system comprised of a UiO-66-NH2 metal–organic framework (MOF) and a 6FDA-Durene polyimide. The UiO-66-NH2 MOF is post-synthetically modified with 6FDA-Durene oligomers through either chemical or thermal imidization routes. The thermal(T)-PSM approach does not require reactive small molecules for synthesis, thereby allowing the cross-linking reaction between the MOF and the polymer to occur only on the MOF surface. Conversely, the chemical(C)-PSM approach requires reactive small molecules for synthesis, thereby altering the internal amine functionality of UiO-66-NH2. Upon embedding these PSM-MOFs in high molecular weight 6FDA-Durene polymers, CO2 and CH4 pure-gas permeation tests reveal that the T-PSM-MOF has synergistic improvements in both permeability and selectivity due to enhanced interfacial compatibility without significant modification to the internal MOF framework. Conversely, MMMs formed with the C-PSM-MOF experienced a significant decrease in permeability due to reactions that occurred inside of the MOF pore structure. These findings suggest that there is a critical need to consider reaction conditions and their corresponding effect on MOF structure when pursuing a PSM approach to form MMMs with desired performance enhancements.

Poster 43 - Nanofiltration Membranes by Salinity Gradient Induced Aqueous Phase Separation

Elif Nur Durmaz* (University of Twente), Muhammad Irshad Baig (University of Twente), Joshua D. Willott (University of Twente), Wiebe M. de Vos (University of Twente).

Current phase separation techniques for the production of polymeric membranes mostly involve the use of toxic solvents. Many alternatives for the solvents were investigated, however, a recent breakthrough has found that membranes can also be produced through an aqueous phase separation (APS) approach, utilizing a pH gradient [1]. A downside of this is that quite extreme pH values are needed. Here, we propose an alternative that requires much milder conditions such as a salinity gradient. Two oppositely charged polyelectrolytes (PEs) are mixed at a high salinity where the PEs cannot complex. Immersion of the mixture in a low salinity bath results in the formation of a water-insoluble polyelectrolyte complex (PEC). The resultant PEC membranes are in the nanofiltration (NF) range with <300 Da of MWCO and ≤1 LMHbar permeability. First, the effects of molecular weight (MW) and total PE concentration were investigated. The optimum casting solution was selected based on membrane water permeability and the simplicity of the membrane formation procedure. Then, the effect of coagulation bath salinity was investigated. While the MW and the polymer concentration do not affect the membrane structure significantly, the coagulation bath salinity affects both the membrane structure and performance. High salinity coagulation baths led to thicker skin layers, lower permeabilities and higher MgSO4 retentions (up to 80%). Besides the separation performance, the stability of the membranes under high pressures and after long-term tests confirmed that these membranes are suitable for NF applications.

References

[1] Baig M.I., Durmaz E.N., Willott J.D., de Vos W.M., Adv. Funct. Mater., 2019, 1907344.

Poster 44 - Co-permeation behavior of methanol and acetate in polyether-based cation exchange membranes

Jung Min Kim* (Auburn University), Bryan Beckingham (Auburn University).

The permeation of multiple solutes in ion exchange membranes is crucial for innovative energy conversion applications including photoelectrochemical CO2 reduction cells, which reduces CO2 to valuable chemicals such as methanol and acetate. One of the primary concerns with such devices is the lack of ion exchange membranes with the requisite conductivity but limited crossover of CO2 reduction products. Towards developing such membranes, understanding the permeation, and co-permeation behavior of these products in ion exchange membranes is critical. In previous work, co-permeation behavior of methanol and acetate across the cation exchange membrane Nafion® 117 was investigated and differences in transport behavior relative to their single solute permeation behavior were observed. Here, to further understand this emergent transport behavior, we investigate a series of polyether-based membranes of varied charge density. Model membranes are synthesized by UV photopolymerization with varying poly(ethylene glycol) diacrylate to 2-acrylamido-2-methylpropanesulfonic acid content and characterized by measuring water uptake and in-plane ionic conductivity. Diffusive permeabilities of these membranes to methanol, acetate, and the mixture of two were measured by a diffusion cell coupled with in situ attenuated total reflectance Fourier transform infrared spectroscopy to monitor the time-resolved concentration of each permeant in the downstream receiver cell. We find the inclusion of AMPS in these membranes has a distinct increase in permeability to acetate in the presence of co-permeating methanol. This behavior is attributed to the screening of electrostatic interactions (Donnan exclusion), in which methanol interferes with the electrostatic repulsion between membrane-bound sulfonate anions and transporting acetate anions.

Poster 45 - Radio frequency induction heated membranes in vacuum membrane distillation

Arezou Anvari* (Temple University), Avner Ronen (Temple University).

Recently, there is a growing interest in membrane distillation (MD) technology, a thermally driven membrane separation, to treat high salinity water or contaminated wastewater. However, MD commercialization is still limited by technical challenges such as temperature polarization (TP), scaling, and high-energy consumption, as the feed solution requires continuous heating to provide an efficient driving force. We suggest and assess the ability and efficiency of self-heating thermally conducting membranes heated by radio frequency induction heating, to overcome distillation limitations. A composite membrane containing iron oxide coated carbon nanotubes was spray-coated on a PTFE membrane and heated by induction heating, using radio frequency (RF) altering magnetic fields. The performance of RF-MD system was evaluated in terms of distillate flux and specific heating energy consumption at optimized operating conditions and results were compared to a conventional MD system. In addition, the impact of RF heating on calcium sulfate (CaSO4) scaling was addressed in terms of distillate flux and crystal formation. Results show the ability of heating water directly on the membrane surface in RF-MD systems, leading to low TP, high distillate flux, and low specific heating energy compared to a conventional MD system. Scaling results showed the impact of RF magnetic field on salts crystallization, leading to smaller crystal size and overall less scaling. Following analysis of the RF-MD membrane surfaces, only sporadic small CaSO4 crystals were detected, while high concentrations of small crystals were detected at the concentrate stream exiting the MD process. The scaling mitigation is hypothesized to be a result of the high-frequency movement and collision of the ions in the solution. These results are promising as they show the RF-MD system has potential to improve MD processes, specifically for high salinity distillation where pressure-based applications cannot be used.

Poster 46 - Using Inorganic Salts in Forward Osmosis to Remove Water from High Osmotic Pressure Mineral Processing Effluents

Noel Devaere* (University of Toronto), Vladimiros Papangelakis (University of Toronto).

Reverse Osmosis (RO) is the technology of choice for extracting pure water at a low cost from high flow, low concentration, ionic salt solutions. However, as the concentration of ionic compounds increases, the osmotic pressure of the solution becomes more elevated; consequently, the operating pressure required for the RO process must be increased to compensate for this. Once the feed goes beyond the osmotic pressure of seawater (0.5 M NaCl), the RO process starts to encounter problems with operating pressures, scaling and membrane fluxes.
The osmotic pressure of certain industrial mineral/mining processing effluents could reach values as high as a 2.5 M NaCl equivalent. Forward Osmosis (FO) offers an alternative approach wherein an engineered high osmotic pressure salt solution is used in order to spontaneously draw water across an asymmetrical polymeric or hollow fiber membrane. We are developing a patented process by which common inorganic salt draw solutions are employed which are regenerated by freezing, wherein clean water is recovered as ice. With this approach, the osmotic pressure ceiling for membrane-based water removal technologies can be increased. In this work, we evaluate the performance of one such forward osmosis process on real industrial mineral processing effluents as well as synthetic NaCl solutions simulating industrial effluents. Comparative results are presented and key challenges related to the nature of the membrane, pH, fouling and draw solution selection are discussed.

Poster 47 - Radio Frequency Induction Heated Membranes in Vacuum Membrane Distillation

Arezou Anvari* (Temple University), Avner Ronen (Temple University).

Recently, there is a growing interest in membrane distillation (MD) technology, a thermally driven membrane separation, to treat high salinity water or contaminated wastewater. However, MD commercialization is still limited by technical challenges such as temperature polarization (TP), scaling, and high-energy consumption, as the feed solution requires continuous heating to provide an efficient driving force. We suggest and assess the ability and efficiency of self-heating thermally conducting membranes heated by radio frequency induction heating, to overcome distillation limitations. A composite membrane containing iron oxide coated carbon nanotubes was spray-coated on a PTFE membrane and heated by induction heating, using radio frequency (RF) altering magnetic fields. The performance of RF-MD system was evaluated in terms of distillate flux and specific heating energy consumption at optimized operating conditions and results were compared to a conventional MD system. In addition, the impact of RF heating on calcium sulfate (CaSO4) scaling was addressed in terms of distillate flux and crystal formation. Results show the ability of heating water directly on the membrane surface in RF-MD systems, leading to low TP, high distillate flux, and low specific heating energy compared to a conventional MD system. Scaling results showed the impact of RF magnetic field on salts crystallization, leading to smaller crystal size and overall less scaling. Following analysis of the RF-MD membrane surfaces, only sporadic small CaSO4 crystals were detected, while high concentrations of small crystals were detected at the concentrate stream exiting the MD process. The scaling mitigation is hypothesized to be a result of the high-frequency movement and collision of the ions in the solution.

These results are promising as they show the RF-MD system has potential to improve MD processes, specifically for high salinity distillation where pressure-based applications cannot be used.

Poster 48 - Membrane Processes for CO2 Removal and H2 Reuse for Solid Oxide Fuel Cells: Process Design and Techno-Economic Analysis

Kai Chen* (The Ohio State University), Yang Han (The Ohio State University), Mike Gasda (Bloom Energy Corporation), Winston Ho (The Ohio State University).

In order to enhance fuel utilization in solid oxide fuel cells (SOFCs), part of the anode exhaust may be recycled to join the anode inlet stream. However, the CO2 present in the exhaust would dilute the fuel concentration and lower the efficiency of the SOFCs. Few papers have considered removing CO2 from the recycle stream, and none have reported on using a membrane process for this task. A CO2-selective facilitated transport membrane was developed with fluoride- and hydroxide-containing species for the CO2/H2 separation, and the membrane exhibited a good stability, showing a CO2 permeance of 108 GPU and a CO2/H2 selectivity of 106 at 120°C. Moreover, the membrane was scaled up to 21" in width, which had consistent transport results as the lab-scale samples.
Herein, a process analysis was done on using the developed membrane for an SOFC system. First, due to its wide availability, air was used as the sweep gas. Furthermore, a vacuum was proposed for the permeate side to further enhance the driving force and allow for CO2 capture. Also, as oxygen is absent, amine-containing membranes with better transport performances could be used for the vacuum process at 120°C. At 99% H2 recovery, the air sweep process was estimated to remove 44.7% of CO2 from the recycled anode exhaust, whereas the vacuum process could boost the removal to 76.9%. A less stringent H2 recovery requirement could yield a higher CO2 removal. For instance, a CO2 removal of over 90% could be achieved when the H2 recovery was relaxed to 96.7% for the vacuum operation. In addition, a high-level techno-economic analysis was conducted, without including the H2 reused value. The CO2 removal costs corresponding to the air sweep and vacuum processes were $62.9/tonne and $53.3/tonne, respectively, even with the added cost of the vacuum pump for the amine membrane process. The cost is even lower at $35.4/tonne if the captured CO2 is utilized for enhanced oil recovery.

Poster 49 - Conducting thermal energy to the membrane/water interface for the enhanced desalination of hypersaline brines using membrane distillation

Jingbo Wang* (University of California, Los Angeles), Yiming Liu (University of California, Los Angeles), Unnati Rao (UCLA), Navid Ebrahimi (University of California, Los Angeles), Eric Hoek (UCLA), Tzahi Cath (Colorado School of Mines), Nils Tilton (Colorado School of Mines), Craig Turchi (NREL), Yongho Sungtaek Ju (University of California, Los Angeles), David Jassby (UCLA).

Membrane distillation (MD) is a membrane-based thermal desalination process capable of desalinating hypersaline brines. Standard MD systems rely on the feed stream as the thermal energy carrier to drive the desalination process. However, the reliance on the feed to carry thermal energy has limitations including thermal polarization and feed stream temperature drop (and subsequent driving force drop). Providing thermal energy to the membrane/water interface has the potential of eliminating thermal polarization, reduce limitations on single-pass recoveries, and decrease the number of heat exchangers in the system. Recent developments in solar collection technologies enabled solar desalination due to its availability and low cost, so long as this heat can be delivered to where it's needed. We developed an MD process using advanced layered composite membrane materials that are thermally and electrically conducting. The integrated polymeric and metallic layers address needs for high-salinity desalination, such as the ability to conduct heat from an outside source to the membrane/water interface and drive electrochemical anti-scaling reactions. The composite membranes were tested in a vacuum MD configuration using feed water containing 100g/L NaCl. The new MD system showed stable performance with water flux up to 9 LMH, and salt rejection >99.9% over hours of desalinating hypersaline feed, with the only heat provided through the thermally conducting membrane layers. The system exhibited >75% energy efficiency in a single pass. System configuration, vacuum levels and crossflow velocities all contribute to the overall system performance. Through capturing solar heat outside of the membrane module and delivering it directly to the membrane/water interface, this process has the potential of solving several problems associated with thermally-driven desalination processes, such as minimizing heat losses, increasing thermal efficiency, and limiting the number of heat exchangers..

Poster 50 - Membrane Surface Characterization Upon Cleaning of a Heavily Fouled Reverse Osmosis Membrane from an Advanced Reclamation Facility

Bilal Abada* (Texas A&M University), Shankar Chellam (Texas A&M University).

Reverse osmosis (RO) membrane from the third stage of a three RO-pass train in an advanced water purification facility (Orange County, CA) was thoroughly examined for foulants identification and chemical cleaning assessment. Silicon-based foulants were the most dominant foulants on the membrane surface and existed in different forms like silica, silicate, and organosilicons. Biological fouling was observed to a lesser extent as compared to silicon foulants. Minor contributions of aluminum and calcium scales accompanied the dominant combined silicon-biological foulants. Membranes were cleaned with either NaOH or a mixture of sodium tripolyphosphate and sodium dodecylbenzene sulfonate.
Neither of the chosen agents successfully recovered the declined flux with a single cleaning cycle. Due to silicon-based fouling dominance, multiple NaOH cleaning cycles were attempted. Ten cleaning cycles effectively restored the flux, with no improvement in salt rejection. Fourier transform infrared (FTIR) spectroscopy showed evidence of near-complete removal of biological fouling. X-ray photoelectron spectroscopy (XPS) scans indicated substantial removal of Ca scales but only negligible removal of Si and Al from the membrane surface. Therefore, this partial removal of foulants did not explain complete flux restoration.
Interestingly, virgin membrane cleaning within the same cleaning protocols also showed a flux increase.
Both XPS and FTIR spectra did not reveal chemical alteration to the membrane active layer. Thus, flux increase could be linked to membrane swelling, rather than a deterioration of the polyamide layer. Hence, we propose that the efficient flux restoration in fouled membranes was due to a combination of partial removal of foulants and swelling of the membranes. These results underline the importance of conducting an in-depth analysis of cleaned membranes that is beyond only flux and salt rejection measurements to assess foulants removal and polyamide layer integrity.

Poster 51 - Biomimetic Analysis of crown ethers as ion channels for rapid and selective Li+ transport

Chenhao Yao* (University of Texas at Austin), Manish Kumar (University of Texas at Austin), Nathaniel Lynd (The University of Texas at Austin).

Lithium recovery from brackish and sea water has drawn great attention in the recent years, as the global demand for lithium is growing rapidly. Ion separation membranes are being proposed to recover lithium from waste water; however, the common processes under study focus solely on either one of the separation parameter: permeability (ion-exchange membranes) or selectivity (membrane absorption), which leaves the question: can we optimize the selectivity and the permeability at the same time? In this study, we learn about the selective interaction between crown ethers (CEs) and alkaline metal ions, and evaluate a 1D hopping model to describe the ion transport property when they form ion channels in a lipid bilayer membrane. Based on the model, we demonstrate the effect of thermodynamic and kinetic properties of CE-ion interaction (equilibrium constant, rate of complex formation), as well as the operating condition (feed concentration), on the single channel ion permeability and selectivity of an ion transport membrane. Among these parameters, we centered our study on the equilibrium constant Kd, because it plays an important role in selective ion transport and can be controlled by modifying CE structure. We measure Kd values of different CE-ion pairs, and compare the permeability/selectivity of each CE ion channel using the 1D hopping model. This study suggests a reliable transport model which can be used in future studies, and shows the possibility of using crown ethers to form selective ion channels in ion transport membranes.

Poster 52 - Multiphysics Simulation of CO2 Removal from Seawater using Hollow Fiber Membrane Contactors

Joanna Rivero (University of Pittsburgh).

Capturing CO2 from the ocean is imperative to protect coral reefs and shellfisheries from the damaging effects of ocean acidification. Very few researchers have attempted to strip CO2 from seawater, and our team is the first to attempt this using a passive membrane device. The purpose of this work is to simulate CO2 removal from seawater in a hollow fiber membrane contactor using sodium hydroxide (NaOH) as the solvent. NaOH has been extensively studied and used for CO2 separation from flue gas, and is a cheap solvent that is harmless to marine life. Our model has been developed in COMSOL Multiphysics 5.4 and represents a 2D-axisymmetric single hollow fiber. This model includes detailed chemical species on the seawater side, and reaction kinetics for CO2 absorption into NaOH based on literature values. A polypropylene membrane is modeled with an overall 1-inch diameter by 5.5-inch length based on parameters provided by 3M. The model will be validated against experimental results collected on a 3M hollow fiber membrane bundle in our lab. Parametric studies will also be performed to determine the impact of flow-rates, temperature, and membrane design on CO2 removal rate. Future work will include modeling facilitated transport hollow fiber membranes to boost the rate of CO2 removal from seawater, which is a low-CO2 source. In particular, carbonic anhydrase and amine group coatings will be modeled on the seawater side of the membrane. This work will be complemented by experimental testing with our collaborator Dr. William Federspiel from the Bioengineering department at the University of Pittsburgh. Finally, a 3-D hollow fiber bundle model will be developed to study flow patterns and species distributions in the bundle.

Poster 53 - Reactive membranes to prevent fouling by generating in situ microbubbles

Weiliang Bai* (University of Texas at Austin), Manish Kumar (University of Texas at Austin), Navid Saleh (Univeristy of Texas at Austin).

Membrane filtration is an effective technology due to the high removal efficiency and mild working conditions, and is thus widely used in water treatment. However, the removal efficiency and flux drops with time because of fouling by organics. This is especially true in nanofiltration and reverse osmosis systems. Hydrogen peroxide which has relatively low price has been employed as an oxidant in drinking water treatment with AOP process. In this project, following our group's previous research, we produced NF and RO membrane coating with polydopamine as well as different types of metal and metal oxide nanoparticles including copper oxide, manganese dioxide, zerovalent iron and silver as the catalysts. Polydopamine was used to increase membrane surface hydrophilicity and also form a cohesive layer to enable further surface metal catalysts coating. Periodically added hydrogen peroxide could be catalytical decomposed by metals to generate oxygen microbubbles on the surface of the synthesized membrane, which could prevent organic fouling deposited on the membrane. Polydopamine layer between metal oxide and membrane surface helped protect membrane from damage by microbubbles and reactive radicals, which enabled long term use in membrane filtration. This research provided a comprehensive evaluation for the application of different metal catalysts antifouling membrane on RO and NF system and provides feasibility assessment for different kinds of organic fouling. Oil is a common contaminants in several wastewater streams from oil and gas industry. It is a problematic foulant for membranes because of its membrane adhesion properties and its difficulty of removal by pretreatment methods. Results will also presented on the efficacy of these catalytic membranes for preventing fouling of water by oily streams.

Poster 54 - The Influence of Ion Association on Ion Solubility in Ion Exchange Membranes

Rahul Sujanani* (The University of Texas at Austin), Oscar Nordness (The University of Texas at Austin), Joshua Saunders (The University of Texas at Austin), Joan Brennecke (The University of Texas at Austin), Lynn Katz (The University of Texas at Austin), Benny Freeman (The University of Texas at Austin).

Ion exchange membranes (IEMs) are widely used in water purification and energy generation systems because the fixed charge moieties on the polymer backbone allow for selective permeation of ionic species. However, a detailed understanding of the impact of polymer structure on ion solubility and diffusivity in IEMs for various electrolytes remains largely incomplete. In prior studies, we developed a thermodynamic modeling framework (i.e., Manning/Donnan model) to predict ion solubility in IEMs. Results for NaCl and MgCl2 solubility in IEMs showed that our framework describes this data remarkably well. However, we recently discovered that our model is unable to predict Na2SO4 solubility in cation exchange membranes (CEMs). Na2SO4 was expected to be less soluble than NaCl, because mobile ions of higher valence are subject to a higher Donnan potential that excludes them more effectively from IEMs. Na2SO4 uptake was higher than expected and was, instead, in quantitative agreement with monovalent salt (i.e., NaCl) uptake. We attribute this to ion association as Na2SO4 forms significant amounts of monovalent [NaSO4]- ion pairs in aqueous solution, that can contribute to enhanced sorption in a CEM. Additionally, we measured the solubility of $MgSO_4$ and results are available for several commercial IEMs. Remarkably, the solubility behavior of $MgSO_4$ in IEMs is similar to that of an electroneutral molecule (i.e., CH_3OH). The extent of ion association for Na_2SO_4 and $MgSO_4$ in aqueous solutions was measured and found to be as high as 50% at higher salt concentrations. These findings and their impact on ion separation using IEMs will be discussed.

Poster 55 - Enhanced water evaporation through graphene nanopores via the Kelvin effect

Yuhang Fang* (Purdue University), Prabudhya Roy Chowdhury (Purdue University), Xiulin Ruan (Purdue University), David Warsinger (Purdue University).

An understanding of the evaporation rate at liquid–vapor interfaces confined by nanopores is significant in optimizing phase-change-related applications. The Kelvin equation theoretically associates a higher vapor pressure to convex, rather than flat, liquid surfaces. To validate the Kelvin effect at the nanoscale, molecular dynamics (MD) methods are used to simulate the evaporation of pressurized water from graphene nanopores (0.6–6 nm). The results suggest that the molecular evaporation rate is inversely proportional to the radius of the liquid-vapor surface. The curved surface geometry reduces the free energy barrier associated with the release of water molecules across the liquid–vapor interface. Our findings demonstrate the possibility of using nanoporous membranes to facilitate evaporation-related applications in heat transfer and energy conversion.

Poster 56 - Batch counterflow reverse osmosis for energy efficient desalination at high recoveries

Abhimanyu Das* (Purdue University), David Warsinger (Purdue University).

Counterflow reverse osmosis (CFRO) is a new RO configuration capable of handling high salinities and high recoveries. Unlike regular RO, in CFRO both sides of the membrane are saline and in counterflow, which lowers the osmotic pressure difference across the membrane. The process produces a concentrated brine stream and a diluted stream which is then desalinated using conventional RO. In this study, we develop a fully transient, system-level mathematical model for RO and extend it to study a novel configuration termed as Batch CFRO. We hypothesize that, by individually recirculating the concentrate and diluate streams of a CFRO system in a transient process, the concentration profiles on either side of the membrane can be flattened out. This would avoid over-pressurizing and maintain a uniform flux across the membrane, bringing down the specific energy consumption. Moreover, for a significant portion of the cycle or batch, permeation can occur at lower pressures not possible with its steady-state variant. It is, however, apparent that for the same membrane flux and recovery, such a system will require higher applied pressure towards the end of the batch than steady-state CFRO. The net energy consumption for a batch can be expressed as a sum of gradually increasing pressurization work and a significant circulation pump work. For the test cases studied, our model shows a 7% reduction in specific energy consumption as compared to steady-state CFRO, which comes at the cost of higher maximum operating pressure.

Poster 57 - The Effect of Microplastic Shape, Size, and Concentration on Membrane Performance in Municipal Wastewater Treatment

Ryan LaRue* (McMaster University), Blake Patterson (McMaster University), Todd Hoare (McMaster University), David Latulippe (McMaster University).

The presence of anthropogenic plastic waste in aquatic environments is widespread and poses significant environmental, human, and economic consequences. Microplastics (MP) are of particular concern due to their small size, highly irregular shapes, and potential ecotoxicity. It is estimated that trillions of MPs are discharged from American wastewater treatment plants (WWTPs) daily. The membrane technologies used in some WWTPs for tertiary wastewater polishing are well-positioned to mitigate this efflux of MPs, however little is known about how the MPs affect the performance of these membranes. Previous filtration studies are limited to regularly-shaped particles in well-defined solutions, and no works exist that document MP fouling in WWT. Here, we present findings using MP particles (MPPs) and fibers (MPFs) suspended in wastewater—representative of real WWTP effluents. It is a challenge to directly obtain MPs with non-spherical shapes, so MPPs were created by processing polyethylene powder via a planetary ball mill that was then sieved to yield non-spherical MPPs (~14 μm). In parallel, polystyrene was electrospun into an aligned mat of fibers and cut using a cryostat to yield MPFs with a diameter on the order of microns and lengths averaging 70 μm. PVDF UF and MF flat-sheet membranes (0.03 and 0.2 μm pore size, respectively) were used to filter the suspensions of MPs in wastewater. Flux decline experiments were conducted at constant pressure to study the effect of MP properties (e.g. shape, concentration) on the fouling propensity of the membranes. While both the MF and UF membranes experience more significant flux decline over time in the presence of MPs, this decline is more pronounced for the MF membrane. For example, the mass of permeate water collected over 24 hours declined by 15% for UF and 33% for MF membranes at a given concentration of MPPs. Particle shape (e.g. MPP vs. MPF) and concentration plays a significant role in severity of membrane fouling.

Poster 58 - Does modifying commercial reverse osmosis membranes with solvent pre-treatment and additional polymerization enhance water permeability and salt rejection?

Mikayla Armstrong* (University of North Carolina at Chapel Hill), Orlando Coronell (University of North Carolina at Chapel Hill).

An important challenge in membrane technology is to advance the theoretical trade-off between water permeability and salt rejection. Solvent pre-treatment or activation has been shown to enhance the water permeability of thin-film composite membranes, but at the expense of salt rejection. We hypothesize that solvent pre-treatment followed by additional polymerization may result in a membrane that exhibits enhanced performance in terms of both water permeability and salt rejection. In our work, commercial membranes were subjected to three sequential modification steps: 1) pre-treatment with organic solvent, 2) exposure to trimesoyl chloride (TMC) in organic phase, and 3) exposure to m-phenylenediamine (MPD) in water. The solvent pre-treatment swells the active layer, while the exposure to TMC impregnates the active layer with TMC, facilitating additional polymerization to occur within the active layer. We are evaluating solvent type used in the solvent pre-treatment, polymerization time, and monomer concentrations. Initial results show that modified membranes exhibited higher water permeance and lower salt rejection (by <4.5 percentage points) than the unmodified commercial membrane. Data shows that exposing the membrane to solvent has a larger effect on membrane performance than polymerization reaction time, where the latter caused slightly lower water flux and slightly better rejection. In general, the membranes modified with a higher TMC concentration exhibited better rejection than the membranes modified with lower TMC concentration. Future experiments will focus on fine tuning the exposure and reaction parameters to amplify the effects of polymerization and reduce the effects of solvent exposure on membrane performance. Our work builds on the concept of solvent pre-treatment with the aim to develop membranes with improved water permeability and salt rejection.

Poster 59 - Thin-film composite gas separation membranes prepared by interfacial polymerization of macrocyclic molecules

Woochul Song* (The University of Texas at Austin), Jaesung Park (The University of Texas at Austin), Benny Freeman (The University of Texas at Austin), Manish Kumar (The University of Texas at Austin).

Interfacial polymerization is facile chemistry that provides industrial scale production of thin-film composite (TFC) membranes. However, tuning the microstructure of these TFC membranes have been challenging due to lack of knowledge that bridge chemistry-to-microstructure relationship. Here, we propose an approach to integrate macrocyclic molecules into TFC membranes to overcome this challenge. Macrocycles are a class of organic molecules that have specific number of aromatic repeating units and, consequently, have precisely defined central cavity structure at angstrom scale. They also maintain versatile processability of organic materials such as solvent compatibility and feasibility of chemical modification. These physical and chemical properties make macrocycles, which are pillar[5]arenes here, promising source materials for TFC membranes that possess designed microstructures at molecular level. As a result, as pillar[5]arenes were integrated into the membrane matrix, the TFC membranes enhanced molecular sieving capability for gas separations. Pure-gas selectivity of representative gas pairs were calculated as 26 (H_2/CO_2), 120 (CO_2/CH_4), 10 (O_2/N_2), and 3200 (H_2/CH_4) based on preliminary pure gas permeation results. In addition, these membranes demonstrate good resistance to CO_2-induced plasticization due to its highly crosslinked chemistry.

Poster 60 - Development of Novel Processes for the Synthesis of Membranes with Special Wettability

Siamak Nejati* (University of Nebraska-Lincoln).

Creating reentrant structures on flexible and porous substrates poses a significant challenge to the scalable fabrication of omniphobic membranes. Designing such membranes requires control over the surface topography and chemistry of the interfacial domains. We developed a semi-continuous vapor phase polymerization approach, based on initiated chemical vapor deposition, to impart the desired chemical and structural properties to a variety of substrates. By adjusting the free radical polymerization rate and controlling the dynamics and growth mechanism in our process, we assembled desired reentrant structures on a variety of porous substrates, making their surface superhydrophobic. Subsequently, by depositing a conformal thin film of low surface-energy polymer on the assembled structures, we further improve the non-wetting properties of the membranes, rendering them omniphobic. The omniphobic properties of these membranes were evaluated using sessile-drop contact angle measurement for liquids with different polarities and surface energies ranging from 72 mN/m (water) to γ~22 mN/m (ethanol). The omniphobic structures hindered intrusion of the large droplets (1.5-1.8 mm) of low surface energy liquids (e.g., Ethanol) impacted the surface with Weber numbers as high as 300. To evaluate the wetting resistance of the developed interfaces in a dynamic process, we constructed an omniphobic layer on macroporous supports, made of polyvinylidene fluoride, and evaluate the performance of these membranes in direct contact membrane distillation (DCMD). A few complex feed compositions, including a synthetic brine with surface tension as low as ~40 mN/m and a municipal brine were utilized for the DCMD experiments. We report on the robust performance of these membranes.

Poster 61 - Dissipative Particle Dynamics Simulation of PVDF Membrane Formation via Non-Solvent Induced Phase Separation Process

Suphanat Aphinyan (University of Arkansas), Eric Ledieu (University of Arkansas), David Ford (University of Arkansas), Xianghong Qian (University of Arkansas).

Polymeric membranes produced by Non-solvent Induced Phase Separation (NIPS) have been extensively used across a wide array of microfiltration and ultrafiltration processes. However, despite their broad applications, current fabrication processes are largely empirical, determined largely by trial and error. Better insights into the factors affecting membrane morphology would be of great use. To that end, Dissipative Particle Dynamics (DPD), a mesoscale simulation tool, was used to simulate membrane formation process involving PVDF in DMF with H2O as a nonsolvent. The DPD simulations were performed for a total of several million beads and over several microseconds. The interactions between these beads are determined according to Flory-Huggins thermodynamic theory and continuum transport equation. Significant insights have been obtained on the effects of polymer concentration and polymer molecular weight on the pore size, pore size distribution and membrane morphology.

Poster 62 - Influences of Microwave Irradiation on Performances of Membrane Filtration and Catalytic Degradation of Perfluorooctanoic Acid (PFOA)

Fangzhou Liu* (NJIT).

Perfluorooctanoic Acid (PFOA), one of the common per- and poly fluorinated alkylated substances (PFASs), is increasingly detected in the environment due to the diverse industrial applications and high resistance to degradation processes. This study evaluated degradation of PFOA in microwave-assistant catalytic membrane filtration, a process that integrates microwave catalytic reactions into a ceramic membrane filtration. First, water permeation of the pristine and catalyst-coated membranes were examined under the influence of microwave irradiation to analyse the impacts of the coating layer and water temperature increase on permeate flux, which were well interpreted by the Carman-Kozeny, Hagen-Posieulle and Boussinesq models. Then, the PFOA removal was first assessed in a continuous filtration model with and without microwave irradiation. Our results show that PFOA adsorbed on membrane and catalyst materials and fully penetrated the membrane filter after reaching adsorption equilibrium. Under microwave irradiation (7.2 watt·cm-2), approximate 65.9% of PFOA (25 μg·L-1) in the feed solution was degraded within a hydraulic time of 2 min (at the permeate flow rate of 43 LMH) due to the microwave-

Fenton like reactions. In addition, low flow rates and moderate catalyst coating densities are critical for optimizing PFOA removal. Finally, potential degradation mechanisms of PFOA were proposed through the analysis of degradation by-products (e.g., PFPeA). The findings may provide new insight into the development of reactive membrane-enabled systems for destruction of refractory PFAS.

Poster 63 - Performance of Combining Ozonation, Ceramic Membrane Filtration with Biological Activated Carbon Filtration for the Advanced Treatment of Hypersaline Petrochemical Wastewater

Jinting Hu* (Tsinghua Shenzhen International Graduate School).

An integrated process combining ozonation, ceramic membrane filtration with biological activated carbon filtration (O3+CMF+BAC process) was designed and evaluated using a pilot scale (10 m3/d) test for the advanced treatment of hypersaline petrochemical wastewater in a coastal sewage plant. The membrane flux and ozone dosage were optimized for the optimal treatment performance of this integrated process. The results showed that this integrated process performed well in pollutant removal, with CODCr, phosphate and color in the effluents being 17.9 mg/L, 0.25 mg/L, and 5 dilution times in average, respectively, meeting the local discharge standard even under a high organic loading (195 mg/L in average). The synergistic effect of the ozonation and ceramic membrane filtration was investigated through the fluorescence characteristics and hydrophobic/hydrophilic properties of organic compounds. It revealed that ozonation mitigated the membrane fouling and the nanopores in the ceramic membranes enhanced the ozonation efficiency. Meanwhile, the Fenton process had a slightly better effluent quality than the integrated process, but Fenton process consumed much more chemicals and required the sludge disposal, resulting in higher cost. The estimated unit cost for this integrated process was only 39% of that for the Fenton process. Overall, the integrated process demonstrated high stability, reliable effluents and low cost, providing a promising and cost-efficient technology for the treatment of hypersaline petrochemical wastewater.

Poster 64 - Cinchona-decorated cyclodextrin organocatalyst for asymmetric synthesis in a continuous-flow membrane reactor

Peter Kisszekelyi* (Budapest University of Technology and Economics) Abdulaziz Alammar (University of Manchester), Jozsef Kupai (Budapest University of Technology and Economics), Peter Huszthy (Budapest University of Technology and Economics) Barabas Julia (Budapest University of Technology and Economics), Tibor Holtzl (Budapest University of Technology and Economics), Lajos Szente (Cyclolab, Ciklodextrin Kutató-Fejlesztő Kft), Carlo Bawn (University of Manchester), Ralph Adams (University of Manchester), Gyorgy Szekely (King Abdullah University of Science and Technology).

In this work a cyclodextrin-enhanced organocatalytic methodology is presented covering molecular design and process engineering.[1] To facilitate the catalyst recovery, cinchona-thiourea and -squaramide catalysts were covalently anchored to an inherently large, stable and well-defined permethyl-β-cyclodextrin. The asymmetric catalysis was successfully demonstrated on the Michael reaction of 1,3-diketones and trans-β-nitrostyrene. Both emerging green and conventional solvents were screened for the asymmetric addition (up to 99% ee). The Kamlet–Taft solvent parameters were correlated to the enantioselectivity. Furthermore, quantum chemical modelling revealed that the catalyst anchoring resulted in favorable structural changes, and stronger intermolecular interactions between the catalyst and the reagents. The size enlargement of the cinchona organocatalyst resulted in its complete recovery and continuous, in-line recycle with nanofiltration membranes, while other solutes showed less than 5% rejection. The continuous organocatalysis was performed in coiled tube flow reactor coupled with a membrane separation unit, which allowed 50% solvent (2-MeTHF) recycling. The full conversion, 98% purity, 99% ee, 80 g L-1 h-1 productivity and in-line catalyst recovery makes the proposed methodology an attractive integrated synthesis-separation catalytic platform.

References

[1] Kisszekelyi, P., Alammar, A.; Kupai, J.; Huszthy, P.; Barabas, J.; Holtzl, T.; Szente, L.; Bawn, C.; Adams, R.; Szekely, G. J. Catal. 2019, 371, 255–261..

Poster 65 - Membrane Separation of Nitrogen and Phosphorus Nutrients with Downstream Recovery as Struvite Fertilizer

Zahra Anari* (University of Arkansas), Lauren Greenlee (University of Arkansas).

Phosphorous is both essential and limited fertilizer for crops. Recovery of phosphorous from municipal and animal wastewater as struvite ($MgNH4PO4.6H2O$) can overcome the shortage of phosphate sources and, at the same time, reduce the wastewater nutrients to a level that meets the discharge regulations. Struvite is a phosphorous mineral with low solubility in the neutral conditions. The low solubility causes the fertilizer to release nutrients slowly, which is a considerable advantage over conventional fertilizers, resulting in less nutrient input to water streams. Also, struvite provides the crops with an equal molar of ammonium, phosphorous, and magnesium, maintaining a sustainable agricultural system.

Electrochemical recovery of struvite has demonstrated that higher aqueous concentrations of ammonium and phosphate yield greater recovery of struvite. This suggests pre-concentration prior to electrochemical precipitation will enhance N and P recovery. Here, Nanofiltarion of ammonium and phosphate is proposed as a concentration step. In this scenario, both filtrate and concentrate are filtration products. The concentrate is directly used at the electrochemical reactor to produce struvite fertilizer, and the filtrate water can be recovered as a source of irrigation water.

In this project, we focused on understanding and quantifying the separation performance of commercially-available polymeric membranes for ammonium and phosphate in municipal and poultry wastewater. The impact of membrane polymer chemistry and wastewater composition are investigated to predict the highest yield in nutrient concentration. Life cycle assessment is performed to determine potential environmental benefits/impacts of membrane-based nutrient recovery from wastewater.

Poster 66 - Effects of resin bead chemistries on the selective removal of industrially relevant metal ions using wafer-enhanced electrodeionization

Humeyra Ulusoy Erol* (University of Arkansas), Jamie Hestekin (University of Arkansas), Christa Hestekin (University of Arkansas).

There is a need for an efficient and economical process of ion-selective separation. In hydraulic fracturing, for instance, several million gallons of water are used in a site, most of which are disposed of by deep well injection taking the water from the ecosystem and causing earthquakes. However, due to the high amount of ions in wells at fracking sites, pressure driven processes for water recovery are impossible to use. This presentation will demonstrate an emerging environmentally friendly Wafer Enhanced Electrodeionization (WE-EDI) technology which attempts to eliminate the shortcomings that come with normal EDI and separates effluent streams into concentrated ions and clean water. In WE-EDI, ion-exchange resins are replaced by wafers that consist of cation and anion exchange beads and a polymer binding agent. In this study, selective separation of these ion exchange wafers have been performed for various anions and cations ($Na+$, $K+$, $Ca2+$, $Cl-$) and improved selectivity and higher water recoveries have been observed. This project is supported by the Membrane Science and Technology (MAST) Center and is conducted with the collaboration and guidance of industry mentors.

Poster 67 - Ceramic membrane centered hybrid processes for water treatment

WANYI FU* (Tsinghua University), Xihui Zhang (Tsinghua University).

Membrane technology is the leading edge water solution of the 21st century. Ceramic membranes have attracted more and more interest because of their high permeability, mechanical strength and excellent chemical resistance. Particularly, ceramic membrane filtration coupled with oxidation demonstrate efficient mitigation of membrane fouling and high removal efficiency for the pollutants, especially for the emerging contaminants that are challenging the conventional treatment processes. In the past decade, our research group has focused on the development of ceramic membrane centered treatment processes. Various lab researches, pilot studies and engineering projects have proven that our processes have excellent treatment performance for the drinking water treatment, wastewater treatment and portable reuse of secondary effluents. The integration of ceramic membrane filtration with ozonation merges five units (including pre-oxidation, sedimentation, filtration, main oxidation and ultrafiltration) into one unit.

Such a hybrid treatment process is much shorter and suitable for the upgrade of conventional water treatment plants for better supply of drinking water. This integrated process also enables the decentralized water supply in rural areas and the portable reuse of secondary effluents in urban areas. Overall, we believe that the development of ceramic membrane filtration will facilitate a game-changing mode for the water treatment and water reuse.

Poster 69 - Membrane charge weakly affects ion transport in reverse osmosis

Mikhail Stolov* (Technion - Israel Institute of Technology).

Polyamide RO membranes have been a "golden standard" in water desalination, with aromatic polyamide layer providing an excellent balance between salt rejection and water permeability. Yet, there still are gaps in understanding the salt transport in RO, closely related to the nature and formation of membrane charge and its effect on ion permeation.

We report here for the first time a systematic investigation of charge properties of polyamide layer at different pH that integrate experimental data - nominal fixed charge, salt diffusion permeability and membrane conductivity. Comparing all 3 types of measurements permits differentiating between the contributions of co- and counter-ions, and allows to distinguish the roles of charge repulsion and dielectric exclusion and effect of proton on salt permeation by looking at the pH dependence.

In the mid-pH range, most important in desalination practice, observed relations between the conductivity, salt permeability and the nominal charge show a much weaker dependence than the Donnan model predicts. This indicates that fixed charged groups inside polyamide films are largely deactivated, thereby the membrane behaves as effectively neutral or weakly charged. However, a substantial charge may form at extreme acidic and basic pH via uptake of $H+$ and $OH-$ ions, which increases ion uptake and conductivity, promoting polymer swelling and weakening salt rejection. Overall, the results strongly suggest that fixed charges weakly affect salt rejection in RO and it is dominated by dielectric and steric mechanisms.

Poster 70 - Experimental measurement of single- and mixed-gas permeation through photo-responsive gas membranes

Bradley Ladewig* (Karlsruhe Institute of Technology), Jinju Zhang (Karlsruhe Institute of Technology), Nicole Jung (Karlsruhe Institute of Technology).

This poster describes recent developments at Imperial College London and Karlsruhe Institute of Technology to develop a robust apparatus for routine measurement of gas permeation through photo-responsive membranes under controlled conditions.

Early experimental membrane work at Imperial College London (as well as extensive work on CO2 sorption in photo-responsive MOF materials) showed how critically important careful control of the membrane temperature is. While consistent illumination (to date with 365nm UV light, but in future a wider ranger of light sources and broad spectrum light will be used) is important to activate the photo-switching phenomenon in the functional materials, over-illumination leads to localised adsorption and heating of the material. The subsequent gas desorption due to an increase in the material temperature is difficult to deconvolute from the photo-induced gas desorption, and we are developing fundamental and applied methods to explore this further.

The current status of a completely new test station, which will allow for single-gas pressure decay measurements, as well as mixed-gas Wicke-Kallenbach type experiments with a sweep gas carrying permeate to an online GC, will be described - in both cases membranes can be tested with constant or dynamic illumination under moderate pressures (up to around 10 bar, this may be increased in future cell designs).

Karlsruhe Institute of Technology is a pioneer in the development of open source, highly robust electronic lab notebooks (https://chemotion.net/), and this membrane test station will be complemented with a world-leading data storage repository for all tested materials. This provides unparalleled opportunities to share research results openly with collaborators both within and outside the institution, and the interactive poster will share some examples of the features of the Chemotion electronic lab notebook and data repository.

Poster 71 - Radical filtration for Water Purification

Shuyana Heredia* (University of Twente).

One of the improvements regarding water purification is to treat the water more effectively and economically, avoiding the production of secondary waste pollutants and ensuring a safe water supply.
Advanced oxidation techniques and membrane filtration have attracted increasing attention to treat and purify water. Among these methods, photocatalytic oxidation with titanium dioxide (TiO2) is widely studied as this treatment avoids the solid waste, and membranes have gained an important place in chemical technology and are used in a broad range of applications. A reactive membrane is expected to reduce the concentration polarization via the chemical conversion of reactants. Thus, the production of purified water can be increased.
This research seeks to provide a novel water treatment method, based on the synergy between photocatalytic oxidation and membrane separation within a single material to remove micropollutants and inactivate microorganisms. To elucidate this synergy, a transport model that contains the membrane function (rejection) and photocatalytic degradation (reaction), including light distribution, has been designed.

Poster 72 - membrane-toolkit: a Python package for fast, accurate, automated experimental data management

Ryan Kingsbury* (University of North Carolina at Chapel Hill).

Processing experimental data to determine membrane properties is a crucial task that provides the foundation for much of membrane science. Such data is often managed manually by, for example, saving to one or more spreadsheets and writing calculation formulas from scratch. This approach is time-consuming and error prone, especially for large quantities of data. Moreover, requiring every researcher to write calculations by hand makes it difficult to compare results across studies due to inconsistencies in assumptions and methods. For example, a cursory search of the literature revealed numerous equations for computing Donnan equilibrium, each making different assumptions that were not always explicitly stated.
This presentation will introduce membrane-toolkit, an open-source software package designed to accelerate membrane research by streamlining and standardizing routine data management tasks. Its three primary goals are to 1) provide standardized, well-tested functions for routine calculations, such as equations for permeability, Donnan exclusion, and Manning theory; 2) automate the processing and aggregation of common types of experimental data (e.g., spreadsheets containing dead-end cell test data) into more powerful data structures; and 3) provide an object-oriented interface to facilitate the construction of continuum transport models.
membrane-toolkit is at an early stage of development, and hence this interactive presentation will provide an excellent opportunity for interested membrane scientists to provide feedback on the goals of the project and learn about how to contribute to its development.

Poster 73 - In-situ measurement of the mechanical properties of a foulant layer at a membrane surface

Jose Agustin Epstein* (Technion - Israel Institute of Technology), Guy Ramon (Technion - Israel Institute of Technology).

Fouling remains a limitation in membrane-based water purification and desalination. Little is known about the mechanical properties of foulant deposits and how these relate to fouling reversibility and resistance to permeation. A better understanding of foulant mechanical properties, in-situ, can provide practical insight on more effective back-pulsing strategies for foulant removal, e.g. selecting the most appropriate frequency. Measurements using bulk rheometry would not represent the realistic structure of the material deposited at a membrane surface.
Here, we demonstrate in-situ measurements of micro-scale rheology of a foulant layer deposited at a membrane surface, with alginate as a model foulant. Fouling experiments were conducted employing a custom-made membrane cell fabricated with an optical window, enabling high-resolution, real-time observation with a confocal microscope. Permeate flux and applied pressure are maintained using a software-controller. A video-based, particle-tracking micro-rheology technique was employed to measure mechanical properties at different depth locations within the foulant layer. Through the analysis of particle trajectory statistics, the mechanical properties of

the alginate layer were characterised in terms of the frequency-dependent storage and loss modulus, G'(ω) and G"(ω).
In particular, a comparison of mechanical properties was made when varying the permeate flux and the pressure. Results demonstrate the elastic-gel characteristics of the layer, and the variations in its storage and loss moduli upon flux increase, indicating micro-scale 'stiffening' of the material. The foulant also shows clear depth variations, where a decrease in its elasticity was observed as the distance from the membrane increased. The presented approach opens new possibilities for studying fouling, particularly soft, gel-like deposits encountered with organic and bio-fouling.

Poster 74 - On the reliability and comparability of organic solvent nanofiltration reports

Hai Anh Le Phuong* (The University of Manchester), Christopher F. Blanford (The University of Manchester), Gyorgy Szekely (King Abdullah University of Science and Technology).

Organic solvent nanofiltration (OSN) is an energy-efficient separation technique that has the potential to improve the sustainability in many industrial sectors, including the production of pharmaceuticals, fine chemicals and petrochemicals and in food processing. Some issues, however, hinder the pace of development of this sustainable separation method that could ultimately provide green manufacturing strategies. These issues include lack of clear experimental designs, explicit experimental protocols, comparable performance data and long-term performance tests of membranes at industrially relevant solute concentrations in OSN studies. We have critically assessed the importance of these issues. We support our assessments with an analysis of the recent academic literature on OSN and a survey of the OSN community on data acquisition and reporting. The literature analysis consists of an informed, data mining of information of methodological and performance data from 177 articles published from 2015 to 2019, and a critical analysis of these findings. Based on the results of our survey and literature analysis, we crafted a series of best-practice recommendations for researchers reporting data on membrane fabrication, membrane materials characterization and filtration performance, process integration and fundamental studies.

Poster 76 - Experimental Design Study of Reinforced Hollow Fiber Membrane Casting

Walter Kosar* (Arkema Inc.), Gregory O'Brien (Arkema Inc.), Michele Crane (Arkema Inc.).

Membrane bioreactors (MBR) are recognized as cutting edge technology for waste water treatment in municipal and industrial applications. The MBR has advantages of compact footprint, high filtered water quality, and rapid cycle time. The predominant membrane type used for MBR systems is reinforced PVDF hollow fiber membranes. These membranes share the same excellent chemical resistance properties of unreinforced PVDF membranes, but with improved mechanical durability due to the reinforcing braid.
There have been numerous articles published on formulating and casting unreinforced hollow fiber membranes. However, there are far fewer reports on preparation of reinforced PVDF hollow fiber membranes. Consequently, there has not been much study on key process parameters affecting reinforced hollow fiber membrane quality.
We recently completed an experimental design study on casting reinforced PVDF hollow fiber membranes. Our goal was to understand the effects of formulation and casting conditions on membrane properties. We looked at pure water permeability, pore size, membrane coating thickness, morphology, penetration depth into the braid, and burst pressure.
Key findings are:
Reinforced membranes have higher water permeability than unreinforced NIPS membranes
Reinforced membranes contain substantially more macrovoids in the coating layer when compared to unreinforced membranes
Dope pump and line speed were significant process variables
The active membrane layer in reinforced fibers is about half the thickness of unreinforced membranes
Formulation additives had little effect on measured properties
We will also discuss a new liquid-liquid porometry method we developed for testing hollow fiber membranes. During this study, we discovered a major problem with the isobutanol-water liquids typically used in this method. We will present these results and introduce our preferred liquid combination for this method.

Poster 77 - Determination of Carbon Nanomembrane Permeability Coefficients via Radioactive Tracer Experiments

Raphael Dalpke (Bielefeld University), Anna Dreyer (Bielefeld University), Riko Korzetz* (Bielefeld University), André Beyer (Bielefeld University), Karl-Josef Dietz (Bielefeld University), Armin Gölzhäuser (Bielefeld University).

In recent years, 2D materials have gained a lot of attention as highly efficient filter materials. One class of materials are carbon nanomembranes (CNM), which are fabricated by self-assembly of organic precursors onto metal substrates and subsequent cross-linking by electron irradiation. They can be released from their original substrate and transferred onto arbitrary supports. Specifically, CNMs made from terphenylthiol (TPT) exhibit a very high selectivity towards water in combination with an outstanding permeance [1].
Here, we present new permeation measurements of TPT-CNMs utilizing radioactively marked water as well as carbonate and phosphate ions. We show that water and carbonate can pass through while phosphate ions are completely blocked by the CNM. The resulting permeability coefficients in combination with recent findings indicate that the permeation across the membrane primarily occurs by transport of neutral species. This implies a fast transport of CO2 in aqueous solution, while dry TPT-CNMs are not permeable to CO2, which suggests a facilitated transport in the presence of water.

References

[1] Yang et.al., ACS Nano, 2018, 12(5), 4695-4701.

Poster 79 - Biocatalytic membrane reactor using immobilized phosphotriesterase for the degradation of pesticides

Giuseppe Vitola (National Research Council of Italy, Insitute on membrane Technology, CNR-ITM), Rosalinda Mazzei (National Research Council of Italy, Insitute on membrane Technology, CNR-ITM), Lidietta GIORNO* (National Research Council of Italy - Institute on Membrane Technology (CNR-ITM)).

Pesticides are a category of recalcitrant micro-contaminants present in the environment, especially due to agriculture cultivations. It seems that chronic exposure to small amount of such components may cause health problems.
Phosphotriesterase enzymes from extremophile microorganisms proved very promising for the biodegradation of organophosphate pesticides.
The covalent bond of phosphotriesterase to porous membranes to develop biocatalytic membranes and membrane reactors was investigated by our group. Among the various materials used, hydrophilic polymeric membranes resulted the most suitable to preserve enzyme activity and stability. Therefore, regenerated cellulose membrane was selected to fabricate the biocatalytic membrane reactor. Flat-sheet membrane disk (Ø 4.7 cm) was used. The void volume of the porous membrane formed the reactor volume (0.3 cm3). The paraoxon pesticide solution (up to 1 mM concentration) was fed through the enzyme-loaded membrane at constant flow rate, each membrane pore worked as a continuous micro-reactor and the biodegraded solution was collected in the permeate side. The volumetric reaction rate was derived by steady-state balance equation assuming no accumulation within the biocatalytic membrane reactor. In fact, both reagent and products have lower size than membrane pores and can freely pass through them transported by convective flow. Under optimal biocatalyst amount and residence time a complete conversion of the pesticide can be achieved. The biocatalytic membrane reactor was also tested using real streams constituted by surface waters or olive mill waste waters added with known amount of pesticide. It was observed that the presence of biophenols, such as hydroxytyrosol, which has a molecular structure similar to paraoxon pesticide, provoked competitive inhibition ($Ki = 2.4 \pm 0.3$ mM). Biocatalytic membranes were monitored for some 4 months.

Poster 80 - Membrane Filtration of Poly(dT60) Single-Stranded DNA

Hossein Nouri Alavijeh* (Clarkson University), Ruth E. Baltus (Clarkson University).

Certain unique features inherent to membrane filtration technology offer several advantages for a variety of applications. Microfiltration (MF) and ultrafiltration (UF) membranes are commonly employed in the food processing, biotechnology, pharmaceutical, and water/wastewater treatment industries. Membrane filtration of flexible particles has sparked a renewed interest in DNA purification in pharmaceutical companies and filamentous virus filtration in water treatment plants. The ability of flexible particles to adopt conformations and enter membrane pores influences their filterability. Understanding the effects of mechanical flexibility on the rejection coefficient is crucial for designing and implementing of membrane separation. The main objective of this study is to obtain quantitative data on the transmission of single-stranded DNA (ssDNA) -which is designed for DNA therapeutics- through track-etched polycarbonate ultrafiltration membrane by considering the factors that govern the transmission of ssDNA through membrane pores. The effects of flexibility, transmembrane pressure (TMP), solution ionic strength, and membrane pore size on the rejection coefficient were investigated using track-etched polycarbonate ultrafiltration membrane in a stirred cell set-up. To characterize flexibility of DNA, Fluorescence Recovery After Photobleaching was integrated with Worm Like Chain polymer model to determine the persistence length. The results demonstrate that increasing TMP results in decreasing the rejection coefficient due to the ability of DNA to deform and pass through the pores easier in higher pressure. In addition, studying the effects of solution ionic strength shows the electrostatic interactions between the same charged DNA and membrane increase DNA rejection. Results generated from this study will be important for the optimal design of micro- and ultrafiltration systems for water and wastewater treatment and DNA purification in biotechnological sectors.

Poster 81 - Scaling Resistance in Nanophotonics-Enabled Solar Membrane Distillation

Francois Perreault* (Arizona State University).

Membrane distillation (MD) is a promising method for treating high salinity waters since water vapor pressure is largely independent of salt concentration, making this process less sensitive to salinity then reverse osmosis. MD systems generally operate at high temperatures (~60□C), which impose a high energy cost for water treatment. Recently, nanophotonics-enabled solar membrane distillation (NEMSD), which uses direct sunlight and carbon black-coated membranes to provide localized surface heating on the membrane, was shown as a promising alternative to conventional MD for energy efficient off-grid desalination. In this project, we investigated the scaling propensity of the NESMD system in order to assess its applicability to brackish water with high scaling potential. Our results show that NESMD can resist membrane scaling due to the lower operating temperature in the feed water, which reduce calcium precipitation on the membrane. Low scale formation was confirmed by scanning electron microscopy, elemental analysis of the fouled membrane, and equilibrium speciation modelling. These results highlight the advantages of NESMD beyond energy efficiency for the treatment of high salinity water.

Poster 82 - Generalized model for the prediction of the permeability of mixed-matrix using impermeable fillers of diverse geometry

Haoyu Wu* (University of Ottawa), Boguslaw Kruczek (University of Ottawa), Jules Thibault (U. Ottawa).

Using mixed-matrix membranes (MMMs) offers the possibility to modify membrane transport properties for specific applications. This paper is concerned with the numerical evaluation and prediction of the enhancement of barrier properties of packaging materials by adding impermeable nanoparticles within the matrix of thin polymer films. Adding impermeable nanoparticles creates a more tortuous pathway for diffusion thereby reducing the effective permeability of the membrane. Various shapes of particles can be used: nanocuboids, nanospheres, and nanorods. Some mathematical models, such as Maxwell's model and Neilson's model, have been proposed to estimate the relative permeability coefficient of MMMs. The accuracy of these models can be assessed by comparing their results with those obtained from numerical experiments. The latter were obtained by solving the three-dimensional Fick's second law of diffusion previously developed by our group.

In this paper, we propose a novel model, which was developed for impermeable nanocuboids dispersed in a continuous polymer phase. The proposed model is surprisingly simple and yet it provides a more accurate prediction of the effective permeability coefficients of MMMs compared to the predictions of existing models. Although our model was originally developed for impermeable cuboids, it can also be easily adapted for fillers of different geometries such as nanospheres, nanorods, and nanotubes. Application of our model to other geometries simply requires the introduction of a shape factor, which transforms any geometry into a cuboid having the same projected area and an equivalent relative thickness.

Poster 83 - Factors controlling the structure formation in isoporous hollow fiber membranes

Kirti Sankhala* (Helmholtz-Zentrum Geesthacht, Geesthacht), D. C. Florian Wieland (Helmholtz-Zentrum Geesthacht, Geesthachtge), Joachim Koll (Helmholtz-Zentrum Geesthacht, Geesthacht), Maryam Radjabian (Helmholtz-Zentrum Geesthacht, Geesthacht), Clarissa Abetz (Helmholtz-Zentrum Geesthacht), Volker Abetz (Helmholtz-Zentrum Geesthacht).

Isoporous block copolymer membranes hold the potential of providing energy-efficient separations.1 The membranes are fabricated via evaporation-induced self-assembly of block copolymers and non-solvent induced phase separation (SNIPS) in flat sheet and hollow fiber (HF) geometries.1,2 The self-supported HF geometry provides a higher active surface area for lesser volume with an additional self-sustaining feature. However, the spinning process complicates the kinetics of microphase separation of block copolymers due to the shear during extrusion and relaxation afterwards.2,3,4

Different solutions of polystyrene-block-poly(4-vinylpyridine) diblock copolymer were used to understand the influence of spinning parameters on the structure formation. In situ small-angle X-ray scattering (SAXS) experiments were conducted during conventional HF spinning.3 In addition, a novel approach of controlling the kinetics of microphase separation of block copolymers by using gas flow is introduced, i.e., so far the only feasible way to achieve evaporation-induced self-assembly in compact geometries.4,5

The structural features obtained from in situ SAXS investigations are correlated with the structures in the solutions and morphologies in the flat sheet and the HF membranes obtained by ex situ SEM.3 Technologically-relevant inside-out isoporous HF as integral asymmetric and composite membranes of inner diameter ca. 1 mm were fabricated by spinning and coating processes, respectively.4,5 The emerging isoporous membranes can be applicable for numerous separation applications.

References

1. V. Abetz, Macromol. Rapid Commun. 2015, 36:10
2. M. Radjabian, J. Koll, K. Buhr, U. Vainio, C. Abetz, U.A. Handge, V. Abetz, Polymer 2014, 55:2986
3. K. Sankhala, D. C. F. Wieland, J. Koll, M. Radjabian, C. Abetz, V. Abetz, Nanoscale 2019, 11:8049
4. K. Sankhala, J. Koll, M. Radjabian, U. A. Handge, V. Abetz, Adv. Mater. Interfaces 2017, 4:1600991
5. K. Sankhala, J. Koll, V. Abetz, ACS Macro Lett. 2018, 7:840.

Poster 84 - Solubility of Rare Earth Sulfates and Chlorined toward Electrochemical Membrane Recovery: Effect of pH

Mohammad U. Shafiq (Texas A&M University-Kingsville), Lucy Camacho* (Texas A&M University-Kingsville).

Currently society relies heavily on technology such as touchscreen, microchips, magnets, lasers, and superconductors. A key component in these high technologies are rare earth elements (REE). Over the past decade REE have shown substantial increases in demand and this trend is expected to continue into the foreseeable future. Although demand is expected to increase, the supply of REE is not expected to meet the demand. China, which currently produces approximately 97% of the world's supply of REE, recently placed restriction on the mining of REE. Recently, great interest has been shown in identifying new sources and methods of obtaining REE. One of the barriers in commercial production of REE is the lack of knowledge on REE. In comparison to other elements, REE are relatively new and are found in varying concentrations and compositions throughout the earths crust, making it difficult for the REE to be studied. A potential source of REE are geothermal waters, which contain these elements at low concentrations. These geothermal waters can vary greatly in pH and composition. Understanding how the REE will behave at different pH values can provide valuable information that can potentially be used to identify new electrochemical membrane methods of recovery and sources of REE. The solubility, mass balance, and complexation of REE at different pH values are being researched. Initial laboratory research has identified a correlation between the solution pH and the solubility of REE, with solubility reducing as the pH increases. The results are promising, and knowledge gained can be used to help increase the overall production of REE and therefore increase the overall global supply. Additionally, the removal of these heavy metals from the geothermal waters would also help the environment.

Poster 142 - Improved batch reverse osmosis configuration for better energy efficiency

Sandra Cordoba* (Purdue University), Abhimanyu Das (Purdue University), David Warsinger (Purdue University).

Recent progress in batch and semi-batch reverse osmosis processes such as CCRO have shown the promise to be the most efficient desalination systems. Despite their progress, it is critical to further increase their efficiencies, and reduce the downtime between cycles that worsens their cost performance. In this study, we model in new detail a further improved batch desalination system that uses a high pressure feed tank with a reciprocating piston. A high-pressure pump fills the inactive side with the following cycle's feedwater, providing two main benefits. First, no tank emptying step is needed because feed is already present, thus reducing downtime. Second, the tank fully empties each cycle, thus avoiding the small energy losses from brine mixing with the new feed that past best designs had. The modeling methodology is the most thorough yet for batch processes, as it uses a discretized module that includes transient mass transport equations for salt boundary layers, membrane permeability effects, and minute salt permeation through the membrane. Comparing the new configuration to standard reverse osmosis with and without energy recovery, the new process vastly outperforms, with the potential to be below 2 kWh/m3 for seawater. The new process has less downtime too, around 2% of cycle time, compared with 10% for CCRO or 16% from past batch studies.
We are now, working in the experimental phase, building the first Batch RO double acting piston prototype where different practical factors affecting the real performance of the system will be tested, such as, salt built-up between cycles, air infiltration in the system or the influence of the reset time in the overall energy consumption.

Poster 143 - Microalgae Filtration using Electrochemically Reactive Ceramic Membrane: Filtration Performances, Fouling Kinetics and Foulant Layer Characteristics

Qingquan Ma* (NJIT).

Electrochemical membrane filtration has proven successful for microbial removal and separation from water. In addition, membrane fouling could be mitigated by electrochemical reactions and electrostatic repulsion on reactive membrane surface. This study assessed the filtration performances and fouling characteristics of electrochemically reactive ceramic membranes (a Magneli phase suboxide of TiO2) when filtering algal suspension under different DC currents to achieve anodic or cathodic polarization. The critical flux results indicate that when applying positive or negative DC currents (e.g., 1.25~ 2.5 mA·cm-2) to the membrane both significantly mitigated membrane fouling and thus maintained higher critical fluxes (up to 14.6×10-5·m3·m-2·s-1 or 526 LMH) compared to the critical flux without DC currents. Moreover, applying DC currents also enhanced membrane defouling processes and recovered high permeate flux better than hydraulic and chemical backwash methods. Moreover, fouling kinetics and the cake layer formation were further analyzed with a resistance-in-series model that revealed many important but underexamined parameters (e.g., cake layer resistance and cake layer thickness). The cake layer structures (e.g., compressibility) were shown to vary with the electrochemical activity, which provide new insight into the biofouling mechanisms. Finally, the algogenic odor, Geosmin, was shown to be effectively removed by this reactive membrane under positive DC currents (2.5 mA·cm-2), which highlights the multifunctional capabilities of electrochemically reactive membrane filtration in biomass separation, fouling prevention and pollutant degradation.

Poster 144 - Green solvent PVC ultrafiltration membrane

Baicang Liu* (Sichuan University), Wancen Xie (Sichuan University).

Membrane fabrication currently relies on the use of traditional solvents, which are harmful to the environment and human health. Replacing traditional solvents with green solvents while maintaining or improving membrane performance is a challenging task at the forefront of research in the field of membrane technology. Here, a green solvent, dimethyl sulfoxide (DMSO) is used to partly replace traditional solvents and prepare high-performance poly (vinyl chloride) ultrafiltration membranes. The amphiphilic copolymer poly (vinyl chloride)-graft-poly (ethylene glycol) methyl ether methacrylate (PVC-g-PEGMA) is blended with PVC to improve the membrane performance. PVC cannot dissolve in DMSO, so based on the Hansen solubility parameter calculation, we investigated the mixed solvents of traditional solvents and DMSO. We found that membranes fabricated by solvent 1-methyl-2-pyrrolidinone (NMP)/N, N-dimethylacetamide (DMAc)/DMSO = 4/3/3 had the high pure water permeability of ~ 900

L m−2 h−1 bar−1 and the high sodium alginate (SA) rejection of nearly 95%. Furthermore, in order to completely replace toxic solvents during membrane synthesis, a novel green solvent, methyl-5-(dimethylamino)-2-methyl-5-oxopentanoate (Rhodiasolv® PolarClean) is used to fabricate PVC/PVC-g-PEGMA membranes. The effects of PVC concentration, addition of amphiphilic copolymer PVC-g-PEGMA concentration, and use of a non-woven polyethylene terephthalate fabric as support layer were investigated systematically. The membrane fabricated with 8% PVC, 5% PVC-g-PEGMA, and non-woven PET fabrics as support layer showed the best overall performance, presenting ultrahigh pure water permeability of >5000 L m−2 h−1 bar−1 and high SA rejection of nearly 98%. Both green solvents are used in PVC membrane system for the first time and demonstrate the feasibility of using green solvents to increase the sustainability and effectiveness of membrane fabrication.

Poster 145 - Porous single-layer graphene membranes for carbon capture

Kuang-Jung Hsu* (Ecole polytechnique fédérale de Lausanne), Shiqi Huang (Ecole polytechnique fédérale de Lausanne), Guangwei He (Ecole polytechnique fédérale de Lausanne), Kumar Varoon Agrawal (Ecole polytechnique fédérale de Lausanne).

Carbon capture by membrane separation is regarded to be one of the potential solutions for reducing greenhouse gas emissions, due to its high energy efficiency, modularity, and ease of retrofit. The performance target for the post-combustion carbon capture is as following: permeance > 1000 GPU with a separation factor exceeding 20. Ideally, the selective layer of the membrane should be ultra-thin to perform beyond the target. Single-layer graphene (SLG) is highly promising for this but has an impermeable lattice [1]. A precise etching of sub-angstrom pores in graphene is regarded as a bottleneck for the development of SLG membranes. Recently, we developed a millisecond ozone etching technique followed by oxygen expansion [2]. By this new approach, nanoporous single-layer graphene (N-SLG), with high flux gas and down to 0.2 Å molecular sieving resolution, has been achieved. The fine-tuned molecular-sieving pores enable CO2/N2 selectivity of 21.7 with corresponding CO2 permeance of 11850 GPU.

To further enhance the gas selectivity, functionalizing a CO2-philic layer on porous graphene is a promising approach. Polymer-functionalized nanoporous graphene (SPONG) membranes offer a platform for selectively adsorbing CO2, increasing the separation factor. Porous graphene serves as a molecular-sieving layer and also as a substrate for an ultra-thin (<20 nm) polymeric layer. The two layers synergistically enhance the performance with CO2 permeance up to 11790 GPU and CO2/N2 selectivity up to 57.2 [3].

Finally, an automated etching reactor has been developed which allows for improved control over the etching kinetics and offers potential for scale-up without compromising the sub-angstrom level etching precision.

References

1. Wang, L., et al.,. Nature Nanotechnology, 2017, 12, 509-522.

2. Shiqi Huang, L.F.V., Shaoxian Li, Deepu J. Babu, Mohammad Tohidi Vahdat, Emad Oveisi, Kumar Varoon Agrawal., Submitted.

3. He, G., et al., Energy & Environmental Science, 2019, 3305-3312.

Poster 146 - Bimodal Reverse Osmosis and Pressure Retarded Osmosis Framework

Akshay Rao* (Purdue University), Stephen Coan (Purdue University), Luke Wrede (Purdue University), Owen Li (Purdue University), George Elias (Purdue University), sandra cordoba (Purdue University), Michael Roggenburg (Purdue University), David Warsinger (Purdue University), Luciano Castillo (Purdue University).

Reverse osmosis (RO) desalination encounters economic challenges in increasingly renewable-dependent grids, where the price of electricity largely fluctuates due to variable power generation. Pressure retarded osmosis (PRO) is a closely-related clean energy technology that, with current materials, struggles to produce enough power to justify capital expense at an industrial scale. In this work, a framework for the integration of RO and PRO using the same parts for both is proposed to maximize profitability through the sale of producing either water or producing electricity depending on electricity peak prices. Compared to the standalone systems, the proposed integrated system is estimated to approximately decrease the efficiency of similar RO systems by 1% and that of PRO by 5%, over a range of salinities. An optimized process would use high-selectivity, high-permeability membranes that combine the selectivity of RO while maintaining the high power density of PRO. A case study of the Ashkelon,

Carlsbad, Ras Al-Khair, and Sorek RO plants shows that the costs associated with implementing the bimodal system could be less than 5% of the initial capital expense. While stand-alone PRO is viewed as too expensive, the results show it can be very competitive when integrated with RO, and can change desalination from a base-load power demand to being highly synergistic with variable energy sources.

Poster 147 - Effect of packing nonuniformity at the fiber bundle-case interface on flow distribution and separation performance for hollow fiber membrane modules

Lili Sun* (University of Toledo), Atabong Etiendem (University of Toledo), Glenn Lipscomb (University of Toledo, Toledo).

Membrane gas separation processes most commonly utilize hollow fiber membrane modules. Experimental performance of these modules can differ dramatically from theoretical predictions based on independent experimental measurements of membrane transport properties. One of the most significant sources of these deviations is flow maldistribution that prevents uniform counter-current contacting between retentate and permeate. Previous work has demonstrated the impact of fiber size uniformity on flow distribution and performance for lumen-fed modules. While poor packing of the fiber bundle adjacent to the enclosing case has been hypothesized as another significant factor for shell-fed modules, the impact has not been investigated thoroughly.

Simulations of flow through fiber bundles with non-uniform wall packing are reported. Results are reported for fiber bundles possessing a regular square or triangular packing as well as randomly packed bundles. The simulation approach relies on determining the length scale over which the case influences flow distribution and the extent to which flow near the case increases or decreases relative to that expected for a large bundle where the wall has no influence.

The length over which the case influences flow increases dramatically with increasing fiber packing fraction. As the packing fraction approaches 0.6, the impact can extend over the entire bundle diameter for small modules (<5,000 fibers). The impact of poor packing is greater for triangular packings than square packings because irregular triangular packing leads to larger flow regions near the case. The impact is even greater for randomly packed bundles.

The flow distribution results are used to evaluate the performance of a shell-fed hollow fiber module. The results clearly demonstrate the detrimental effect of poor packing along the case especially for small, randomly packed modules.

Poster 148 - Techniques for Measuring Gas and Vapor Solubility and Transport in Membranes

Mark Roper* (Hiden Isochema Ltd), Darren Broom (Hiden Isochema Ltd), Michael Benham (Hiden Isochema Ltd).

The accurate characterization of membrane materials is essential to their development for practical applications in fields including gas and vapor separation and purification, and as barrier materials. Key parameters include solubility, diffusivity and permeability (transport) properties.

We present an overview of lab scale techniques for the measurement of gas and vapor solubility and transport in membrane materials. Examples of analyses from literature, using Hiden Isochema sorption analyzers, are presented including polymeric, inorganic and mixed-matrix membrane materials. Conditions range from high pressure gas solubility to vapor diffusivity under elevated temperature and humidity conditions.

Poster Session 3: Applications

Poster 85 - Super water permeable PVDF UF membrane for fine separation

Shun Shimura* (Toray Industries, Inc.), Hiroki Eimura (Toray Industries, Inc.), Masayuki Hanakawa (Toray Industries, Inc.), Masahiro Kimura (Toray Industries, Inc.).

Membrane technology is regarded as indispensable in this century for various separation processes related to water treatment, production of foods or chemical product. Poly (vinylidene fluoride) (PVDF) Ultrafiltration (UF) membranes have been widely used for separation processes in various fields due to high chemical resistance and good filtrate water quality and hence ease of operation and maintenance. While there has been growing expectations for small substrates separation (fine separation) and high water permeability from the perspective of energy saving and cost reduction, it is generally difficult to achieve both fine separation and high water permeability at the same time, as resistance in the UF membrane structure increases as the pore diameter becomes smaller and water permeability declines.

In order to overcome this problem, we pursued the limits in terms of refinement of the size of pores on the surface of a membrane, which is necessary for fine separation, and coarsening of pores inside the membrane, necessary for high water permeability. Based on the in-situ observation of phase separation process for forming membrane carried out at SPring-8, a large synchrotron radiation facility, we got relationships between various process conditions and membrane structures. By leveraging the proprietary membrane process technology and precise structural design using phase separation simulation based on computational chemistry combined with the in-situ observation, we succeeded in controlling the phase separation speed and growth speed of the membrane. This made it possible to overcome the existing trade-off between pore size and permeability, and enabled the realization of nano-metric sized fine pore structure that has six times the permeability of existing membranes.

Poster 86 - Highly Permeable Polyethersulfone Substrates with Bicontinuous Structure for CO2-Selective Composite Membranes

Ruizhi Pang* (The Ohio State University), Kai Chen (The Ohio State University), Yang Han (The Ohio State University), Winston Ho (The Ohio State University).

The substrate pore structure has a significant effect on the CO2 transport performance of a thin-film-composite (TFC) membrane. In this study, a novel solvent system comprising 2-pyrrolidone (2PD) and 2-methoxyethanol (2-ME) was used to prepare polyethersulfone (PES) substrates with bicontinuous structure by vapor-induced phase separation, followed by immersion in a nonsolvent, water. Compared with the common solvent, e.g., N-methyl-2-pyrrolidone (NMP), the 2PD/2-ME system is more hydrophilic and can decrease the thermodynamic stability of the casting solution significantly. As a result, the phase separation during the membrane formation process was induced via the spinodal decomposition mechanism. The process parameters, including PES concentration, 2PD/2-ME weight ratio, water vapor exposure time, and water bath temperature, were optimized by characterizing the membrane morphologies and the CO2 transport performance. Compared with the baseline PES substrate optimized previously with NMP, the new substrate showed a high surface porosity of 19.9%, indicating a 48.5% increase. Also, the new substrate exhibited a very high CO2 permeance of 133,226 GPU at 57oC, which was 5 times more permeable than the baseline substrate. This high permeance was contributed mainly by the absence of a top dense layer and the presence of highly interconnected pores in the substrate. By using this substrate, the prepared TFC membrane containing amines showed a CO2 permeance of 908 GPU, which was 48 GPU higher than that with the baseline substrate at 57oC. This improvement can be explained well using the resistance-in-series model. The improved CO2 permeance was attributed to both the reduced lateral diffusion and substrate transport resistances. In addition, the scalability of the PES substrate with the bicontinuous structure was demonstrated by the successful roll-to-roll fabrication of a 50-foot long scale-up PES substrate with a width of 21".

Poster 87 - Nutrient recovery from synthetic livestock wastewater effluent using electroactive membranes

Kartikeya Kekre* (Temple University), Arezou Anvari (Temple University), Avner Ronen (Temple University).

Nitrogen (N) and phosphorus (P) are essential elements needed for agricultural crop fertilization. The production of both nitrogen and phosphorus fertilizers is energy-intensive, and their demand is expected to increase in the coming decades. Meanwhile, agricultural wastewater from livestock (LWW) farms contains high concentrations of nitrogen and phosphorus in addition to organic loads. Runoff of Livestock wastewater from farms to aquatic ecosystems leads to contamination and severe environmental problems. Currently, conventional wastewater treatment plants are unable to remove such high concentrations, which require complex anaerobic/anoxic treatment stages and

reduce the efficiency of the treatment. An important strategy to mitigate phosphorus and nitrogen scarcity is to recycle them from waste streams back into agricultural production.
We present a novel 'polishing' filtration process based on electroactive membranes to simultaneously produce a solid fertilizer and high-quality water from LWW effluents following anaerobic treatment, thereby closing resource cycles in agriculture. The suggested process uses low-voltage electrically-charged membranes to recover and separate Struvite, an eco-friendly fertilizer from LWW effluents. The recovery process is based on local hydrolysis and pH increase near the membrane surface and does not require chemical additives apart from magnesium. We evaluated permeate flux and change in pH of the permeate as a function of applied electrical potential, time, and solution's ionic strength. Results indicate rapid and continuous removal rates of Phosphorus and Nitrogen. In addition, membrane flux is easily restored to its initial value, and therefore, the recovery process can be repeated following the collection of the solid fertilizer.

Poster 88 - High Aspect Ratio Nanoparticles with Controlled Interfacial Transport for Improved Reverse Osmosis Performance

Ethan Smith* (Virginia Tech), Stephen Martin (Virginia Tech).

Thin film nanocomposite membranes (TFNs) are a class of materials that use nanoparticles to provide improvements over traditional thin film composite (TFC) reverse osmosis membranes by addressing design challenges such as low flux for brackish water sources, biofouling, etc. In our research group, various types of nanoparticles have been investigated as possible TFN additives, including TEMPO-oxidized cellulose nanocrystals (TOCNs), zwitterion-functionalized single-walled carbon nanotubes (Z-SWNTs), and metal-organic framework (MOF) nanoparticles. When these nanoparticles are incorporated during membrane fabrication, interactions at the polymer-nanoparticle interfaces create nanochannels which act as rapid water transport pathways through the membrane. This can lead to improved membrane flux while maintaining selectivity. While each class of nanoparticle possesses unique and interesting properties, they have been primarily chosen due to their high aspect ratios, which facilitate the formation of linear nanochannels more readily than other nanoparticle geometries. The presence of high aspect ratio nanoparticles in TFN membranes has been shown to improve water flux by up to 260% compared to a control polyamide membrane while maintaining high salt rejection, prompting further study into nanoparticle loading and potential variations in surface moieties to explore and control these nanochannels.

Poster 89 - GO-based Membrane as Barrier against Toxic Vapors/Gases vs. Membrane-Supported MOFs

Cheng Peng (New Jersey Institute of Technology), Zafar Iqbal (New Jersey Institute of Technology), Kamalesh Sirkar* (New Jersey Institute of Technology), Gregory Peterson (U.S. Army Edgewood Chemical Biological Center).

Traditional protective garments loaded with activated carbons to remove toxic gases are bulky. Novel graphene oxide (GO) flake-based composite lamellar membrane was developed for a garment for protection against chemical warfare agent (CWA) simulants: dimethyl methyl phosphonate (DMMP) (a sarin-simulant); 2-chloroethyl ethyl sulfide (CEES) (a simulant for sulfur mustard); yet it has a high moisture transmission rate. GO flakes (300-800 nm, 0.7-1.2 nm thick) dispersed in water were formed into a membrane by vacuum filtration on porous polyethersulfone (PES) or poly (ether-ether-ketone) (PEEK) support membrane for noncovalent π-π interactions with GO flakes. After physical compression, Upright Cup tests indicated: it can block toluene for 3-4 days and DMMP for 5 days while exhibiting excellent water vapor permeation. Further, they have low permeances for small molecule gases/vapors. The GO flakes underwent crosslinking later with ethylenediamine (EDA) during vacuum filtration followed by physical compression and heating. With a further spray-coating of polyurethane (PU), these membranes could be bent without losing barrier properties vis-à-vis DMMP for 5 days; a membrane not subjected to bending blocked DMMP for 15 days. For the PEEK-EDA-GO-PU compressed membranes after bending, the separation factors of H2O over others for low gas flow rates in the dynamic moisture permeation cell (DMPC) were: αH2O-He, 42.3; αH2O-N2, 110; αH2O-ethane, 1800. At higher gas flow rates in the DMPC, moisture transport rate goes up considerably due to reduced boundary layer resistances and exceeds the breathable fabric threshold water vapor flux, 2000 g/m2-day. The membrane displayed considerable resistance to permeation by CEES as well. This behavior has been compared with that of a nanopacked bed of membrane supported nanocrystals of UiO-66-NH2 MOF. The PES-EDA-GO-PU compressed membrane shows good mechanical property under tensile strength tests. (In press, ACS Appl. Mat. & Interf.).

Poster 90 - Sterically Hindered Poly(N-methyl-N-vinylamine) Membranes for CO2 Capture from Flue Gas

Ting-Yu Chen* (The Ohio State University), Xuepeng Deng (The Ohio State University), Li-Chiang Lin (The Ohio State University), Winston Ho (The Ohio State University).

Membrane separation technology requires highly CO2-selective and permeable membranes for a cost-effective post-combustion CO2 capture from flue gas. Amine-containing facilitated transport membranes are able to realize an increased permeability due to the reversible reaction between CO2 and amino groups. Here, we have developed an improved method for the synthesis of a high-molecular-weight sterically hindered polyvinylamine (SH-PVAm) as the new fixed-site carrier in our ultra-thin CO2 selective membranes.
Commercial polyvinylamine (PVAm) was mono-methylated into SH-PVAm using the stepwise reductive amination with a highly polar fluorinated alcohol as solvent to enhance the equilibrium shift to the imine formation. The method was found to prevent the over-alkylation, resulting in the increased yield of the target product. The synthesized SH-PVAm was characterized and confirmed by FTIR and NMR. The SH-PVAm solution retained a sufficiently high viscosity after incorporating the aminoacid salt, 2-(1-piperazinyl)ethylamine sarcosinate, as the mobile carrier for the membrane coating without any penetration issue. The SH-PVAm exhibited a significantly enhanced performance in the blended membrane of about 170 nm thickness over the unmodified PVAm for CO2/N2 separation at 57oC. The former membrane demonstrated a CO2 permeance of 1071 GPU and a CO2/N2 selectivity of 183 whereas the latter showed 975 GPU CO2 permeance and 163 selectivity. The improved membrane performance corresponded well with our computational results on the reaction chemistry of CO2 between the sterically hindered vinylamine monomer (SH-VAm) and the pristine vinylamine monomer (VAm) by density functional theory (DFT). SH-VAm showed a preference for the more efficient bicarbonate formation pathway and contributed to a higher CO2 loading capacity. Therefore, the steric hindrance effect of SH-PVAm enhanced the solubility of CO2 in the polymer matrix and resulted in the higher CO2 permeance.

Poster 91 - The impact of pre-chlorination on microfiltration fouling during municipal wastewater effluent filtration for water reuse

Kunal Gupta* (Texas A&M University), Shankar Chellam (Texas A&M University).

Microfiltration (MF) is an integral step of the water reclamation process; however, it's fouling continues to be an impediment. In this study, we looked at chlorination, a commonly employed step before MF in water reuse treatment trains, as a pretreatment step. That is pre-chlorination to control fouling during MF of municipal wastewater effluent. We evaluated the effect of NaOCl dosage (0 to 15 ppm, 90% effective chlorine) on the fouling of 0.1 μm hollow fiber polyvinylidene fluoride (PVDF) membranes. The experiments were performed with disinfected effluent from a local wastewater treatment plant on a bench scale constant flux system for five filtration-backwashing cycles. Fouling after pre-chlorination exhibited a dose-dependent behavior. Higher doses (7.5-15 ppm) employed in this study effectively reduced fouling; however, lower doses (1.5-5 ppm) resulted in increased fouling.
The dose-dependent alteration in fouling was analyzed in terms of interactions of effluent organic matter (EfOM) with the membrane. We investigated (i) sorption tendency of raw and chlorinated EfOM with colloidal PVDF particles (DOC removal during batch adsorption) and membranes (streaming potential variation) and (ii) free energy of adhesion between EfOM and membrane using extended Derjaguin-Landau-Verwey-Overbeek theory. Both sorption tendency and free energy of adhesion varied as a function of chlorine dose and showed a strong correlation to the observed dose-dependent fouling behavior. Also, pre-chlorination effectively changed the dominant fouling mechanism from cake filtration during raw effluent filtration to standard and/or intermediate blocking as revealed by constant flux blocking laws analysis. This change in the fouling mechanism was ascribed to the reduction in particle size with chlorination, which was evinced by particle size distribution measurements. The results from this research suggest the importance of careful consideration while choosing chlorine dosage before MF.

Poster 92 - Rapid Concentration and Isotopic Screening of Plutonium from Aqueous Systems Using Functionalized Membranes

James Foster* (Clemson University), Scott Husson (Clemson University), Timothy DeVol (Clemson University), Brian Powell (Clemson University).

The development of rapid screening methods for plutonium and special nuclear materials remains a crucial component for nonproliferation efforts. The determination of Pu isotopes at trace-levels in water requires tedious off-site techniques to prepare samples within detectable limits; these methods can be time-consuming when combined with analytical techniques like mass spectrometry. Therefore, the capability to combine on-site concentration with spectroscopic screening would prove to be an invaluable nuclear forensics tool.
This work describes the development and evaluation of a reactive polymer thin-film composite membrane to isolate and concentrate Pu from water at concentrations as low as 10-12 M. After filtration, the membranes are used directly as samples for alpha spectroscopy to perform isotopic screening. Four polymer-ligand films types were studied for their ability to recover plutonium from solutions of low (2.3) and circumneutral (6.3) pH, as well as from a low-activity solution composed of mostly Pu(V). High alpha spectroscopy peak energy resolutions were achieved for films prepared with the Pu-selective ligand bis(2-ethylhexyl) phosphoric acid (HDEHP) over a range of film thicknesses from 30 to 250 nm. XPS analysis revealed a surface excess of HDEHP in the films, which is advantageous for Pu recovery and detection. However, it also revealed that some ligand leached from the surface after contact with water, which may require that samples be acidified to low pH prior to filtration. Commercial ultrafiltration membranes coated with the reactive polymer thin films were tested for water permeability. The reactive thin-film composite membranes that yielded the highest water permeabilities were evaluated for recovery and isotopic screening of Pu at trace-levels in water. Alpha spectroscopy detection efficiencies and peak energy resolutions for membrane samples will be presented.

Poster 93 - Amine-Containing CO2-Selective Membrane and Process for Carbon Capture from Flue Gas

Yang Han* (The Ohio State University), Kai Chen (The Ohio State University), witopo salim (The Ohio State University), dongzhu wu (The Ohio State University), Winston Ho (The Ohio State University).

Large-scale application of membrane in post-combustion carbon capture has been limited by the trade-off between CO2 permeance and CO2/N2 selectivity of most polymeric membrane materials. In order to overcome this limitation, research efforts on facilitated transport membrane (FTM) have been devised with the objectives of (1) developing carriers with high CO2 loading capacity and reactive diffusivity, and (2) designing membrane processes that can capitalize on the outstanding selectivity shown by FTM. In this presentation, a novel FTM was synthesized in a composite membrane configuration with a 170-nm selective layer coated on a nanoporous support. In the selective layer, polyvinylamine was used as fixed-site carrier and an amino acid salt, synthesized by deprotonating sarcosine with 2-(1-piperazinyl)ethylamine, was blended as mobile carrier. The membrane was used to fabricate 1.4-m2 spiral-wound modules, which exhibited a CO2 permeance of 1450 GPU and a CO2/N2 selectivity of 185 at 67°C with actual flue gas at the National Carbon Capture Center in Wilsonville, AL, USA.
In addition, for the range of CO2 partial pressure relevant to carbon capture from coal-derived flue gas, significant increase in CO2 permeance can be achieved upon the bulk removal of CO2. For instance, at 67°C, the CO2 permeance can increase from 1464 to 1918 GPU when the CO2 partial pressure reduces from 74.1 to 3.9 kPa. Such a carrier saturation phenomenon has been modeled and incorporated into a two-stage membrane process featuring partial retentate recycle. It has been demonstrated that the bulk CO2 removal reduces the CO2 partial pressure gradually in the membrane module. This feature mitigates the carrier saturation and results in an uprising CO2 permeance upon the CO2 removal. For the membrane performance at 67°C, an attractive capture cost of $41.75/tonne can be achieved for 90% CO2 capture in addition to offering the reduced system footprint.

Poster 94 - Facilitated Transport Membranes with Tunable Amine–CO2 Chemistry for Hydrogen Purification

Yang Han* (The Ohio State University), Xuepeng Deng (The Ohio State University), Li-Chiang Lin (The Ohio State University), Winston Ho (The Ohio State University).

CO_2-selective, amine-containing facilitated transport membranes (FTMs) are of great interest for syngas purification since high-pressure H_2 can be retained upon CO_2 removal. Various FTMs have shown decent chemical and thermal stability at aggressive conditions, but their CO_2/H_2 separation properties are largely limited by the severe carrier saturation at high syngas pressure. Herein, we report a new approach to enhance the CO_2 permeability by manipulating the steric hindrance of the amine carrier. A series of α-aminoacids with different alkyl or hydroxyethyl substituents were deprotonated by 2-(1-piperazinyl)ethylamine, resulting in nonvolatile amine carriers with different degrees of steric hindrance. For hosting the low MW amine carriers, a water-swellable polymer network was synthesized from poly(vinyl alcohol) crosslinked by a bidentate tertiary aminosilane. In the presence of moisture, a bulkier alkyl substituent increased the steric hindrance and hence destabilized the carbamate adduct to afford bicarbonate through hydrolysis. Thus, this drastically increased the chemisorption of CO_2. Further, density functional theory (DFT) calculations were conducted to study the function of the hydroxyethyl substituent, which indicated that the hydroxyl group stabilized the bicarbonate through strong hydrogen bonding, thus further improving the CO_2 sorption. The enhanced CO_2 solubility significantly mitigated the carrier saturation, and an unprecedented CO_2/H_2 selectivity greater than 130 was demonstrated at 107°C and 12.5 atm of CO_2 partial pressure. As the CO_2 partial pressure reduced to 0.4 atm, a less hindered amine yielded a higher reactive diffusivity of CO_2, resulting in a CO_2 permeance of 435 GPU with a selectivity greater than 500. The CO_2/H_2 separation performance of these reaction-mediated polymeric membranes is well above the theoretical upper bound, and they open up a new avenue for designing a highly selective membrane process for syngas purification.

Poster 95 - High-Performance Gas Separation Membranes Based on Poly(benzimidazole)

Alexander Bridge* (The University of Texas at Austin), Joshua Moon (California Nanosystems Institute, The University of California Santa Barbara), Joan Brennecke (The University of Texas at Austin), Benny Freeman (The University of Texas at Austin).

Polybenzimidazole (PBI), commercially produced as Celazole®, is a glassy polymer with high potential to improve and expand applications for commercial gas separation membranes. Unlike current commercial membrane materials which are limited to gas separations below 100°C, PBI exhibits thermal and chemical stability at temperatures up to 400°C. PBI also exhibits size sieving properties that are attractive for dehydrogenation and pre-/post-combustion carbon capture. Challenges preventing integration of PBI membranes into commercial separations include addressing PBI's low gas permeability/permeance and a lack of progress in developing asymmetric membranes formed via phase inversion. In this project, composite membranes were developed by blending PBI with highly permeable thermally rearrangeable polyimides (PI). Phase separated PBI/PI blends exhibited enhanced H_2 permeability at 35°C (7.4 Barrer at 10 atmospheres) while maintaining the native H_2/CO_2 selectivity of PBI (~15). In parallel with this work, concentrated solutions of PBI dissolved in solvent systems based on N,N-dimethylacetamide, tetrahydrofuran, and ionic solids/liquids were optimized for flat sheet asymmetric membrane fabrication via aqueous phase inversion. Resulting membranes exhibited desirable morphologies previously only attainable via phase inversion in toxic coagulants such as acetonitrile, outlining a feasible approach for developing PBI membrane technology at a commercially relevant scale.

Poster 96 - Development and Testing of Pervaporation Desalination Membranes

Elisabeth Thomas* (Arizona State University), Mary Laura Lind (Arizona State University).

A critical challenge facing all water desalination, regardless of feedwater type, is management of the produced concentrated waste solutions. Reverse osmosis (RO) is the most commonly used desalination process because it has the lowest energy consumption compared to alternative thermal technologies and operated near its theoretical thermodynamic floor. RO typically achieves 40-90% recovery of the feed water (dependent on concentration and composition), leaving up to 30% of the water as concentrated waste for disposal. For brackish feedwaters (total dissolved solids < 5 g/L) the maximum recovery is typically limited by sparingly soluble salts or minerals (e.g., calcium sulfate or silica, respectively) and, ultimately, depends on the feed water composition. However, for high salinity waters, such as seawater (total dissolved solids ~ 32g/L) or produced waters from hydraulic fracturing (total dissolved solids 30-200 g/L) the maximum recovery is limited when the osmotic pressure of water exceeds the maximum achievable pressure within the system. Pervaporation is a membrane process that operates based on a vapor pressure difference across a dense, selective membrane. Pervaporation is more energy intensive than RO, therefore ultimately it is best suited for applications that exceed the limitations of RO (such as concentrate management or treatment of waters with TDS > 60-80 g/L).

Our team has been working collaboratively with researchers from at Rice University and University of Texas El Paso and we have developed a facile solution casting method for freestanding pervaporation membranes. These polymeric membranes show incredible promise for use in pervaporation desalination due to ease of casting and excellent performance parameters, with permeance of 135.5 ± 29 kg m-2 hr-1 bar-1 and salt rejection consistently greater than or equal to 99.5%. This reliably outperforms commercial pervaporation membranes.

Poster 97 - Synthesis of charged PEG macromonomers and their application towards antifouling thin-film composite membranes

Swati Sundararajan* (Ben-Gurion University of the Negev), Ron Kasher (Ben Gurion University of the Negev).

The degradation of membrane performance with time is inevitable and is a result of interaction of membrane surface with organic, inorganic or biological substances resulting in clogging of membrane which is known as membrane fouling. Many studies have focused on membrane surface hydrophilicity, surface roughness and surface chemistry as antifouling strategies, but the effect of surface electrostatic charge remains unexplored. The aim of this study was to design series of poly(ethylene glycol) (PEG)-containing surfaces with varied electrostatic charge. We have synthesized positively (quaternary ammonium) and negatively (sulfonate and phosphonate) charged PEG methacrylate macromonomers to study the effect the surface charge on antifouling behavior. The preparation of these PEG derivatives was confirmed by ATR-FTIR, NMR and HPLC. The synthesized macromonomers were grafted on nanofiltration NF90 membrane surface by redox-initiated polymerization. Successful grafting was confirmed by FTIR which indicates the presence of new carbonyl peak at ~1715 cm-1 and CH2 twisting peak at ~945 cm-1. XPS indicated increase in oxygen content due to the grafting of PEG chains. Zeta potential analysis showed increase in streaming potential with increasing negative charge on the surface. The water permeability and salt rejection studies were carried out in dead-end filtration cells. Further, fouling experiments have been carried out in crossflow system and showed antifouling effects of PEG-modified surfaces coupled with surface electrostatic charge. This study will provide a detailed picture about the effects of surface electrostatic charge on protein fouling, which is crucial for designing novel antifouling membranes.

This study was supported by Israel Science Foundation (ISF) in joint program with National Science Foundation of China (NSFC). S.S. would like to thank Blaustein Center for Scientific Cooperation (BCSC), Ben-Gurion University of the Negev, Israel for postdoc fellowship.

Poster 99 - Membrane Adsorbers for Medical Isotope Production

Maura Sepesy* (Case Western Reserve University), Benjamin Fugate (Case Western Reserve University), Christine Duval (Case Western Reserve University).

Radiopharmaceuticals offer promising new approaches for imaging and treatment for cancer patients; however, the current supply of these novel drugs is too low to support clinical trials. Copper-67 (Cu-67), can provide treatment (via beta decay) and imaging (via gamma rays). Cu-67 is purified using resin-packed columns, which are known to suffer from diffusion-limited transport. The practical implications of these limitations are (1) low column flow rates, (2) high elution volumes and (3) long purification times—all of which contribute to the bottleneck in the sustained supply of Cu-67. Due to the short half-life (2.58 days) of Cu-67, rapid purification is key for increasing the availability of the isotope for clinic trials and use.

A promising alternative to resin-packed columns are membrane adsorbers, which have seen recent success in the purification of biologics. This contribution describes our efforts to graft glycidyl methacrylate (GMA) from polyvinylidene fluoride (PVDF) membranes through activator generated by electron transfer, a type of, atom transfer radical polymerization (AGET ATRP). After grafting poly(GMA), diamine ligands of varying chain length (putrescine and ethylene diamine) are attached to the brushes through an epoxide ring opening reaction. Membranes are characterized by attenuated total reflection Fourier-Transform infrared spectroscopy throughout the synthesis process (grafting and ring-opening) to support functionalization. Membrane permeability was calculated from pure-water flux experiments in a dead-end filtration cell. The number of binding sites was quantified through acid-base titration and the binding capacity was calculated by modeling equilibrium batch adsorption data with the Langmuir isotherm. The results of this research are laying the groundwork for implementing membrane absorbers as a new separation material for medical isotope purification.

Poster 100 - Controlling Membrane Pore Structure of Polyelectrolyte Multilayer Nanofiltration Membranes for Selective Ion Removal

Ryan DuChanois* (Yale University).

Nanofiltration (NF) membranes have limited solute-solute selectivity in water treatment applications, especially when considering ions with similar size and charge. To selectively separate contaminants using NF, fabrication techniques with control of membrane pore size are clearly needed, but the role of membrane thickness in selectivity is not well understood. We demonstrate how to tailor membrane pore structure using layer-by-layer assembly of two strong polyelectrolytes, PDADMAC and PSS, on a polysulfone ultrafiltration substrate by systematically controlling the polyelectrolyte and salt concentrations in the deposition solution. Results suggest that membrane pore size and thickness can be simultaneously controlled in layer-by-layer assembly by tuning polyelectrolyte-salt interactions, which provides a tool for studying selectivity mechanisms in membrane separation processes. We subsequently investigated whether apparent energy barriers for anion transport—as an indicator of anion selectivity—primarily arise due to partitioning into the membrane phase (by altering membrane pore size and ion hydration energy) or intra-pore diffusion (by altering membrane thickness). Our findings suggest that anion flux and energy barriers are significantly affected by membrane pore size and ion hydration energy, but negligibly influenced by membrane thickness. Overall, we provide evidence that ion dehydration at the solution-membrane interface, not intra-pore diffusion, controls ion selectivity through NF membranes and will be central to the design of selective membrane materials.

Poster 101 - Uranium concentration using reactive polymer thin films and thin-film composite membranes for spectroscopic analyses

Abenazer Darge* (Clemson University), Yugantar Gera (Clemson University), Timothy DeVol (Clemson University), Scott Husson (Clemson University).

P This contribution describes the development of reactive polymer films for the concentration of uranium from circumneutral pH solutions for spectroscopic analyses. These films were prepared by grafting uranium-selective polymers from polyethersulfone (PES) films and membranes via UV-initiated polymerization, and adding uranium-selective functional groups to polyacrylonitrile (PAN) films and membranes by chemical reaction. Ellipsometry was used to study poly(phosphoric acid 2-hydroxyethyl methacrylate ester) film growth kinetics on PES films. X-ray photoelectron spectroscopy of modified PAN films revealed conversion of nitrile groups to amidoxime groups to be as high as 40% and showed that the extent and depth of reaction could be varied precisely. Static uptake experiments with solutions of depleted uranium spiked with 233U were conducted. An optimum sorption capacity of $1.09 \times 10\text{-}2 \pm 1.03 \times 10\text{-}3$ mmol/m2 and $1.15 \times 10\text{-}2 \pm 1.14 \times 10\text{-}3$ mmol/m2 were obtained at pH 6 for M-PAN and M-PES films, respectively. Uptake kinetics studies revealed that the maximum binding capacity of the films was reached after 2 h of batch contact with uranium solution. Alpha spectroscopy pulse height spectra were analyzed to study the role of selective layer film thickness on peak energy resolution. With the understanding of the physicochemical properties of thin film coatings and their influence on radioanalytical detection and uranium characterization, PES and PAN ultrafiltration membranes were modified with the same experimental conditions to develop uranium selective membranes. Permeability experiments were conducted to evaluate water transport before and after modification. Flow through uptake experiments were performed to study the effect of residence time on detection efficiency. Results from this work are being used to guide the development of thin-film composite membrane-based detection methods for the rapid, fieldable analysis of radionuclides in water for nuclear forensics investigations.

Poster 102 - Macrocyclic Nanofiltration Membranes for the Removal of Per- and Polyfluoroalkyl Substances in Drinking Water

Elham Abaie (Texas Tech University), Limeimei XU (Texas Tech University), Yuexiao Shen (Texas Tech University).

Because of the diverse functionality of per- and polyfluoroalkyl substances (PFAS), they have been mass-produced and applied in a variety of industries and our daily life globally. The abuse of these chemicals without regulations results in the omnipresence in the environment, eventually in drinking water resources, and thereafter environmental concerns[1]. Conventional drinking water treatment technologies are ineffective in removing or destroying chemically diverse and stable PFAS[2]. In the present research, we propose a combination of macrocyclic-based host-guest chemistry with membrane separations as a potential approach to remove PFAS effectively from drinking water resources. Macrocycles such as cyclodextrin containing nano-sized hydrophobic cavity can selectively bind PFAS molecules[3]. Different types of macrocycles such as pillar[5]arenes and cyclodextrins with different cavity sizes, shapes and hydrophobicity will be selected in membrane synthesis using interfacial polymerization techniques and the resulting thin composite membranes will be investigated for PFAS treatment. Different monomer concentrations and trimesoyl chloride concentrations will be employed to investigate the optimized conditions for the best treatment performances in terms of permeability and PFAS rejection. Simulated PFAS samples such as perfluorooctanoic acid (PFOA), perfluorooctane sulfonic acid (PFOS), and Pentafluorobenzoic acid (PFBA) will be used in the study. We expect that the combination of membrane filtration with macrocyclic-based molecular host-guest properties could provide a rapid, universal PFAS removal solution.

References

[1] I. Ross et al., Remediation, vol. 28, no. 2, pp. 101–126, 2018.
[2] M. F. Rahman et al., Water Res., vol. 50, pp. 318–340, 2014.
[3] A. Alsbaiee et al., Nature, vol. 529, no. 7585, pp. 190–194, 2016.

Poster 103 - Ultrafiltration for the purification of nucleic acid-based therapeutics

Ivan Manzano* (Penn State University), Andrew Zydney (Penn State University).

Nucleic acid therapies have the potential to treat a wide range of diseases through their ability to silence and regulate gene expression. This includes the use of short interfering and micro RNA for gene silencing or post-transcriptional regulation of gene expression. More recently, CRISPR technologies have created exciting opportunities for gene editing and novel biotherapeutics, using an engineered sgRNA that binds to specific DNA sites enabling modification / regulation of targeted genes.

Although the development of large-scale processes for the purification of therapeutic proteins like monoclonal antibodies is well-established, most work on nucleic acid purification has focused on laboratory-scale methods, many of which would be very difficult to employ for commercial manufacturing. The objective of this work was to evaluate RNA transmission through a series of composite regenerated cellulose and polyethersulfone ultrafiltration membranes with different NWCO (30 – 300 kDa). Data were obtained using a series of synthetic RNA with well-defined size and tertiary structure over a range of filtrate flux and solution conditions (i.e., added salt and urea). Additional insights into the ultrafiltration behavior were obtained using DLS and AGE to evaluate the effective size of the nucleic acids.

Results with dilute RNA solutions were strongly influenced by RNA adsorption to the RC membrane; this effect was not seen with PES membranes. Sieving coefficients also showed a strong dependence on the ionic strength due to significant changes in RNA structure and effective size. Data obtained with CRISPR-Cas9 complexes indicate the role of protein fouling on the retention of RNA, complicating the identification of appropriate conditions for the removal of free RNA from the ribonucleoprotein complex. These results provide important insights into the factors controlling RNA transmission and the potential for using ultrafiltration for the purification of nucleic acid therapeutics.

Poster 105 - UF membranes modified by covalently grafted graphene oxide with improved antifouling properties

Xiaoyi Chen* (State University of New York at Buffalo), Erda Deng (State University of New York at Buffalo), Dongwon Park (State University of New York at Buffalo), Blaine Pfeifer (State University of New York at Buffalo), Haiqing Lin (The State University of New York at Buffalo).

Graphene oxide (GO) is negatively charged and has good antifouling properties. However, its good dispersibility in water makes it challenging for membrane surface modification to improve antifouling properties. Herein, we demonstrate a versatile approach to covalently graft GO onto UF membrane surfaces with different hydrophilicities in aqueous solutions at room temperature. Specifically, we first functionalize the PSf membrane surface with polydopamine (PDA), a bio-adhesive that can deposit on membranes with a variety of surface characteristics. Second, the GO with carboxyl acid groups on the edges can be activated by 1-ethyl-3-(3-dimethylaminopropyl) carbodiimide (EDC) and N-hydroxysuccinimide (NHS). Finally, the activated GO can be covalently grafted onto the PDA layer through the reaction of the carboxyl groups and amines in the PDA. The GO was prepared using the modified Hummers' method and then etched using H2O2 to reduce the size and maximize the grafting efficiency. The GO and modified membranes were thoroughly characterized using SEM, AFM, Raman spectroscopy, and zeta-potential analyzer. GO grafting improves the surface hydrophilicity without decreasing the water permeance and decreases the bacteria adhesion by 82%. The GO-modified membranes exhibit 63% less fouling rate and 20% greater critical flux than the pristine one using 1 g/L sodium alginate in a constant-flux crossflow system. The modified membranes demonstrate stability for continuous 48-h operation and interval cleanings using NaOH solutions. Given the simplicity of the GO grafting and the versatility of the PDA for various surfaces, our approach can also be used as post-modification of modules for industrial use.

Poster 106 - Defining new performance metrics for the application of NF membranes in 'tough-to-treat' industrial wastewater applications

Abhishek Premachandra* (McMaster University), Nicole Perna (McMaster University), Susan O'Brien (McMaster University), Jacob McGivern (McMaster University), David Latulippe (McMaster University).

Nanofiltration (NF) membranes were developed for water softening and are rated based on their water permeance and salt removal. These metrics do not explain how NF membranes behave in 'tough-to-treat' applications, such as the treatment of industrial wastewater (IWW) containing organic compounds. In collaboration with a local IWW treatment plant, four poly-amide NF membranes (NF90 [Dow], TS80 [TriSep], NFX & NFS [Synder]) were evaluated in a cross-flow flat sheet configuration. Significant variability was observed in the total reduction of organic compounds (measured via chemical oxygen demand [COD]) and individual compounds removed (observed via gas chromatography-mass spectrometry [GC-MS]). For example, a 59±9% reduction in COD was obtained with NFS, while an 83±3% reduction was obtained with NF90. GC-MS characterization of the filtrate showed that compounds such as benzoic acid were completely rejected by NF90 but not NFX. Measuring membrane flux highlighted differences in performance during filtration. For example, NFS & NFX exhibited an increase in flux, while decreases were observed for NF90 & TS80. These results match previous studies on 'solvent activation', where exposure to certain solvents can significantly alter membrane performance. In this application, solvent exposure comes through constituents present in the IWW, and an understanding of how repeated exposure to IWW affects long-term performance is needed. The current phase of this project involves evaluating the best performing membranes (NFX & NFS) in an 1812 spiral-wound (SW) geometry with repeated exposure to IWW of varying concentration and composition to simulate their industrial application. Simultaneously, a pilot-scale system containing a SW NFX element is being installed at the treatment plant to serve as a proof-of-concept and validate the results from the lab-scale studies. This study lays the groundwork for characterizing the performance of polymeric NF membranes in IWW applications.

Poster 107 - Printing polyepoxyether thin film composite (TFC) membranes: Achieving membrane chemical robustness with tunable permeance for Nanofiltration Applications

Xin Qian* (University of Connecticut), Jeffrey McCutcheon (University of Connecticut), Rhea Verbeke (KU Leuven), Ivo Vankelecom (KU Leuven).

Membrane stability in oxidative environments has been regarded as a significant challenge in treating drinking water and reusing wastewater. Compared with conventional polyamides, which typically comprise today's reverse osmosis (RO) membranes, polyepoxyether has exhibited superior chemical stability in such environments. Previously developed crosslinking techniques for polyepoxyether membranes result in relatively low permeance membranes. In this work we evaluate the use of electrospray to print thin layers of polyepoxyether selective layer on a ultrafiltration support layer. The membranes exhibit over 50 LMH/bar water permeance and over 90% Rose Bengal rejection. Printed membranes were found to maintain a water permeance of 35 LMH/bar with a 99% Rose Bengal rejection after strong NaClO attack. Interestingly, interfacial polymerized TFC membranes that were made by the same recipe as printed membranes failed to maintain selectivity after the NaClO attack. Additionally, by post-curing the polyepoxyether membranes, we are able to greatly enhance the Methyl Orange rejection (97.6%). These results suggest that electrospray is able to generate a thin, permeable, chlorine tolerant selective layer.

Poster 108 - A New Concept for Generating Mechanical Work from Gas Permeation

Sarah Moussaddy (Oakland University), Jonathan Maisonneuve* (Oakland University).

Large amounts of energy are wasted from thermal power plants when exhaust is released into the environment. We estimate that the energy potential available from carbon dioxide and water vapor gradients of flue gas and cooling towers of a typical coal-fired power plant represent 1-2 % of the plant's overall capacity. In this study, we introduce a novel process for generating useful work from gas mixture concentration gradients, using membrane-based gas permeation. The theoretical limits to energy recovery from common exhaust streams are established, fundamental transport dynamics are introduced for the proposed process, and experimental validation is provided using a simple binary gas mixture of nitrogen and water vapor with commercially-available polydimethyl siloxane membranes. We observe power density of up 60 mW/m2 under relatively conservative conditions, and estimate that on the order of 1 W/m2 could be reasonably obtained under the right conditions. Implications for improving power plant efficiency and processing exhaust gases for carbon capture are discussed.

Poster 109 - Influences of Membrane Morphology and Biocompatibility on Hemodialysis Treatment Outcomes

Heloisa Westphalen (University of Saskatchewan), Amira abdelrasoul* (University of Saskatchewan).

A large population worldwide currently rely on hemodialysis (HD) for managing chronic and acute kidney diseases. HD is a membrane-based therapy through which toxins and excess fluids are removed from the blood stream compensating for the loss of kidney function. Although HD treatment is fundamental for patients' survival, this treatment option brings severe, life-threatening side effects associated with the contact between patient's blood and the membrane surface. The membrane properties play an important role in understanding and controlling protein adsorption and clearance of toxins in HD. The goal of the present study is to obtain a full characterization of two membranes currently in Canadian hospitals for HD treatment. The study also aimed to compare and explore how the membrane properties can affect protein adsorption and blood activation and the unstable cytokine levels and its consequences. In the present study, membrane morphology, hydrophilicity and its chemical composition using were analyzed using Atomic Force Microscopy (AFM), X-ray Photoelectron Spectroscopy (XPS), Raman mapping, Fourier Transform Infra-Red (FTIR) Spectroscopy, Nuclear Magnetic Resonance (NMR), Brunauer, Emmett and Teller (BET) particle sizer, Zeta potential, Scanning Electron Microscopy (SEM) and Transmission Electron Microscopy (TEM). The interaction between membrane materials and uremic blood after dialysis treatment was investigated using FTIR, Raman, and NMR in order to understand membrane fibers and how they reveal changes in chemistry. An in-vitro study was conducted to analyze the adsorption of fibrinogen using both membranes in a simulated HD session. The results obtained indicated that the morphology, hydrophilicity and chemistry of HD membranes have a significant impact on protein adsorption and lead to problematic outcomes for the patients.

Poster 110 - Highly permeable polymeric membrane fabrication through interfacial polymerization of pillar[5]arene artificial water channels

Matthew Skiles (University of Texas at Austin), Woochul Song (University of Texas at Austin), Diana Cintron (University of Texas at Austin), Manish Kumar (University of Texas at Austin).

We have previously reported that pillar[5]arene (P5A) based artificial water channels (AWCs) efficiently mimic the structure and function of biological membrane proteins, which have uniform pore size, as well as water transport efficiency and selective rejection properties which far exceed that of commercial membranes.(1) Previous research has incorporated contorted monomers within interfacial polymerization (IP) membrane fabrication to enhance microporosity and void interconnectivity.(2) The goal of this project is to incorporate P5A channels as monomers in IP fabrication. We hypothesize that the benefits realized with the IP process in previous studies (2) could be enhanced when the cavity sizes of the monomers are uniform and similar to those seen in biological water channel proteins.

The widely used IP membrane fabrication process was adapted to include P5A channels in the solvent, where the channels became crosslinked within membrane matrices. In order to maximize performance, membranes synthesized with varying solvent composition and monomer ratios were analyzed, and the optimal parameters for fabrication were defined. Optimized P5A channel based membranes demonstrated removal efficiencies >91% for relevant contaminants and water permeance of 3.13±0.35 (L m-2 h-1 bar-1), which was a 6-fold increase over the control with no AWCs.

References

1. Y. X. Shen et al., Achieving high permeability and enhanced selectivity for Angstrom-scale separations using artificial water channel membranes. Nature Communications 9, 11 (2018).

2. M. F. Jimenez-Solomon, Q. L. Song, K. E. Jelfs, M. Munoz-Ibanez, A. G. Livingston, Polymer nanofilms with enhanced microporosity by interfacial polymerization. Nature Materials 15, 760 (2016).

Poster 111 - Sorption of ternary gas mixtures of CO2, CH4 and C2H6 in PIM-1, and role of solubility in multicomponent gas separation

Francesco Maria Benedetti* (University of Bologna), Eleonora Ricci (University of Bologna), Antonella Noto (University of Bologna), Maria Grazia De Angelis (University of Bologna), Tim Merkel (Membrane Technology & Research Inc.), Jianyong Jin (The University of Auckland).

In gas separation performed with dense membranes, permeability is the product or solubility and diffusivity. The extent of each of these contributions significantly changes between pure-gas and mixed-gas experiments, because of competitive sorption and other non-idealities. In particular, when glassy polymers like PIM-1 and gases with significant solubility differences like CO2 and CH4 are involved, the solubility factor has been shown to have a predominant role over the diffusivity component. However, so far, only measurements for the case of sorption of binary mixtures were reported, and there is a lack of characterization of more complex mixed-gas data, which is of remarkable importance to understand transport mechanisms in glassy polymers at relevant industrial conditions.

To fill this gap, we measured the sorption isotherms of C2H6, CO2, and CH4 in PIM-1, and their binary and, for the first time, ternary gas mixture combinations. The development of a new measurement protocol allowed to perform this experimental investigation at constant-composition variable-pressure of the gas phase, using a pressure decay apparatus. For the ternary mixtures, the tests were performed at 15/15/70 mol% and 5/25/70 mol% of C2H6/CO2/CH4, respectively. Sorption isotherms relative to C2H6/CH4 and C2H6/CO2 binary mixtures were measured as well, in order to identify third component effects on the solubility and selectivity of the material.

The results obtained show that C2H6 has a strong exclusion power over CH4 and CO2, both in the binary and the ternary mixture scenarios. The case of C2H6/CO2 mixtures was particularly surprising because the two gases have very similar condensability and pure-gas sorption in PIM-1. The NELF model predictions for binary and ternary sorption were performed using only pure-gas parameters as input, and results are in good agreement with the experimental data. Finally, general considerations on the role of sorption in multicomponent conditions can be drawn.

Poster 112 - Tailoring structural and functional features of high-performance ionenes and ionic composites designed for membrane-based gas separations

Kathryn O'Harra* (University of Alabama), Irshad Kammakakam (University of Alabama), Jason Bara (University of Alabama).

We have developed a library of high-performance (HP) ionenes, or polymers which contain ionic groups along the backbone chain rather than as pendants. These materials combine structural elements typically associated with state-of-the-art gas-separation membranes with functionality utilized in ultra-high-performance engineering polymers (i.e. imide, amide linkages). This work probes the effects of spacing and sequencing of functional features and ionic groups incorporated along the main chain, specifically imidazolium cations paired with fluorinated anions (Tf2N, OTf). Our methods take advantage of established chemistries that impart great control over repeat unit structure and regiochemistry. These ionenes are robust, exhibiting good thermal and mechanical properties, but extremely tailorable based on the modular design and synthetic methods which allow for innumerable potential derivatives. Thus, the thermophysical properties of these materials are highly tunable based on methodical combination of diverse monomers and linkages. These polymeric materials exhibit additional self-assembly and nanostructuring when impregnated with "free" imidazolium-based ionic liquids (IL), which contributes another degree of tunability and increases stability as the IL serves as a non-covalent cross-link between the polymer chains. We are investigating the effects of altering charge density as well as adding functional small molecules and ionic moieties into the polymeric matrix which alter intramolecular interactions and promote further structuring within the framework. These HP-ionenes are thoroughly characterized in order to develop and study structure-property relationships. Composites are similarly analyzed for comparison with the corresponding neat polymers, to gain a better understanding of the coordination between polymer chains and dispersed additives. This presentation focuses on the design and performance of these novel ionenes and composites as gas separation membranes.

Poster 113 - High-temperature H2 purification from atom-thick g-C3N4 nanosheets hosting ordered nanopores

Luis Francisco Villalobos* (Ecole polytechnique fédérale de Lausanne), Kumar Varoon Agrawal (Ecole polytechnique fédérale de Lausanne), Mostapha Dakhchoune (Ecole polytechnique fédérale de Lausanne).

Attractive 2D materials for membrane-based separations should have a high pore density, a narrow pore size distribution, and be as thin as possible. Poly(triazine imide) or PTI, an emerging class of material belonging to the crystalline g-C3N4 framework, is one of the most promising nanoporous atom-thick nanosheets for gas separation [1, 2]. Single-layer PTI lattice hosts unique triangular nanopores formed by imide-bridged triazine groups. The electron-density-gap in these nanopores is ca. 0.34 nm, making PTI highly attractive for selective sieving of light gases (He, H2, and CO2) from larger gas molecules (N2, CH4, larger hydrocarbons, etc.), as shown in Figure 1B. PTI is an extremely stable material that can withstand strong acidic environments, as well as temperatures up to 600 °C in non-oxidizing conditions [3, 4]. This opens up the possibility of using PTI films for H+ transport in electrolytic cells, and to recover H2 directly from hot streams without having to cool them.

In the presentation I will discuss our recent success in i) synthesizing high-crystalline layered-PTI in a scalable manner, ii) an efficient route for exfoliation of single-layer PTI nanosheets, and iii) fabrication of exfoliated PTI-based membranes for hydrogen sieving [5].

Gas permeation studies and first-principles simulations demonstrating the ability of PTI nanopores to sieve H2 from larger molecules will be discussed. Briefly, PTI nanosheet were incorporated in a polybenzimidazole matrix leading to membranes with H2-permeance reaching 1500 GPU, and H2/CO2, H2/N2, and H2/CH4 selectivities reaching 10, 50, and 60, respectively, at 250°C.

References

[1] Bojdys et al., Chem. - A Eur. J. 14, 8177–8182 (2008); [2] Wirnhier et al., Chem. - A Eur. J. 17, 3213–3221 (2011); [3] Schwinghammer et al., Angew. Chemie 52, 2435–2439 (2013); [4] Miller et al., Nano Lett. 17, 5891–5896 (2017); [5] Villalobos et al., Sci. Adv. 6, eaay9851 (2020).

Poster 114 - Interpenetrating Polymer Networks for Organic Solvent Nanofiltration Membranes

Gergo Ignacz* (King Abdullah University of Science and Technology), Gyorgy Szekely (King Abdullah University of Science and Technology).

Marine mussel inspired polydopamine (PDA) has received increased attention due to its good thermal and chemical stability as well as strong adhesion on most materials. PDA-coated membranes have received attention due to their excellent features and potential ability to improve the stability and rejection of membranes. In this work, high-performance nanofiltration membranes based on interpenetrating polymer networks (IPN) incorporating PDA and polybenzimidazole (PBI) were developed for organic solvent nanofiltration (OSN)[1]. Generally, in order to obtain solvent stability, polymers need to be covalently cross-linked under harsh conditions, which inevitably leads to losses in permeability and mechanical flexibility. Surprisingly, by in situ polymerization of dopamine within a PBI support, excellent solvent resistance and permeance of polar aprotic solvents were obtained without covalent cross-linking of the PBI backbone due to the formation of an IPN. The molecular weight cutoff and permeance of the membranes can be fine-tuned by changing the polymerization time. Robust membrane performance was achieved in conventional and emerging green polar aprotic solvents in a wide temperature range covering −10 °C to +100 °C. It was successfully demonstrated that the in situ polymerization of PDA – creating an IPN – can provide a simple and green alternative to covalent cross-linking of membranes. To elucidate the nature of the solvent stability, a detailed analysis was performed that revealed that physical entanglement along with strong secondary interaction synergistically enables solvent resistance with as low as 1–3% PDA content.

References

[1] Zhao, D.; Kim, J. F.; Ignacz, G.; Pogany, P.; Lee, Y. M.; Szekely, G. Bio-Inspired Robust Membranes Nanoengineered from Interpenetrating Polymer Networks of Polybenzimidazole/Polydopamine. ACS Nano 2019, 13, 1, 125-133. https://doi.org/10.1021/acsnano.8b04123.

Poster 115 - Enhancing antimicrobial properties of Chitosan/ Graphene Oxide membrane for waste water treatment application

Raheleh Daneshpour* (University of Arkansas), Lauren Greenlee (University of Arkansas).

Biofouling is one of the major challenges in wastewater treatment by membranes. So, enhancing antimicrobial properties of the membrane is critically needed to avoid biofilm growth which leads to flux reduction and more energy consumption. Graphene oxide (GO) and Chitosan (CS) are promising candidate materials for antimicrobial membranes due to their abundancy, low cost, scalability, stability in water and their oxygen-containing functional group.

The ultimate goal of this project is to achieve an optimum composition of CS/GO and PVA (Polyvinyl alcohol) to create a membrane with a broad chemical and biochemical contaminant rejection. The anti-microbial properties of the CS/GO-PVA membrane allow this type of membrane to be used broadly for disinfection in wastewater treatment processes and demonstrate contaminant rejection performance that is comparable to RO/NF rejection and we will demonstrate the resulting membrane effectively decreased the biofilm formation in cross-flow systems.

Poster 116 - Hindering Lattice Flexibility Of Metal Organic Frameworks For CO2-sieving

Deepu Babu* (Ecole polytechnique fédérale de Lausanne), Kumar Varoon Agrawal (Ecole polytechnique fédérale de Lausanne).

Metal-organic framework (MOF) membranes are advantageous to their nanoporous counterparts owing to a high degree of structural tunability, relatively fast crystallization kinetics, and higher reproducibility [1-3]. Although MOF membranes were first reported almost a decade ago, so far, an attractive CO2/CH4 or CO2/N2 separation performance has remained elusive. This is because MOFs possessing a pore aperture suitable for CO2 sieving typically suffer from lattice flexibility and as a result, a sharp molecular cut-off is not obtained. For example, ZIF-8, has a crystallographically-determined pore aperture of 0.34 nm, ideal for separating CO2 (0.33 nm) from N2 (0.36 nm) and CH4 (0.38 nm). However, CO2/CH4 or CO2/N2 selectivities greater-than 5 have not been reported from ZIF-8 membranes.

Here we report a post-synthetic rapid heat treatment (RHT) method, implemented in a few seconds between 320-400 °C, which drastically improves the carbon capture performance of the ZIF-8 membranes. Lattice stiffening is confirmed by the appearance of a temperature-activated transport, attributed to a stronger interaction of gas molecules with the pore aperture. Spectroscopic and X-ray diffraction studies confirm that while the coordination environment and crystallinity are unaffected, lattice distortion and strain are incorporated in the ZIF-8 lattice, increasing the lattice stiffness. Unprecedented CO2/CH4, CO2/N2, and H2/CH4 selectivities of up to 34, 37, and 200 respectively, which is an order of magnitude higher than the values reported so far for ZIFs, and a complete blockage of C3H6 is achieved [4]. Furthermore, the results from our latest studies describing the shrinkage of MOF crystals at elevated temperature will also be presented.

References

1. Kang et al. J. Mater. Chem. A 2017, 5 (21), 10073–10091.
2. Babu et al. ACS Sustain. Chem. Eng. 2019, 7 (1), 49–69.
3. Hao et al. J. Mater. Chem. A 2020, DOI: 10.1039/c9ta12027k
4. Babu et al. Adv. Mater. 2019, 31 (28), 1900855.

Poster 117 - Graphene-based nanocomposite membranes with a mussel inspired polydopamine coating for produced water treatment

Gyorgy Szekely (King Abdullah University of Science and Technology), Abdulaziz Alammar* (University of Manchester).

The treatment of wastewater from the oil and gas industry presents a very specific problem because the wastewater produced is comprised of a complex mixture of oil and water that can be difficult to treat. In this work, polybenzimidazole (PBI), graphene oxide (GO) and reduced GO (rGO) nanocomposite membranes were developed via the common blade coating and phase inversion technique for the treatment of produced water from the oil and gas industry. The nanocomposite membranes were dip-coated by polydopamine (PDA), which is known for its antifouling properties. For the industrially relevant produced water, stable emulsions with high salinity, sharp unimodal size distribution and average oil droplet size of less than 500 nm were prepared. The incorporation of just a few weight percent GO into the PBI matrix resulted in superior oil-removal efficiency up to 99.9%, while maintaining permeance as high as 91.3 ± 3.4 L m−2 h−1 bar−1. The presence of GO also increased the mechanical stability of the membrane. The biofouling test of the nanocomposite membrane over 180 days showed remarkable improvement compared to the pristine PBI membrane. The nanocomposite membranes described in this work demonstrated promising long-term performance for oil-in-water emulsion separation as well as antifouling and antimicrobial properties without any alkaline or acidic cleaning. The membranes were capable of de-oiling high salinity emulsions with excellent reusability, highlighting that these membranes are promising for produced water treatment under harsh industrial conditions.[1]

References

[1] A. Alammar, S.H. Park, C.J. Williams, B. Derby, G. Szekely, Oil-in-water separation with graphene-based nanocomposite membranes for produced water treatment, J. Membr. Sci., 603 (2020), 118007, https://doi.org/10.1016/j.memsci.2020.118007.

Poster 118 - Enhanced Degradation of 1,4-dioxane by Photo-Fenton Reactive Ceramic Membrane

SHAN XUE* (NJIT).

1,4-dioxane is a potential carcinogen to human and has been an emerging environmental concern due to its wide presence in various water sources. The present study evaluated a photo-Fenton reactive membrane which simultaneously allowed high efficient photo-Fenton oxidation and catalyst retention for enhanced 1,4-dioxane removal performance. The membrane was prepared by covalently coating photocatalyst of goethite (α-FeOOH) on a robust ceramic porous membrane as we reported previously. The effects of UV irradiation, H2O2 and catalyst dosage on the 1,4-dioxane removal efficiency in batch reactors were first evaluated for optimized reaction conditions, followed by a systematical investigation of photo-Fenton degradation-filtration for 1,4-dioxane removal in the photo-Fenton reactive membrane reactor. We show under optimized conditions, the 1,4-dioxane removal rate reached 21% with combination of H2O2 dosage and UV365 irradiation in the photo-Fenton reactive membrane reactor.

Moreover, the total organic carbon, hydroxyl radicals and degradation by-products were also assessed to reveal deeper insight into the degradation mechanisms of 1,4-dioxane. This study demonstrated enhanced 1,4-dioxane degradation processes via synergistic incorporation of photo-Fenton reactions into ceramic membrane filtration.

Poster 119 - Zwitterionic-Containing Ultra-thin Hydrogel Selective Layer for Fouling-Resistant Ultrafiltration Membranes

Alice Oliveira Aguiar* (Tufts University), Hyunmin Yi (Tufts University), Ayse Asatekin (Tufts University).

Zwitterions are defined as neutrally charged molecules with equal number of cationic and anionic groups. They possess several interesting characteristics, such as high dipole moments, super-hydrophilicity, and exceptional resistance to fouling. These properties make zwitterions excellent materials for membranes applications, since membrane fouling by organic compounds is still one of the main obstacles to more extensive use of membrane separation processes. Zwitterion-containing hydrogel networks can potentially be used as selective layers of thin-film composite membranes to create highly fouling-resistant ultrafiltration membranes. We have recently demonstrated a new method, Interfacially Initiated Free Radical Polymerization (IIFRP), that allowed for the formation of an ultra-thin, defect-free hydrogel layer on top of a porous support. First, an aqueous phase with hydrophilic monomers saturates a porous support. Then, the membrane surface is covered with an organic phase with photoinitiator and exposed to UV light. The hydrogel selective layer is formed at the interface of these two phases. The present study aims to evaluate the influence of the zwitterionic monomers sulfobetaine methacrylate (SBMA) and 2-methacryloyloxyethyl phosphorylcholine (MPC) on the membrane's properties. Poly(ethylene glycol) diacrylate (PEGDA) was used as crosslinker, and membranes with poly(ethylene glycol) acrylate (PEGA) were used as control. SBMA membranes had much higher permeance than the other membranes, and MPC the second highest. Moreover, the increase in permeance was not accompanied by a decrease in rejection of myoglobin. Lastly, excellent oil fouling resistance was observed after 24-hour fouling experiments. Flux recovery rates (FRR) were above 90%, except for the MPC-containing membrane (FRR ~ 86%). These results indicate that zwitterion-containing ultra-thin hydrogel selective layers are very promising materials for fouling-resistant ultrafiltration membranes.

Poster 120 - Amphiphilic tercopolymer membranes for water vapor removal in air dehumidification

Faheem Akhtar* (KAUST), Kim Choon Ng (King Abdullah University of Science and Technology (KAUST)), Klaus-Viktor Peinemann (KAUST).

Selective removal of water vapor from humid streams is an important industrial process that is currently dominated by energy-intensive processes like dew-point dehumidification or desiccant-based air dehumidification. Moreover, in tropical countries, the energy consumed for HVAC comprises a significant portion of total energy consumption in a building. There is a need for energy-efficient dehumidification processes. Engineering membranes suitable for dehumidification offer several advantages like a non-regenerative nature (i.e., continuous operation), close control of water vapor, and, most importantly, the advantage of a high energy efficacy. Polymeric membranes especially, PEG-based membranes present an efficient solution to this problem; however, the mechanical stability issues and loss in vapor permeance over time lessen their use in such applications.

Herein, we demonstrate high-performance membranes for water vapor and CO2 separation based on novel amphiphilic copolymers. We have synthesized amphiphilic tercopolymers comprising PAN, PEGMA, and PDMAEMA segments via an economical and facile free radical polymerization. These membranes are mechanically-robust, and water vapor permeability and selectivity are enhanced by the chain-length and content of the PEGMA segments. In all the varying concentrations of PEGMA segments, a substantive improvement in water vapor permeability is observed with very high vapor/N2 selectivities. Optimized PEGMA segments yield very high CO2/N2 and CO2/CH4 selectivities with enhanced CO2 permeability.1 The reason for this superior gas and water vapor separation performance will be discussed in detail during my presentation.

Reference

1. Akhtar, F. H.; Kumar, M.; Vovusha, H.; Shevate, R.; Villalobos, L. F.; Schwingenschlögl, U.; Peinemann, K.-V. J. M., Scalable Synthesis of Amphiphilic Copolymers for CO2-and Water-Selective Membranes: Effect of Copolymer Composition and Chain Length. Macromolecules 2019, 52 (16), 6213-6226.

Poster 121 - Synthesis of sodalite precursor nanosheets and facile assembly for hydrogen purification

Mostapha Dakhchoune* (Ecole polytechnique fédérale de Lausanne), Kumar Varoon Agrawal (Ecole polytechnique fédérale de Lausanne), Luis Francisco Villalobos (Ecole polytechnique fédérale de Lausanne).

Zeolite membranes have been studied for energy-efficient gas separation under thermally and chemically harsh conditions for several decades. However, their implementation in the gas and vapor separations has been hampered because of the reproducibility issue arising from the complex hydrothermal synthesis route and the defect formation during the activation step.

Herein, we report the exfoliation of the layered zeolite precursor of sodalite, RUB-15, into single 0.8 nm-thick nanosheets hosting hydrogen-sieving six-membered (6MR) rings of SiO_4 tetrahedra and their assembly by simple filtration into thin films for H_2 sieving yielding H_2/CO_2 over 100. The hydrothermally synthesized material was confirmed to be RUB-15 using X-ray diffraction, 29Si MAS NMR and SAED. Swelling of RUB-15 with a C16 cationic surfactant was performed to weaken the interlayer interactions, and melt compounding technique was performed to exfoliate the layers. Exfoliated nanosheets crystallinity was confirmed by TEM.

Membrane fabrication was performed by filtration of the nanosheets dispersed in ethanol. As-filtered membranes showed a periodical arrangement of the nanosheets along the z-axis with a d-spacing of 11.4 Å which translates in a gallery spacing of 3.4 Å. Intersheet gaps dominated the overall transport leading to a cut-off in the kinetic dimeter of 3.6 Å yielding H_2/N_2 selectivities over 20, while CO_2 was able to pass between the gallery spacings. The presence of reactive terminal silanol groups in the RUB-15 nanosheets presented a unique opportunity for the elimination of the nanosheets gaps through covalent Si-O-Si linkages. Indeed, calcination of the as-filtered nanosheets film led to a decrease of periodic interlayer spacing to 7.4 Å preserving the nanosheets crystallinity. Upon calcination, the dominated transport was through the 6-MR yielding H_2/CO_2 selectivity in the range 20-100 and H_2 permeance in the range 41-424 GPU at 250-300 ºC.

Poster 122 - Novel Electrospun Nanoparticle Nanofiber Composites for CO2 separation

Amogh Meshram* (Arizona State University), Bin Mu (Arizona State University).

The purpose of this research is to develop novel Electrospun Nanoparticle Nanofiber Composites (ENNCs) using hydrophobic polymers to create a selective coating around commonly used adsorbents such as zeolites and protect them from moisture and corrosive gases like H_2S. The separation principle of these composites is at the interface of membrane separation and high-pressure adsorption. The thin polymer layer around the adsorbents will act as membrane barriers and selectively facilitate the adsorption onto these composites. A recent study by our research group has shown that these electrospun composites can withstand extreme hydrothermal conditions along with a decrease in degradation rates [1]. Hence, this research will mainly focus on optimizing the contents of the electrospinning dope solution to maximize the adsorbent loading in order to obtain the highest separation performance along with minimum degradation of the adsorbents under extreme hydrothermal conditions. These composites will be characterized and compared using BET surface area, CO_2 adsorption isotherms, water contact angle, TGA, FTIR and SEM images.

References

1. Armstrong, M., Sirous, P., Shan, B., Wang, R., Zhong, C., Liu, J., Mu, B. Microporous Mesoporous Mater., 2018, 270, 34–39.

Poster 123 - Novel Nanofiltration Membranes for Isolation of Pharmaceutical Compounds

Gazelle Vaseghi* (Compact Membrane Systems), Kenneth Pennisi (Compact Membrane Systems), Hannah Murnen (Compact Membrane Systems), Sudip Majumdar (Compact Membrane Systems).

Syntheses of pharmaceutical compounds are often carried out in organic solvents and involve intermediates and active pharmaceutical ingredients (APIs). The organic solvents must then be separated in order to produce the final product medications. APIs and the intermediates are often thermally labile; thereby thermal separations are not preferred. Moreover, due to the latent heat associated with solvent vaporization, thermal separations are energy intensive and costly. Crystallization by the addition of anti-solvents is frequently used as an alternative, however, it requires significant optimization to generate acceptable process yields. Additionally, in large scale pharmaceutical processes the isolation of non-thermally labile compounds by crystallization and distillation can be difficult to achieve while increasing the amount of solvent use and waste generated. Hence, pharmaceutical companies are considering alternative methods for recovery of organic solvents.
Organic solvent nanofiltration (OSN) is a superior substitute to phase change separation processes. It is a molecular level size-sieving process operable at ambient or sub-ambient temperatures, providing a considerable advantage for thermally labile compounds. Despite potential opportunities, nanofiltration (NF) has not been widely adopted in pharmaceutical industry due to a lack of robust NF membranes. Hydrocarbon-based NF membranes are either not compatible with common organic solvents or dramatically lose performance at high operating pressures due to the interactions of solvent with the membrane material and subsequent swelling of the membrane. This work describes the development of custom fluoropolymer-based OSN membranes for concentration of APIs and intermediates and recovery of high purity solvents. The proposed membranes show superior chemical stability and performance with great economic potential in pharmaceutical processes. The durability of the membranes over long-term operation is examined.

Poster 124 - Nexar Coated Hollow fibers for membrane dehumidification

Lakshmeesha Upadhyaya* (KAUST).

The removal of water vapor from humid air is one of the energy-intensive processes and consumes more than 60% of overall daily use energy in tropical and Middle Eastern countries. Air condition provides comfortable conditions indoor (25°C and around 45% humidity) for which the outdoor air has to be cooled and then dehumidified. Membrane technology becomes an alternative to traditional dehumidification systems like desiccation & direct chillers due to lower energy consumption, ease of operation, and miniaturization of the order by replacing flat sheets over hollow fibers. In our current work, hydrophobic ULTEM® polyetherimide is selected for the fabrication of hollow fibers using ethylene glycol (EG) as bore fluid using a wet/dry spinning procedure. EG inherits a more finger-like structure with double symmetry due to the difference in viscosity of bore fluid and water during the phase separation process. The vapor permeance was enhanced through the membrane by coating them with commercial sulfonated pentablock copolymer called Nexar®, which is both hydrophilic and hydrophobic. In house dehumidification set up with a modified permeability cup method was used for vapor permeation, which is open to both flat sheet and hollow fibers configuration. Liquid desiccants, air, and vacuum were used in the lumen side to create the driving force for transport of water vapor in the humidified stream, which is present in the shell side. The vapor permeability drastically increased due to the coating with Nexar®. For 2 wt.% Nexar® in tetrahydrofuran coating, about 5000-12000 GPU of water vapor permeance was achieved as the mode of operations were varied with high selectivity (water /nitrogen). This significant increase was mainly due to the formation of hydrophilic channels on the surface of the membrane due to the presence of sulfonated blocks during microphase separation.

Poster 126 - A novel RGO-CCNT-based catalytic membrane with co-enhanced permeability and catalytic activity for high-efficiency degradation of charged antibiotics

Wenli Jiang* (University of California Berkeley), Sunxiang Zheng (University of California, Berkeley), Monong Wang (UC Berkeley), Baoxia Mi (UC Berkeley).

In this work, we demonstrate a high-performance RGO-CCNT modified electro-Fenton catalytic membrane toward degrade low-concentration refractory antibiotics during the electrocatalytic filtration process, which afford a promising approach to effective blocking of antibiotic resistance bacteria and gene generation or transmission in water treatment process. Besides, we analyzed the correlation between antibiotic removal efficiency and its physicochemical parameters, with the conclusion that acid dissociation co-efficient (charge performance), water solubility and bimolecular reaction rate constants of antibiotic with hydroxyl radicals were the three most influential factors. The antibiotic removal efficiency was also influenced by its molecular weight even though the correlation is not very obvious. Likewise, two regression analyzation methods (stepwise regression, stepwise regression of multiple factors and interaction terms) were applied to further clarify the relationship between antibiotic removal efficiency and these four physicochemical parameters. Correspondingly, two regression equations were obtained and the reliability was ulteriorly verified by comparing the experimental and estimated results. This could be used to forecast whether a kind of refractory organics could be effectively removed by this RGO-CCTN modified membrane during the catalytic filtration process, which supply a simple idea and method to assess the universality of this technique.

The results from the present study revealed that the separation and removal performance of RGO-CCNT modified membranes was greatly influenced by the membrane surface charge, pore size and content of catalysts. Overall, this study could provide a new insight in future for the design of catalytic membranes with both improved water permeability and rejection or removal efficiency of pollutants, which might greatly facilitate their development in water treatment and low-concentration refractory organics removal.

Poster 127 - Energy and performance optimization for electrodialysis water treatment

Zahra Abbasian Chaleshtari (New Mexico State University), Abdulhameed Alalwani (New Mexico State University), Reza Foudazi* (New Mexico State University).

The main goal of water treatment is to decrease the energy and cost, and to increase the water recovery rate. The recovery rate of a desalination system is defined as the ratio of volume flow rate of treated water to the input feed water. Electrodialysis (ED) has greater recovery (more than 80%) compared to RO process, with low chemical pretreatment. Additionally, ED units need lower capital costs and less specific energy compared to RO. They are resistant to scaling of membranes and are able to remove different cations and anions regardless of their size. In the present study, a laboratory scale ED system with the nominal capacity of 4-8 L/h per dilute and concrete cells is used for treatment of water with total dissolved solid (TDS) level of 1000 ppm to drinking water TDS level. Three parameters of applied voltage, feed flow rate, and temperature are varied in the range so that the maximum ion removal will be achieved with the minimum energy consumption. During the experiments, water samples will be collected and analyzed using a Dionex ICS-5000 Dual Channel IC System. The results of the present study have pivotal implications for desalination technologies.

Poster 128 - Enhancement of Nanofiltration Membranes through Chemical Bonding of Lignosulfonic Acid

Phillip Sandman* (University of Kentucky).

Nanofiltration (NF) has recently emerged as an important technology in industries including oil and gas, textiles, and water filtration. Although NF is not designed to remove all monovalent ions from water, it performs well in water softening and uses less energy since fewer solutes are separated. To apply green chemistry concepts and increase membrane permeability, rejection, and anti-fouling properties, an attempt was made to bond lignosulfonic acid (LSA) in a NF membrane through an esterification reaction to introduce greater negative charge. NF synthesis occurred via interfacial polymerization on a PS-35 ultrafiltration (UF) membrane with an aqueous solution of piperazine (PIP) and LSA and an organic solution of trimesoyl chloride (TMC). Bonding of LSA to TMC was evaluated using FT-IR. Permeability and MgCl2 and Na2SO4 rejection measurements were recorded in a dead-end stirred cell, and zeta potential measurements were taken using an electrokinetic analyzer. To model these phenomena, the Donnan Steric Pore Model with dielectric exclusion (DSPM-DE) will be employed. For a NF membrane synthesized using 1.3% LSA and 0.1% PIP in the aqueous solution and 0.14% TMC in the organic solution, MgCl2 and Na2SO4 rejections were 94% and 4%, respectively, at 6.9 bar. For a membrane synthesized using the same solutions without LSA, MgCl2 and Na2SO4 rejections were 82% and 1.6%, respectively, at 6.9 bar, indicating a slight change in rejection properties. Permeabilities were similar, with values of 25.5 LMH/bar and 27 LMH/bar for the LSA and non-LSA membranes, respectively. However, there are concerns concentration polarization may be introducing error into these values. Additionally, because the UF membrane dominated the signal in FT-IR, LSA bonding could not be verified since the signal for an ester bond could not be discerned. By enhancing membrane properties with LSA, potential for improved, greener water filtration methods exists.

Poster 129 - TiO2 Nanowires Based System for Urea Photodecomposition and Dialysate Regeneration

Guozheng Shao* (Univ. of Wash.), Yushi Zang (University of Washington), Bruce Hinds (Univ. of Washington).

More than 2 million End Stage Renal Disease patients receive dialysis to sustain life. Conventional hemodialysis removes urea and other metabolic waste from the body by running ~120 L of dialysate over hollow fiber dialysis membranes each session, which is typically 3-4 hours and 3 times a week. A portable dialysis device, that is working continuously, would bring significant health, quality of life and economic benefits to patients. The key technological challenge to portable devices is to recycle the dialysate to reduce its usage from 120 L to under 1L. We have developed an efficient photooxidation system based on hydrothermally grown TiO2 nanowires, UV LEDs, and catalytic gas diffusion barriers to decompose urea from the dialysate at rates sufficient to remove daily production of urea at 14 g/day. We show that a photoelectrochemical cell with TiO2/FTO anode, 10 mM urea/0.15 M NaCl electrolyte, and 4 mg/cm2 Pt black loaded carbon paper cathode can remove urea efficiently. Under 4 mW/cm2 illumination using 365 nm LED with 40% quantum efficiency, the device yielded a photocurrent density of ~ 1 mA/cm2, corresponding to 40% quantum efficiency in urea decomposition per incident photon. A device with 0.21 m2 area and a current draw of 11 A is able to decompose a daily 14 g urea production sufficient to regenerate dialysate. We have further demonstrated the removal of 14 g of urea in 24 h in a full scale device.

Poster 130 - Improving Membrane Performance for CO2 Capture With Ultrahigh MW Polyvinylamine

Kai Chen* (The Ohio State University), Yang Han (The Ohio State University), Winston Ho (The Ohio State University).

Membranes for post-combustion CO2 capture are required to have a high CO2 permeance due to the limited driving force. For facilitated transport membranes formed by polyvinylamine (PVAm), a defect-free selective layer of about 100 nm is usually required to render sufficient permeance with high CO2/N2 selectivity. In order to meet such a demand through the knife-coating process, the coating solution needs to have a high viscosity at a relatively low concentration to minimize its penetration into the substrate.

In this study, the challenge was overcome by synthesizing PVAm with an ultrahigh molecular weight (MW) via inverse emulsion polymerization (IEP). Compared to solution polymerization, IEP isolates the reaction in inverse micelles suspended in a continuous organic phase, which allows excellent dissipation of the heat generated by the reaction and reduces gel formation drastically. Hence, the monomer concentration used for the polymerization could be increased to 42.5 wt.%, and the MW, measured by static light scattering technique, was enhanced to as high as 13.5 million. For comparison, the PVAm synthesized via solution polymerization had a MW of only 1.2 million. As a result, the viscosity of a 3 wt.% PVAm solution was improved from 2,000 cp to 30,000 cp, which allowed a lower concentration to formulate the coating solution. Subsequently, a reduction in the selective layer thickness from 170 nm to 120 nm was achieved in a roll-to-roll continuous knife coating process. Besides the PVAm, small molecule amino acid salts were also blended into the membrane to increase the amine content, which enhanced the facilitated transport of CO2. With the ultrahigh MW PVAm to strengthen the polymer matrix, the content of the amino acid salts could be raised from 65 wt.% to 85 wt.%. These two factors led to a ca. 25% increase in the CO2 permeance, i.e., 1077 GPU with >140 selectivity at 57°C.

Poster 131 - Haloorganic Sorption and Degradation by Temperature Responsive Membranes

Rollie Mills* (University of Kentucky).

This research is aimed at evaluating the effect of temperature on the removal of perfluorochemicals and on the degradation of polychlorinated biphenyl (PCB) compounds in water filtration. Polyvinylidene difluoride (PVDF) microfiltration membranes are functionalized with poly-N-isopropylacrylamide (PNIPAm), and its temperature responsive behavior is studied as it relates to water flux and adsorption/desorption of perfluorochemicals. PNIPAm has a relatively low Lower Critical Solution Temperature (LCST) of around 32°C, which makes it appealing for industrial use. PNIPAm has a sharp transition from hydrophilic behavior below its LCST to hydrophobic behavior above its LCST, thus, when the PNIPAm is exhibiting hydrophobic behavior, it retracts/collapses upon itself in an aqueous environment. In doing so, the isopropyl groups dehydrate first, causing the polymer chain to collapse, thus causing hydrophobic contaminants to partition into that domain. Reversely, bringing the temperature down past the LCST will rehydrate the isopropyl groups, which reduces adsorption and allows desorption to occur. Additionally, iron-palladium nanoparticles are immobilized on the cross-linked polymer matrix in the membrane pores to degrade PCBs via catalytic dechlorination. Specifically, this study's goal is to evaluate the feasibility of temperature swing adsorption/desorption of Perfluorooctanoic acid (PFOA) and of using these temperature-responsive membranes for degrading PCBs in water filtration. Full-scale PVDF membranes were developed through collaborative work with Nanostone-Sepro (Oceanside, CA, USA).

Poster 132 - Polymer Membrane-based Liquid Hydrocarbon Fractionation

Ronita Mathias* (Georgia Institute of Technology), Kirstie Thompson (Georgia Institute of Technology), Daeok Kim (Imperial College London), Jihoon Kim (Imperial College London), Neel Rangnekar (ExxonMobil), JR Johnson (ExxonMobil), Scott Hoy (ExxonMobil Research and Engineering), Irene Bechis (Imperial College London), Andrew Tarzia (Imperial College London), Kim Jelfs (Imperial College London), Benjamin McCool (ExxonMobil Research and Engineering), Andrew Livingston (Imperial College London), M.G. Finn (Georgia Institute of Technology), Ryan Lively (Georgia Tech).

The separation of multicomponent, complex liquid mixtures, which are mixtures without a clear singular solvent, is an important, emerging area of membrane science. In this work, novel spirocyclic polymers were incorporated into thin-film composite membranes with the aim of separating liquid hydrocarbons that are typically found in the light fractions of crude oil. These cover a range of classes such as alkanes, cycloalkanes, alkyl aromatics, and polyaromatics. These hydrocarbon species were utilized not only to create a molecular weight cut-off (MWCO) curve but also to assess the performance of this class of membranes in the fractionation (as opposed to purification) of complex mixtures. A MWCO of ~ 250 Da was obtained for dilute mixtures, accompanied by high selectivities amongst different classes of hydrocarbons as well as molecules within the same class but with sub-nm differences in size. The polymer membranes also demonstrated the fractionation of light crude oil through a combination of class- and size-based "sorting" of molecules, where the permeate is enriched with molecules lighter than 170 Da equivalent to a boiling point less than 200°C. Such scalable, selective membranes offer potential for the hybridization of energy-efficient technology with conventional processes like distillation.

Poster 133 - The impact of monochloramines and dichloramines on reverse osmosis membranes in wastewater potable reuse process trains: A pilot-scale study

HYEJIN LEE (McMaster University), Amin Halali (McMaster University), Siva Sarathy (Trojan Technologies), Charles-François de Lannoy* (McMaster).

Water reuse is an important strategy to optimize water management. Microfiltration (MF) followed by reverse osmosis (RO) and ultraviolet (UV) based advanced oxidation process (AOP) is a proven treatment train for delivering potable recycled water. Chloramines are typically applied as disinfectants to mitigate biofouling of MF and RO membranes. While monochloramine is the intended oxidant, dichloramine is also produced. This study investigated mono- and dichloramine permeability across pilot-scale RO membranes, their oxidative effects on polyamide RO membranes, and their implications for downstream UV treatment. Permeability of mono- and dichloramine was 91% and 96% in the pilot system, respectively. The conversion of dichloramine to monochloramine was observed across the RO membrane suggesting that membrane-rejected ammonia promoted monochloramine speciation. At the RO membrane surface, dichloramine exhibited stronger oxidation of membranes as compared with monochloramine. Interestingly, when exposing polyamide membranes to dichloramine, flux decreased by 36%, suggesting that strong oxidation causes structural collapse of the polyamide thin film. Finally, a greater reduction of downstream UV transmittance (UVT) was observed by the permeation of monochloramine through RO membranes. This study suggests that any technoeconomic analyses of potable reuse processes should consider the impacts of both monochloramine and dichloramine on the capital cost associated with RO membrane lifetime and the operating costs associated with the UVT of UV processes.

Poster 134 - Synchrotron-based X-ray Spectroscopy to Investigate the Fouling Mechanism on Reverse Osmosis Membranes

Valerie Niemann* (Stanford University), Hans-Georg Steinrück (Paderborn University), Michael Toney (SLAC National Accelerator Laboratory), William Tarpeh (Stanford University), Sharon Bone (SLAC National Accelerator Laboratory).

Reverse osmosis (RO) is the leader in water desalination and purification technologies, but its major challenge lies in membrane fouling. In order to effectively mitigate fouling, a more mechanistic understanding is necessary. Calcium may play a major role in the fouling mechanism, as it can independently contribute to both scaling and fouling and interact with natural organic matter. It has been shown that calcium can form a bridge between negatively-charged ligands. Calcium may contribute to the fouling layer for its ability to sorb organic matter to the membrane surface and enhance the aggregation of organic and inorganic species.

In this work, we examined the calcium speciation on RO polyamide membranes using a set of synchrotron-based X-ray spectroscopic tools. We harvested membranes from the Orange County Water District pilot plant, which features RO with a pretreatment step and anti-scalant to purify wastewater to potable water. X-ray fluorescence mapping provided information on particle composition, size, and spatial distribution, on membranes at the lead and tail of the plant. Calcium K-edge X-ray absorption near-edge structure (XANES) spectroscopy provided definitive speciation information. These experiments were complemented with thermodynamic modeling.

Major findings included that the tail module had a higher number of particles, with larger particles on average and a larger particle size distribution, than the lead module. Calcium had the highest concentration on the membrane surfaces among a range of elements. XANES analysis showed that no crystalline, inorganic scales formed. Meanwhile, thermodynamic calculations showed that scaling was expected at feed conditions. We have identified likely calcium-organic complexes as foulants and concluded that calcium can connect organic ligands to negatively-charged functional groups on the polyamide membrane. These mechanistic findings help inform effective fouling-mitigation strategies.

Poster 135 - Detection of biofouling on gold-coated MF membranes by in-situ electrical impedance spectroscopy

Nan Zhang* (McMaster University), HYEJIN LEE (McMaster University), Charles-François de Lannoy (McMaster University).

Detection of biofouling in microfiltration processes enables early and effective strategies for fouling prevention. This work investigates the use of electrical impedance spectroscopy (EIS) is used for online monitoring the biofilm development on the surface of electrically conductive membranes (ECMs). A simple technology to fabricate ECMs is introduced through sputter deposition of the ultrathin gold layer (30 nm) on microfiltration polyethersulfone (PES) membranes. Gold-coated membranes exhibit extremely high electrical conductivity (~ 5 × 105 S/m) and high water permeability. SEM and AFM images of modified membranes show a homogeneous and flat coating surface.

To monitor the onset and development of biofouling in situ, a two-electrode crossflow filtration cell is customized in which a gold-coated membrane acted as the working electrode. Fouling experiments with a bacterial culture harvested from tap water were conducted over a duration of 12 h. Biofilm-induced permeate flux decline showed two stages of biofouling development. The initial stage was related to attachment and deposition of bacteria on the membrane surface. The second stage was associated with the accumulation of extracellular polymeric substances (EPS). Correspondingly, impedance spectra indicated the impedance at low frequency region (< 10 Hz) sharply decreased with fouling early on, and gradually decreased to the end of the experiments. Further, the measured impedance was modeled by an equivalent circuit from which a EIS-derived parameter, the normalized impedance related to diffusion was achieved. It is observed that the impedance-based detection is more sensitive to changes as compared to the decline of permeate flux during the early stage of biofouling. Hence EIS applied to ECMs has the potential to be used as canary cells installed on a side-stream of commercial membrane modules for fouling detection as well as fouling control in industrial applications.

Poster 150 - Highly efficient plasmonic membrane activation of peroxide for alcohol oxidation

Hao Tang* (University of Washington), Guozheng Shao (Univ. of Wash.) Bruce Hinds (Univ. of Washington).

Many industrial oxidation processes based on peroxide, have a difficulty of over oxidation in homogeneous solution principally due to faster oxidation rates of already oxidized species and high statistical probability of secondary oxidation near the completion of reaction. Ideal for stepwise oxidation is to limit the residence time of target molecule in a reaction zone to allow for single (or quantized) reaction events. This can be achieved in a membrane geometry where catalyst along a pore length and set flow velocity can precisely control residence time for oxidation. It is also known that Au nanoparticles can catalytically activate peroxide under light irradiation due to the formation of concentrated surface plasmon electromagnetic fields and thus hypothesized to be present in nanoporous planes. A plasmonic membrane was synthesized by evaporation of 25nm thick Au films onto pore entrances of anodized aluminum oxide membranes (AAO) with pore diameters of 20-200nm. This allowed solutions of peroxide, benzyl alcohol, sec-phenethyl alcohol to flow through membrane and interact with Au surface plasmon upon exit of the membrane. Under light illumination of 10-100 mW/cm2, quantum efficiencies (photon/peroxide radical) above 100% were seen (up to 350%), indicating a combination of both field induced activation of peroxide as well as a hot electron injection mechanism. Flow-rate controlled peroxide-assisted oxidation of benzyl alcohol and sec-phenethyl alcohol are demonstrated.

Poster 151 - Synthesis of γ-alumina supported ZIF-8 membranes for enhanced propylene/propane separation

Seyede Fateme Banihashemi* (Arizona State University), Jerry Lin (Arizona State University).

Zeolitic imidazolate framework-8 (ZIF-8) membranes have shown promising potential for propylene/propane separation, which is considered one of the most energy-consuming processes in the petrochemical industry. It is desirable to synthesize thinner ZIF-8 membranes with good crystal integrity to enhance their separation performance. In this work, we report the synthesis of thin (~3 μm) and high-quality ZIF-8 membranes on top of α-alumina substrates coated with a thin mesoporous γ-alumina layer. Permeation of pure gases and separation of equimolar (propylene/propane) mixture experiments were conducted to test the performance of membranes. Compared to bare macroporous α-alumina supported ZIF-8 membranes, membranes prepared on γ-alumina show much higher gas permeance and separation characteristics, with a propylene/propene separation factor of 218 and propylene permeance of 9.5×10-9 mol/(m2sPa). These ZIF-8 membranes on γ-alumina support are favorable compare with the best-performing ZIF-8 membranes which require complex synthesis procedures.

Poster Session 4: Undergraduate Competition

UG 1 - Enhancing the Energy Efficiency of Desalination via Batch Reverse Osmosis

Katie Brodersen* (Purdue University).

As the demand for water security expands globally, sustainable desalination technologies become more critical. Traditionally, continuous reverse osmosis (RO) requires that constant, high pressure must be applied to the system to overcome the osmotic pressure. Closed-circuit reverse osmosis (CCRO) recirculates brine into the feed line, increasing the pressure of the system with each pass. This project investigates batch reverse osmosis, which varies the applied pressure over time to closely follow the osmotic curve, conserving the excess energy that is used to maintain a continuous RO system. Unlike CCRO, batch RO avoids the exergy generated when brine and feed are mixed by using a variable displacement cylinder (VDC), into which brine output from the RO membranes recirculates. As the system reacts to increasing brine concentration with each "batch", the pressure of the system increases. Batch RO has been modeled to achieve the lowest specific energy consumption per volume of water desalinated. This team is currently integrating the batch RO model as a hydraulic system to demonstrate these expected results. The hydraulic system is operated by a control system using pH and conductivity sensors to track the salinity over time, as well as pressure and flow sensors which interface with a variable frequency drive that sets the applied flow rate of the high-pressure pump. Solenoid valves are used to switch direction of the recirculating brine stream from one side of the VDC to the other with each pass. The goal of this research project is to improve the sustainability of desalination and the economic viability of its use globally.

UG 2 - Continuous covalent organic framework membranes for dye/salt separation

Miguel Jaimes* (UW-Milwaukee).

Covalent organic frameworks (COFs) are an emerging class of porous membrane materials constructed from organic building blocks via strong covalent bonds. They possess a wide range of unique properties including high crystallinity, ordered two or three-dimensional pore structure, tunable pore size and functionality, and excellent thermal and chemical stability. With a pore size in the range of 1-3 nm, COF membranes are well suited for nanofiltration applications such as the separation of dye molecules. In this work, we will present the synthesis and dye/salt separation properties of COF membranes based on ketoenamine-linked two dimensional (2D) COFs. Their pore structures are constructed by aldehydes and diamines through a Schiff base reaction followed by irreversible enol-keto tautomerism. These 2D COFs can be synthesized as continuous membranes on polymeric substrates using a facile and scalable interfacial crystallization method. In a cross-flow filtration setup, measurements using single dye or salt solution, the membranes showed a high water permeance of >50 L m-2 h-1 bar-1 with a high rejection of dye molecules (e.g., 100% rejection of Congo Red) and low rejection rates (<5%) for mono- and divalent salts (e.g., NaCl and MgSO4). The dye/salt mixture separation properties of these membranes at different separation conditions will also be discussed in this poster presentation.

UG 3 - Functionalization of Polyvinylidene Fluoride Membranes for Ion Separation Applications

Ronald Vogler* (University of Kentucky), Saiful Islam (University of Kentucky), Dibakar Bhattacharyya (U. Kentucky).

The removal of toxic heavy-metal ions from industrial waste streams and the recovery of valuable lanthanides for use in electronics are prominent challenges that make the development of metal ion- separation technologies a critical task. This work involved the incorporation of polyacrylic acid (PAA) into commercial polyvinylidene fluoride (PVDF) membranes for metal ion capture as well as further modification of these membranes to add thiol groups for heavy-metal adsorption. The PAA-functionalized membranes were fabricated by polymerizing PAA within the membrane pores and were used for the capture of lanthanum (III). These membranes were also functionalized with thiol groups via the linking of cysteamine to PAA. The adsorption of mercury (II) from industrial effluent water by the resulting thiol-functionalized membrane was assessed. PVDF-PAA membranes demonstrated a convective-flow adsorption capacity of 3.7 g La3+/m2 membrane and 73% of the captured La3+ was leached upon regeneration with an acidic solution. Fourier-transform infrared spectra verified the functionalization steps performed prior to obtaining PVDF-PAA Cysteamine membranes. With the addition of thiol groups to a PVDF-PAA membrane, the water permeability corresponding to a feed pH of 5.9 decreased from 80.9 ± 2.3 LMH/bar to 67.7 ± 0.8LMH/bar (16.4% decrease). When used to treat effluent water containing 40 - 50 ppb of Hg2+ and~70 ppm of Ca2+, the PVDF-PAA-Cysteamine membrane adsorbed 82% of the Hg2+ in a single pass. Overall, the lanthanide capture by PVDF-PAA membranes demonstrated their potential in this area. Also, the addition of thiol groups to these membranes indicated the potential for further membrane modification for heavy-metal capture applications. Together, these results demonstrated the versatility of the PVDF-PAA membrane with respect to separation applications involving metal capture.

UG 4 - Ammonia and Water Flux in Membrane Distillation Using Spacers

Nobuyo Watanabe* (Barnard College of Columbia University).

The Haber-Bosch Process converts atmospheric nitrogen into ammonia, an essential element in synthetic fertilizer. However, this process demands 1-2% of the world's annual energy. Conversely, excess nitrogen from Wastewater Treatment Plants (WWTP) causes eutrophication and harmful algal blooms posing an environmental and public health threat. To address these problems, nitrogen recovery from source-separated urine, before WWTP, was investigated via membrane distillation. Ammonia permeation across a hydrophobic PVDF membrane is driven by a vapor pressure gradient of volatile ammonium from the aqueous feed to the acidic collector. The temperature difference of 40°C and 20°C between feed and collector facilitates the driving force, yet it is hindered due to temperature polarization. Spacers were used in this direct contact membrane distillation to improve ammonia flux by mitigating temperature and concentration polarization through promoted mixing at the membrane surface. Two commercials spacer, diamond, and square configurations, and a 3D printed spacer were tested for optimal performance. Based on the result, the diamond configured commercial spacer gives the best performance with 105% higher ammonia flux than without a spacer.

UG 5 - A Liquid-liquid Hollow-Fiber Membrane contactor device to remove oceanic CO2

Christopher Snodgrass* (University of Pittsburgh).

The purpose of this project is to investigate the viability of a liquid-liquid Hollow-Fiber Membrane contactor device to remove oceanic CO2. This technology would indirectly reduce atmospheric CO2 and counteract ocean acidification in at-risk regions like coral reefs. Reefs are especially susceptible to ocean acidification, which harms life by holding less oxygen and affects available carbonate species, both of which result in coral bleaching. This work is being done under Prof. Katherine Hornbostel (MEMS) at the University of Pittsburgh. A Liquid-Cel mini-module consisting of polypropylene hollow fibers arranged in a cylindrical housing was used. We constructed a test set-up consisting of peristaltic pump driven countercurrent flow of synthetic seawater and NaOH solution through the bundle. Samples were taken periodically from the inlet and outlet of each stream and run through an Orion Carbon Dioxide electrode to measure PH and dissolved CO2 levels vs. time. The CO2 flux across the membrane was then calculated. Using a liquid-liquid contactor allows CO2 flux to be passively driven by the NaOH solvent on the other side. Using a CO2 selective solvent prevents other gases from crossing over, resulting in a pure CO2 stream once stripped from the solvent. This design could therefore be implemented as a carbon capture alternative to direct air capture systems. The high surface area of the hollow fiber membranes and high CO2 flux due to the permeability of the membrane allows for effective CO2 transport. Initial results demonstrate an average loss of 19.7% (73.5 ppm) of dissolved CO2 from seawater during a single pass through the system. These preliminary results demonstrate that CO2 can be passively removed from seawater using a polymer hollow fiber membrane contactor. Future tests will investigate other solvents (e.g. Na2CO3 solution, MEA solution), gas sweeps (e.g. air, oxygen), and surface coatings (e.g. carbonic anhydrase, amine groups) to optimize CO2 removal.

UG 6 - Predicting Octanol/Water Partition Coefficients from Molecular Structure

Vy (Jasmine) Tran* (Miami University.

Knowledge of the partitioning of a solute between two immiscible phases is of central importance for a wide range of environmental, biological, and physical systems. Furthermore, the partition can be used to characterize the intermolecular interactions of a system. The octanol/water partition coefficient has long been used to characterize the hydrophilic/lipophilic nature of a solute. This molecular-level connection has undoubtedly led to the use of octanol/water partition coefficients in the correlation of a wide range of physical properties, including membrane permeability. For early stage processes conceptualization and design applications, the measurement of octanol/water partition coefficients for a novel solute may not be feasible, due to limitations of the material, cost, time, etc. The development of predictive methods is therefore of utmost importance. The development of methods to predict octanol/water partition coefficients remains a contemporary area of research. In this work, we will demonstrate the use of state-of-the-art electronic structure calculations, and structural-based machine learning methods to predict octanol/water partition coefficients.

UG 7 - Evaluating Electrocoagulation Process for Treating Produced Water

Alix-Cynthia-Ineza Karangwa* (University of Arkansas).

In this study, both batch and continuous electrocoagulation (EC) reactors are employed to investigate the feasibility of removing suspended solids and organic matters from produced water. The effects of current, pH and reaction time are studied to evaluate the working parameters of electrocoagulation. Different pH values (5, 7, 9) are evaluated with respect to TSS, TOC and turbidity removal efficiency. A wide range of current (1 to 8 A) and time (1 to 90 min) is tested to reach a high removal efficiency of TSS, TOC and turbidity. Using EC system as a pretreatment step presents an effective technique to eliminate TSS and TOC from produced water. Moreover, a high removal efficiency of TSS and turbidity is achieved at different pH values, while the highest removal efficiency of TOC is obtained at pH 7. By changing the pH from 5 to 9 there is no significant enhance in the removal efficiency of TOC. The highest removal efficiency of TOC is obtained at 90 min reaction time. Increasing the reaction time from 20 to 60 min can increase the removal efficiency of TOC from 55% to 67%, while increasing the time further to 90 min can lead to a slight increase in the removal efficiency of TOC to about 68%. The electrical energy consumption increases from 8.9 to 103 KW h / m3 when increasing the reaction time from 20 to 60. At 90 min, the highest electrical energy consumption is recorded about 217 KW h / m3. The results obtained here indicate that high removal efficiency of TOC can be achieved by working at optimized EC parameters.

UG 8 - Influence of molecular interactions, membrane swelling and plasticization on pure and mixed fluid transport in OSN membranes

Kelly Bye* (University of Oklahoma).

The vast majority of industrial chemical synthesis occurs in organic solution. Solute concentration and solvent recovery consume approximately 50% of the energy required to produce chemicals and pose problems that are as relevant as the synthesis process itself. Novel, energy-efficient technologies based on polymer membranes are emerging as a viable alternative to distillation. Despite organic solvent nanofiltration (OSN) could revolutionize the chemical industry, its development is still in its infancy for two reasons: i) the instability of traditional polymer materials in chemically challenging environments, and ii) the lack of fundamental knowledge of elemental transport phenomena in OSN membranes. Most of available transport data refer to composite membranes, where the presence of a fabric backing makes it difficult to provide a fundamental description of solvent and solute sorption and transport in the active layer. The lack of fundamental information has hampered the development of rational methods to design better materials for OSN application.

In this study, the solubility of several pure and mixed organic species in Celazole®, a commercial polybenzimidazole, has been investigated. Polymer volumetric dilation upon liquid sorption has been investigated as well, using the optical method. The role of polymer-penetrant interactions, membrane degree of swelling, and penetrant clustering on small molecule sorption and transport in Celazole® has been discussed and several structure-property correlations were identified. Methanol was selected as a model penetrant to run sorption and diffusion experiments in the activity range 0-1. Methanol and other polar lower alcohols cause severe matrix plasticization. We have hypothesized the possible mechanism of Celazole® plasticization. Specifically, methanol (as well as other polar penetrants) likely break the inter-chain hydrogen bonds in favor of polymer-penetrant hydrogen bonds, according to a mechanism that we could define as competitive hydrogen bonding. The breaking of the original polymeric network would increase, in turn, the distance between polymer chains, thus enhancing their mobility. This picture helps to rationalize mixed methanol-PEG400 sorption data. In contrast, Celazole® is very stable in non-polar aliphatic and aromatic hydrocarbons.

Finally, the polymer mechanical properties were measured before and after soaking in liquid water and methanol. The results provide further proof that polar liquids severely plasticize Celazole®.

UG 9 - Reduced graphene oxide membranes: properties and potential water applications

Trisha Nickerson* (University of Kentucky), Ashish Aher (U. Kentucky), Dibakar Bhattacharyya (U. Kentucky).

Graphene oxide (GO) membranes are comprised of stacked 2D carbon lattice sheets with oxygen-containing functional groups that create interlayer channels for water transport. GO exhibits moderate hydrophilicity and a negative surface charge in neutral pH water, giving it the potential to achieve high retention of ions while maintaining high water flux. In this study, reduced graphene oxide (rGO) membranes were synthesized by depositing aqueous graphene oxide onto commercial membranes via vacuum filtration and film casting methods, followed by thermal reduction. Rejection of PFOA (perfluorooctanoic acid) by an rGO membrane was tested in a dead-end mode of operation in a variety of conditions, achieving the highest rejection (near 80%) for a low concentration PFOA solution in neutral pH water. Lower rejection values were observed for an increased PFOA concentration and adsorption was observed at pH 3 where zeta potential is significantly reduced, indicating charge to be a major factor in exclusion. Rejection studies were also conducted for ionic salt solutions using both crossflow and pure diffusion modes of operation. Identical experiments were performed using a rGO membrane and commercial DOW NF270 membrane for comparison. Rejection trends obtained from crossflow experiments were consistent with permeability trends from diffusion experiments, with the highest rejection values corresponding to the lowest permeabilities. NF270 membranes displayed trends based on size exclusion whereas rGO membranes appeared to reject based on charge. As the material properties of rGO are better understood, large-scale synthesis will become possible. In addition, upon further study of the primarily charge-based exclusion mechanisms, rGO has the potential to serve a unique purpose in the membrane community for applications such as removing a range of PFAS (perfluoroalkyl substances) and partial water softening.

UG 11 - Perylene-Polyimide membranes

Aryan Louise Tan* (UCLA).

In the past decade there has been a lot of interest in using graphene and its derivatives as the selective layer in thin-film composite (TFC) membranes. Taking into account the unique properties of pi-stacked systems and pathways through graphene networks, this work is a study of "graphitic" polymers: their synthesis, characterization, and the feasibility of making them into thin-film composite (TFC) membranes. In particular, our end objective is to evaluate the aqueous separation capability of perylene-polyimides, how their diffusive mechanisms may differ from traditional polymeric or inorganic systems, and how this may be applied to removal of trace contaminants in recycled water. Major strides have been made in understanding transport properties and fabrication of these systems, including recent work using Thin-Film Lift Off (TFLO). Using T-FLO, we can use very brittle and otherwise difficult-to-process materials as selective layers. Due to the periodicity of polymeric systems, we can influence the points of rotation about graphitic regions and their functionality through monomer design in an attempt to deterministically design of diffusive pathways. T-FLO, unlike other techniques, allows the specification of the thickness and chemistry of a membrane's selective layer, as it is formed first on a substrate and the support is cured atop the thin film's surface. Reaction of epoxy groups in the support-layer pre-polymer with nitrogen groups in the polyimide thin-film, enabledelamination from the film's substrate. Monomers are formed through the reaction of perylene tetracarboxylic dianhydride with diamines of different structure and electron density including 3,5-diaminobenzoic acid, 4,4-aminophenyl sulfone, and 4,4-oxydianiline. Polymerization can then be done under ambient conditions by reaction of perylene-diamines with pyromellitic dianhydride. Monomers and polymer structures are confirmed using NMR, ATR-FTIR, and Xray photoelectron spectroscopy. Crystallinity, relative surface tension components, and physical structure/morphology of membranes are measured using X-Ray Diffraction, sessile-drop contact angle, and scanning electron microscopy respectively. Complete TFC membrane performance is tested using a dead-end cell apparatus, beginning with small-molecule dyes and divalent salts.

UG 12 - Effect of mechanical strain on the transport properties of thin-film composite membranes used in osmotic processes

Alton O'Neal* (Clemson University).

The goal of this work was to study water and sodium chloride transport through TFC membranes that were subjected to known degrees of strain. We used two methods to evaluate this effect. In one method, we prepared membrane samples that were preconditioned by applying a defined strain and measured their transport properties. In the other method, we proposed an osmotically-driven burst pressure test for flat sheet membranes, which allowed us to evaluate the pressure and strain when the membrane failed. We define failure of the membrane as the loss of selectivity, rather than irreversible mechanical deformation. We demonstrate the importance of knowing the stress-strain curve of the membrane and highlight that stiffer membrane structures are desirable to avoid reaching a strain above the reported onset fracture strain of the selective layer. Since membrane deformation has been reported regardless of the feed spacer (or permeate carrier) used, we assumed that the stress on the membranes, during operation in osmotic processes and reverse osmosis, is above the yield strength of the membrane supporting structure. With this information, we introduce a deformability coefficient and our solution diffusion model with defects to guide the design of membranes and modules for pressurized osmotic processes.

UG 13 - Advanced Water Treatment for Direct Potable Reuse

Ciara Lugo* (University of Arizona).

Direct potable reuse (DPR) is becoming more reliable water resource in areas with water supply stress, groundwater withdrawal impacts, water quality challenges associated with conventional water sources, and where non-potable reuse are limited. DPR is the treatment of wastewater to drinking water quality without an environmental buffer, such as a river or groundwater aquifer.1 Conventional potable reuse trains consist of microfiltration (MF), reverse osmosis (RO) and advanced oxidation processes (AOPs).2 However, there are many limitations to RO-based treatment trains – including membrane fouling, concentrate disposal, operation costs and the necessary pre-treatment of the influent water. Therefore, there has been discussion of alternative treatment technologies, such as ozonation and biologically active filter (BAF).2 The proposed treatment train that will be evaluated consists of ozone/granular activated carbon (GAC)-BAF. Ozone oxidizes organic compounds, so they are readily biodegradable for BAF, and granular activated carbon (GAC). is utilized as a final polishing step to absorbs the remaining organic carbon.2 The difference between GAC and BAF is that BAF is seeded and aerated with microbial cultures to promote adsorption and microbial degradation of contaminants, whereas GAC only provides absorption. The main objective of this research project is to evaluate and compare two different biologically active filters (BAFs) based on their efficiency and effectiveness on removing organics compared to conventional membrane-based DPR treatment trains. Two different wastewater effluents were evaluated for organic carbon and nutrient removal. Furthermore, their energy and carbon footprints were investigated. Results indicated that selection of advanced treatment processes are highly dependent on source water quality and effluent standards.